AGRICULTURAL AND FOOD MARKETING IN DEVELOPING COUNTRIES
Selected Readings

AGRICULTURAL AND FOOD MARKETING IN DEVELOPING COUNTRIES
Selected Readings

Edited by

John Abbott

Formerly Chief, Marketing Service, FAO, Rome

C·A·B International

in association with the

Technical Centre for Agricultural and Rural Co-operation ACP–EEC

C·A·B International
Wallingford
Oxon OX10 8DE
UK

Tel: Wallingford (0491) 832111
Telex: 847964 (COMAGG G)
Telecom Gold/Dialcom: 84: CAU001
Fax: (0491) 833508

© C·A·B International 1993, unless otherwise indicated.
All rights reserved. No part of this publication
may be reproduced in any form or by any means, electronically,
mechanically, by photocopying, recording or otherwise,
without the prior permission of the copyright owners.

The Technical Centre for Agricultural and Rural Co-operation

The Technical Centre for Agricultural and Rural Co-operation (CTA) operates under the Lomé Convention between member States of the European Community and the African, Caribbean and Pacific (ACP) States.

The aim of CTA is to collect, disseminate and facilitate the exchange of information on research, training and innovations in the spheres of agricultural and rural development and extension for the benefit of the ACP States.

Headquarters: 'De Rietkampen', Galvanistraat 9, Ede, Netherlands

Postal address: Postbus 380, 6700 AJ Wageningen, Netherlands
Tel: (31)(0)(8380) – 60400
Telex: (44) 30169 CTA NL
Telefax: (31)(0)(8380) – 31052

A catalogue entry for this book is available from the British Library

ISBN 0 85198 804 0

Typeset by Leaper & Gard Ltd, Bristol
Printed in Great Britain by Redwood Books Limited, Trowbridge, Wiltshire

CONTENTS

Preface ix

Introduction xi

Abbreviations xiv

PART I: DIAGNOSIS, INNOVATION AND EXTERNAL ASSISTANCE 1

   1 The Economics of Marketing Reform 3
       P.T. Bauer and B.S. Yamey

   2 Market Channel Coordination and Economic Development 25
       C.J. Slater

   3 There is Method in My Madness: or is it Vice Versa?
      Measuring Agricultural Market Performance 34
       B. Harriss

   4 Markets and States in Tropical Africa: The Political Bases
      of Agricultural Policies 41
       R.H. Bates

   5 Food System Organization Problems in Developing
      Countries 51
       H.M. Riley and J.M. Staatz

   6 Marketing, The Rural Poor and Sustainability 65
       J.C. Abbott

PART II: MARKETING ENTERPRISES   93

  7 The Lebanese Traders in Sierra Leone   95
     H.L. van der Laan

  8 Farmers and Middlemen: Aspects of Agricultural Marketing in Thailand   104
     A. Siamwalla

  9 Women's Activities in Food and Agricultural Marketing   112
     FAO

 10 New Types of Multinational Firms in the Agribusiness Sector: Implications of their Emergence in the Least Industrialized World   117
     J.C. Dufour, G. Ghersi and R. Saint-Louis

 11 Cooperatives – Effects of the Social Matrix   127
     G. Hunter

 12 Commodity Marketing Through Cooperatives: Some Experiences From Africa and Asia and Some Lessons for the Future   136
     COPAC

 13 Economic Tasks for Food Marketing Boards in Tropical Africa   144
     W.O. Jones

PART III: PHYSICAL INFRASTRUCTURE FOR MARKETING   165

 14 Marketing Aspects in Planning Agricultural Processing Enterprises in Developing Countries   167
     H.J. Mittendorf

 15 A Spatial Equilibrium Model for Plant Location and Interregional Trade   182
     M. von Oppen and J.T. Scott

 16 Infrastructure for Food Marketing: Some Investment Issues   197
     B.W. Berman

17	Improving Physical Marketing Infrastructure in Africa Through More Self-help *H.J. Mittendorf*	208
18	A Market-oriented Approach to Postharvest Management *A.W. Shepherd*	216

PART IV: INSTITUTIONS AND POLICIES FOR MARKETING   223

19	Regulatory Uncertainty, Government Objectives and Grain Market Organization and Performance: Senegal *M. Newman, P.A. Sow and O. Ndoye*	225
20	Domestic Price Stabilization Schemes in Developing Countries *O. Knudsen and J. Nash*	236
21	Training Programmes for Human Resource Development in Agricultural Marketing *L.A. De Andrade and A. Scherer*	254
22	Agricultural Marketing Extension *C.Y. Lee*	262
23	Agricultural Market Information Services *B. Schubert*	268
24	The Marketing Development Bureau in Tanzania *E. Seidler*	273

PART V: PROVISION OF SEEDS AND FERTILIZERS   277

25	Financing Fertilizer Distribution Networks *J.C. Abbott*	279
26	Problems of Marketing and Input Supply *World Bank*	291
27	A Framework for Seed Policy Analysis in Developing Countries *C.E. Pray and B. Ramaswami*	306

PART VI: FOOD GRAIN SUPPLY MANAGEMENT 323

28 The Nonanswers to Food Problems 325
C.P. Timmer, W.P. Falcon and S.R. Pearson

29 Agricultural Marketing and Price Incentives:
a Comparative Study of African and Asian Countries 333
R. Ahmed and N. Rustagi

30 Assessment of Food Subsidy and Direct
Distribution Programmes 341
United Nations World Food Council

PART VII: EXPORTS 355

31 Responsiveness of a Nomadic Livestock Economy to
a Profitable Export Opportunity: The Case of Somalia 357
E. Reusse

32 Contract Farming and its Impact on Small Farmers in
Less Developed Countries 369
N.W. Minot

33 Nontraditional Export Crops in Traditional Smallholder
Agriculture: Effects on Production, Consumption and
Nutrition in Guatemala 378
J. von Braun, D. Hotchkiss and M. Immink

34 Kenya's Horticultural Export Marketing: a Transaction
Cost Perspective 388
S. Jaffee

Index 404

PREFACE

This book is intended for students of food and agricultural marketing in the developing countries. Its intention is to bring together some of the most significant writings in this area over the last 40 years. To meet students' needs within a compact volume the editor has selected strategic sections of influential books and papers and set them in a context of professional interest over the period.

Government officers concerned with the formulation of marketing policies and their implementation will also find this book useful and convenient. It should also be helpful to aid agencies concerned with marketing and to their advisors working directly with governments and on training programmes.

Collections of papers by different authors published in one volume do not always achieve the convenience to which they aspire. They may be too diverse in subject area and quality. Titles chosen for their sales appeal can be directly misleading on content. Sets of papers presented at a conference can jar the reader with their duplications. In this collection we have kept closely to our central theme and cut down the materials included to their essential contribution.

Necessarily this book provides only a very limited selection of what is available. Leads to additional materials appear in the references provided by many of the authors. Readers may also refer to several bibliographical sources. They are reviewed summary in the introduction which follows.

The good will of the many authors, publishers and editors of professional journals who have kindly agreed to the republication of material in this book is hereby acknowledged. Their generosity is most appreciated. Their names are listed on the opening pages. The project to prepare such a book was the outcome of suggestions from people engaged on training programmes for developing countries. For the selection the editor takes full responsibility. The comment of a former head of agriculture in FAO was 'if you still have friends in marketing, Mr Abbott, you will lose them now'.

There has been valuable assistance from several well-informed quarters; their names are withheld so that they do not lose their friends also.

INTRODUCTION

In accordance with current academic practice, marketing is defined as the business activities associated with the flow of goods and services from production to consumption. The marketing of agricultural products begins on the farm, with the planning of production to meet specific demands and market prospects. It is completed with the sale of the fresh or processed product to consumers, or to manufacturers in the case of raw materials for industry. Agricultural marketing also includes the supply, to farmers, of fertilizers and other inputs for production.

SOURCES OF MARKETING LITERATURE

Most of the literature on agricultural marketing in the developing countries has been commissioned by their governments, has been the outcome of aid agency activity or has been undertaken on university research and degree programmes.

The first major sponsor was the Government of India. It set up an agricultural marketing department in its ministry of agriculture. From this came a series of detailed country-wide surveys. Their publication promoted awareness of a wide range of deficiencies in the conditions and procedures of rural marketing. They led directly to remedial legislation. It provided for regulated rural assembly markets managed by committees including farmers' representatives, and for a system of public warehouses issuing certificates that could be the basis of credit from a bank; it defined weights, measures and quality grades that would be clear to all users and established an export quality grading system.

Behind this lay a clear concept of marketing undertaken by private enterprise with a government providing support services adapted to conditions where farms were small, manpower abundant, capital scarce and government resources limited.

The Indian studies were carried out by local personnel following a model originally established under the aegis of a British advisor. For British controlled territories in Africa, marketing studies often began with a report by a government commission of enquiry. The work would be done by an experienced administrator in consultation with the agricultural and trading interests concerned. These reports were also influential. Often they led to legislation setting up marketing boards with monopoly trading rights over a specified commodity, or assigning a protected role to cooperatives.

It was in protest against such government intervention that a traders' association in Cyprus commissioned Peter Bauer (with B.S. Yamey) to make an independent study: *Aspects of Government Intervention in Cooperation and Agricultural Marketing in Cyprus*, Cyprus Federation of Trade and Industry, Nicosia, 1954. A professorship in the economics of developing countries was created for him at the London School of Economics. Systematic academic interest in the subject area, however, awaited the adoption by the Government of the USA of continued funding of university research concentration on a particular country or region. This was implemented under United States Agency for International Development (AID) contracts. Graduate students from that area could be put to work with others from the USA under the guidance of a professor who made this a continuing interest. Associations as of W.O. Jones of Stanford with Africa, J. Mellor of Cornell with India, and H.M. Riley of Michigan State with Latin America then became the fount of a rich body of marketing research.

While such a degree of university specialization was not emulated in Europe, the British Overseas Development Administration has supported Barbara Harriss on in-depth marketing studies in India. Public finance has been available for German graduate students to take up marketing research topics in developing countries. African studies centres at Leiden in the Netherlands and Uppsala in Sweden have been active in this field.

Studies on marketing in the developing countries now appear regularly in the proceedings of national and the international agricultural economics association, and in the professional journals *Food Policy, International Food and Agribusiness Marketing, Development Studies* and *World Development*. The Birla Institute of Scientific Research, a private foundation in India, has sponsored a series of economic studies; several are on marketing.

Specialized work on food and agricultural marketing was taken up by the Food and Agriculture Organization of the United Nations (FAO) in the mid-1950s. It produced a series of marketing guides, which did much to extend appreciation of marketing issues, and also a continuing flow of advisory papers and commodity market situation studies. The International Trade Centre in Geneva has provided specific information on export markets for products of developing countries. Detailed monographs on the preparation of products for export have also come from the Tropical Products Institute in the UK, now incorporated in the Natural Resources Institute at Chatham.

Much of the marketing research of the World Bank has remained confidential, carried out in the course of loan preparation. However, it reviewed its marketing projects in *Agricultural Marketing: the World Bank's Experience 1974–85*, Washington, 1990. It has also published, through its Economic Development Institute and the Johns Hopkins University Press, books on marketing and related issues intended to guide policy formation and implementation.

BIBLIOGRAPHIES

A continuing bibliographical source for agricultural marketing has been the *World Agricultural Economies and Rural Sociology Abstracts* (*WAERSA*) issued by CAB International. Significant publications and reports are listed together with abstracts of their contents in volumes issued monthly. There is good developing country coverage. The Tropical Institute of the Netherlands has a similar annotated listing concentrating on the developing countries, covering the technical side of marketing. From 1975 the International Information System for Agricultural Sciences and Technology, known as AGRIS, has provided computer print-outs of marketing titles with a covering of developing countries that has improved steadily. The FAO Marketing Service set up a specialized bibliography for use in briefing its own advisors, available to libraries and outside researchers on request. This also provides notes on contents. It was kept up to date through the 1980s by the issuance of supplements. The relative convenience for users of WAERSA, AGRIS and the FAO marketing bibliography was appraised in a paper by J.C. Abbott in the *Journal of Agricultural Economics* **23**(2), May 1982.

ABBREVIATIONS

AID	Agency for International Development, Government of the USA, Washington
COPAC	Committee for the Promotion of Agricultural Cooperatives, c/o FAO, Rome
ESCAP	UN Economic and Social Commission for Asia and the Pacific, Bangkok
FAO	Food and Agriculture Organization of the United Nations, Rome
ICRISAT	International Crops Research Institute for Semi-Arid Tropics, Hyderabad, India
IFDC	International Fertilizer Development Center, Muscle Shoals, Alabama, USA
IFPRI	International Food Policy Research Institute, Washington, DC, USA
IMF	International Monetary Fund, Washington, DC, USA
ITC	International Trade Centre, Geneva
LIC	Less industrialized country
MNC	Multinational corporation
MNFFC	Multinational food and feed corporation

NGO	Non-governmental organization
OECD	Organization for Economic Cooperation and Development, Paris
UN	United Nations, New York
$	United States dollars

DIAGNOSIS, INNOVATION AND EXTERNAL ASSISTANCE

Peter Bauer's critical stance on compulsory marketing 'improvement' became known widely with the publication of *West African Trade* (Cambridge University Press, 1954). In the paper presented here, he and Professor Yamey of London University argue their case more generally. Bauer remained sceptical on aid as a whole, with the dictum 'If a country cannot progress without aid, it certainly will not do so with it'. His papers have been published under various titles, most recently in *The Development Frontier* (Harvard University Press, 1991). Looking back in *Pioneers in Development* (Oxford University Press, 1984), edited by G.M. Meier and D. Sears, Bauer said he had misjudged how far economic life would become politicized, i.e. how the new governments of developing countries would distort marketing systems to retain power. M. Lipton's comment was that Bauer had been influenced by the more dynamic groups in the developing countries; there were others unlikely to progress without a push.

Systematic review of the nature and volume of international aid for marketing was initiated at a 1976 OECD/FAO meeting reported as *Critical Issues on Food Marketing Systems in Developing Countries* (OECD, Paris). It was continued in J.C. Abbott's 'Technical assistance in marketing; a view over time' (*Proceedings of the 17th International Conference of Agricultural Economists*, Banff 1979, Gower, Aldershot).

Research and teaching in food and agricultural marketing has long been handicapped by restricted views on the part of both practitioners and their sponsors. Those approaching the subject from an agricultural economic background tended to concentrate on the farmers' interest and see little beyond the wholesaler. Another set of marketing teaching and research has been concerned with the whole range of products offered to consumers, with demand niches and sales promotion a major interest and no particular concern for food.

Dying in mid career, 'Chuck' Slater became almost a cult figure. He was a leader in the marketing research teams fielded by Michigan State University for AID in Latin America in the 1960s. His experience there illuminated for him the critical importance of seeing the marketing channel as a whole.

Analysis of marketing efficiency in terms of its structure, conduct and performance has been a favoured approach since it was formulated by J.S. Bain in the 1950s (*Industrial Organization*, Wiley, New York, 1959). It was brought into agricultural marketing by R.L. Clodins and W.E. Mueller ('Market structure analysis as an orientation for research in agricultural economics' in *Journal of Farm Economics* **17**(3), 513–553, 1961). His approach was used in a number of AID sponsored studies of marketing organization in developing countries in the 1960s and 1970s. They are listed in the paper by Barbara Harriss (Queen Elizabeth House, Oxford) summarized here. She questions their conclusions. More accurate assessment of marketing efficiency is now being sought through econometric models.

Writing from the Institute of African studies, Los Angeles, R. H. Bates attracted considerable attention. In a sense he opened the way to IMF/World Bank use of financial leverage to obtain changes in a government policy frame that was thought negative to development. Recognition that many African governments were prisoners of their own armies and civil servants, and of the potential for riot in their capital cities, provided a rationale for applying a countervailing pressure. Bates has had his impact, though Barbara Harriss saw his thesis as essentially assertive. (Review in *Journal of Development Studies* **19**(3), April 1983).

From time to time meetings of aid organizers and teachers in marketing have been called to review the state of their art. The report of one such workshop is reproduced here. It was sponsored jointly by the American Agricultural Economics Association and the Agricultural Development Council. The senior editor, Harold Riley, was for many years chairman of the department of agricultural economics of Michigan State University. He had a leading role in securing support for its work on marketing in Latin America and the designation of that university as a centre of excellence in the subject.

Aid in practice is appraised for marketing in *Marketing Improvement in the Developing World* (FAO, 1986) and more broadly in *Does Aid Work?* (R. Cassen et al., 1986, Clarendon Press, Oxford).

The final paper in Part I, the editor's 'Marketing, the rural poor and sustainability', brings into a marketing context the dominant concerns of the early 1990s – widespread continuance of poverty and the issue of sustainability. It is based on a policy paper prepared for the International Fund for Agricultural Development (IFAD), Rome, Italy. This organization was established in 1977, it has been said, as a penance by the petroleum exporting countries for the damage done to the developing countries by the sharp rise in its price. It focuses specifically on poverty alleviation among rural people. How such projects can be continued after external financing ceases is a crucial issue for IFAD. The paper also addresses the policy issues arising from the reduced role of government urged by the IMF and World Bank and concern for the welfare of those disadvantaged in consequence.

1
THE ECONOMICS OF MARKETING REFORM

P.T. Bauer and B.S. Yamey

Statutory measures designed to control or to modify the processes of agricultural marketing or to reshape the structure of trade in agricultural produce are in force in many parts of the world. They are of three broad types. First, there are measures designed primarily to raise the returns of certain classes of producers: monopolistic restriction of supply, differential prices, and subsidies fall in this group. Second, there are various measures designed, at least ostensibly, to stabilize prices or incomes. These two types of measure have received considerable attention in the literature of our subject. In addition, a miscellany of measures designed to improve agricultural marketing have been introduced in recent decades in many countries; these include reduction in the number of intermediaries, control of the channels of marketing, delimitation of the places where transactions may take place and elimination of inferior grades of products. Advocates of these miscellaneous measures contend that they are in the interests of producers and of consumers as well. These measures have received little attention in the literature. It is with their implications that this paper is concerned.

It is a common complaint of marketing reformers that unnecessary categories of middlemen are able to interpose themselves between producers, on the one hand, and 'necessary' middlemen and dealers, on the other, and that thereby the costs of marketing are raised. As a corollary, reformers advocate compulsory elimination of the supposedly redundant links in the chain of distribution.

The advocates of such measures generally fail to ask the relevant question concerning why the so-called 'redundant' intermediaries are not by-passed by those with whom they deal. For the services of an inter-

Source: *Journal of Political Economy* **62**(3), 210–235, June 1954 (footnotes omitted). Copyright University of Chicago Press.

mediary will be used only if the price (margin) he asks is less than the value his customers set on the services he performs for them. They will by-pass him if he provides no services at all (i.e. if he is redundant) or if his charges for his services are excessive in comparison with the costs his customers will incur if they provided these services for themselves. Thus redundant intermediaries and intermediaries charging excessive prices will be eliminated without official intervention. This result must follow unless the parties served by them are unaware that it is cheaper to by-pass them or unless institutional arrangements prevent the more economical direct method from being followed.

The agricultural producer rarely sells his output directly to consumers. Usually there are several stages in the marketing process, and it is generally conceded that the middlemen are necessary in some of the stages. The supposedly redundant middleman must stand between another middleman and the producer, between a middleman and the final consumer, or between two middlemen. In each case, at least one of the parties served by an allegedly redundant middleman is a middleman himself. Now, even if it were true that the average farmer or peasant is unaware of other marketing alternatives or is unable to perform simple commercial calculations, a redundant middleman would not be used so long as his middleman-customer was able to see a profit or a saving in direct dealing. It is unlikely that dealers will fail to see an economic opportunity within their field of business or fail to take advantage of it. Hence the knowledge or capacity of the producer is largely irrelevant.

However, it may be doubted whether the general assumption, which underlies so much of the demand for marketing reforms, of the commercial ignorance and lack of wisdom of the average agricultural producer agrees with the facts. Where economic activities are largely specialized, as in western Europe or North America, the farmer earns the bulk of his income from the sale of farm produce. Hence he has a direct, significant, and continuing interest in opportunities in the market and will try to choose the most suitable channel of distribution. Moreover, in such economies it is usually not difficult for groups of farmers to set up cooperative marketing agencies if they are dissatisfied with the services or margins of the available middlemen.

In the underdeveloped areas, where economic activities are not generally so specialized, the average producer perhaps does not secure the bulk of his income (cash *and* other) from the sale of one or a few farm products. On the other hand, the low cash incomes and the often very low opportunity costs (which are low because of the absence of other profitable opportunities) give the producer a strong incentive to market his output advantageously and keep low the cost to him of discovering and examining available alternative opportunities. This clearly applies in West Africa, where, no doubt as a result of the low level of incomes and the comparative lack of other more productive occupations, Africans will spend much time and effort to secure price advantages in selling their produce or in purchasing merchandise. Sellers of produce are particularly sensitive to price differences. In the Eastern Provinces of Nigeria, for example, women selling palm oil will walk or cycle several miles to secure another penny or twopence on the sale of a beer bottle of palm oil or another shilling on the sale of a four-gallon tin of palm oil. In these circumstances a redundant intermediary (i.e. a dealer whose margin is higher than the value of his services to the parties served by him) would certainly be by-passed.

In these circumstances the compulsory elimination of any class of intermediaries (e.g. itinerant buyers in country districts, market commission agents, or touts) will mean that their services have to be performed on more onerous terms by one or another of the parties between whom they stand. It will involve an otherwise uneconomical measure of vertical integration and, at the same time, a reduction in the number of marketing alternatives open to the parties concerned. Both these consequences are likely to be particularly serious in underdeveloped areas.

Institutional arrangements may sometimes deny customers the right to use certain channels or methods of marketing. Restrictive practices may prevent customers from by-passing middlemen whose services are redundant or, more frequently, whose charges are higher than the costs of the prohibited alternatives. The pressure of trade interests has led to such arrangements in the United Kingdom and in North America. There are also examples, though they are less numerous, in underdeveloped economies. Such practices are against the interests of producers and consumers because they impede the competitive elimination of redundant middlemen and the competitive moderation of excessive charges. But improvement is not to be found either in the compulsory removal of the class of middlemen concerned or in the restriction of their numbers. Both these courses merely reduce the alternatives open to producers and consumers.

It is sometimes contended that the producer in underdeveloped areas is not always free to use the most advantageous method or channel for marketing his produce because he may be forced to employ the services of a particular intermediary to whom he is financially indebted. But where the producer has a choice among a number of would-be lenders and trader-lenders, he will choose to borrow where the terms are most advantageous to him. The terms of loans from trader-lenders may be a combination of interest payments and the obligation to sell the produce to the lender, possibly on favourable terms to the latter; what in isolation appears to be a forced sale at a low price may simply represent an indirect part of a payment of interest on the loan. Compulsory reduction in the number of trader-lenders is likely to reduce the sources of funds and to make the terms of loans more onerous to borrowers. Any weakness in the position of the borrower flows from his need for a loan; compulsory reduction in the number of would-be lenders would aggravate the borrower's weakness. Moreover, any attempt to prevent traders from lending to producers on condition that the crop is sold to them would not improve the net position of the borrower; the latter might secure an apparently better price on the sale of his produce, but only by being required to make correspondingly higher direct interest payments to the lender.

The tacit assumption that producers are so ignorant of their own interests and of the marketing opportunities open to them that they cannot be relied upon to sell their produce as advantageously as possible also underlies occasional proposals that producers be forced to take or to send their produce to specified markets or along specified routes. It is to be inferred from these proposals that, without compulsion, many producers would drive their cattle along unprofitable routes or send their produce to the wrong markets. Again, the proposals do not seem to take into account the fact that, even if the producers were ignorant and undiscriminating, better-informed and specialized middlemen would buy up the produce and reroute it. This process insures both that erroneous decisions of producers do not remain effective for long (that is, for no longer than the time necessary to enable middlemen to redirect the misdirected produce) and that the cost of such decisions is minimized. Further, if compulsory marketing routes are laid down by authority, there is a danger that the route, once chosen (even if correctly chosen in the first place), will not easily be changed even if supply and demand conditions change, with resulting costs (losses) to producers. In underdeveloped territories the prescription of permitted market places or of marketing routes may seriously retard the opening-up of new areas of production and the spread of the exchange economy, which in the past has often followed the activities of traders pioneering in new regions and along new routes.

Critics of agricultural marketing often state or imply that competition among middlemen works against the interests of producers. The argument takes different forms and gives rise to proposals for a variety of remedial measures. Sometimes it is said that there are too few competitors; in other cases it is said that there are too many.

In some markets or areas there are only a few buyers of agricultural produce. Almost necessarily, the number of operators declines as the distance from the main assembly, transport, or consuming centres increases. Hence some producers are confronted with only a handful of buyers in the immediate neighbourhood of their farms or holdings. It is argued that such conditions of oligopsony place the producers at the mercy of a few dealers. One type of proposal to improve the position of the producers requires the compulsory centralization of market transactions. The underlying idea is that the producer will be forced to take his produce to one of a limited number of markets, at each of which there will be a larger number of buyers than would otherwise visit his farm or buy in his locality. In this way competition among buyers is increased for the benefit of the producer.

The essence of this type of proposal is to foster one kind of competition (i.e. the competition of more buyers in one place) by denying to producers access to other opportunities of selling their produce. The idea that competition is stimulated by reducing the alternatives open to producers and consumers is here driven almost to its logical but absurd conclusion.

The proposal for compulsory centralization of transactions (i.e. for standardization of one condition of selling) is only one among several suggestions intended to promote competition by compulsory standardization, i.e. by reducing the number of alternatives open to customers. Such proposals rest upon a misconception of the nature of competition and of competitive forces. From the narrow definition of perfect competition in static economic analysis, it is inferred that competition prevails only where the customers can choose among a number of *identical* alternatives, that is, among alternatives which are identical as regards quality, place, and time of supply. In fact, where *demand* is not identical or uniform in these respects, competition is not promoted by forcing demand into a common mould by denying to suppliers the opportunity of adapting their product or service to the varying requirements and unstandardized desires of customers. Geographical differences are generally significant in bringing about unstandardized demands. In such circumstances competition tends to take the form of making available to customers a range of alternatives adjusted to individual circumstances. Moreover, the offering of *new* services is a form of competition which may improve the terms upon which the customary services are made available.

Exaggerated emphasis on the dangers of local buying monopolies in any area neglects the fact that such buyers, while appearing to have no competitors, nevertheless have to set their buying prices in competition with other buyers elsewhere; they cannot depress their own buying prices so low that producers (or other intermediaries) would be better off by taking their produce to other more distant buyers. Accordingly, where producers are able to get their produce to other markets without great sacrifice of time, effort, or resources, the prices they receive locally cannot be far below those obtaining in the more important market centres. In West Africa, for example, producers are sensitive to price differences, knowledge of which spreads quickly, and they are generally prepared to cover long distances to secure attractive prices. On the other hand, where producers are not in a position to undertake long journeys or to hold produce so as to reduce the number of journeys, the compulsory closure of local buying would cause otherwise avoidable loss and hardship. This measure would either deprive them of markets altogether or require them to undertake costly journeys to markets, when previously less costly methods of disposing of their produce has been available to them. This reasoning is borne out by the fact that buyers in outlying areas are voluntarily supported by sufficient producers to keep them in business. Even if there is only one buyer in a particular locality, the margin for his services is limited by the ease of entry of competitors into small-scale trading.

The compulsory centralization of trading may itself promote monopoly practices. Where all dealers and transactions are concentrated, it becomes easier for dealers to form market-sharing or price agreements, provided that they are not too numerous and that entry of new competitors is not easy. It is significant that in some industries effective price agreements require a high degree of standardization of products and conditions of sale.

The reform of marketing is often urged on the grounds that there are too many competing middlemen and that their competition is not to the best advantage of the producers. This complaint is made not only against the competition of traders but also against that of the first processors of agricultural produce, e.g. the owners of slaughterhouses or of cotton ginneries. The argument takes two forms: first, if there are many competitors, competition is too severe, and this leads to the payment of excessively or uneconomically high prices to producers; second, if there are many competitors, competition is wasteful, and this leads to the payment of unnecessarily low prices to producers. The two forms of the argument are often combined; moreover, each gives rise to the same proposals for reform, namely, that the number of intermediaries (dealers or processors) should be reduced in the interests of producers and that the margins for their services should be controlled.

It has been frequently and influentially argued that severe competition among intermediaries results, or may result, in bankruptcies among them, with adverse effects on the producers. This argument has been advanced in support of compulsory limitation of the number of intermediaries and/or prescription of *minimum* margins for the performance of their services. Yet producers can only benefit from increased competition for their crop. If some buyers go bankrupt through paying excessively high prices, this cannot harm the producers who have received these prices. Further, as long as entry is free, there is no danger that the surviving merchants or processors will subsequently be able to exploit producers. There is no need to insure a supply of the services of the intermediaries either by protecting a number of them by means of statutory control of numbers or by prescribing *minimum* margins for their services. On the other hand, if the entry of new firms is difficult because of high costs of establishment or similar obstacles, competition is not likely to be severe; and there is no need for official measures to protect the established firms by statutory barriers to new entry.

Compulsory reduction in the number of intermediaries to raise their returns increases the supply prices of their services and thus harms the producers whose services are in joint demand with those of the intermediaries. In contrast, a reduction in the number of intermediaries as a result of competition implies the survival of those who offer the services at a lower supply price, which is to the advantage of producers.

The more common version of the case against so-called 'excessive competition' is that each competitor is unable to obtain sufficient supplies to keep his plant operating at lowest average cost. This, it is argued, means that competitors are forced to pay producers less than the economic price, so that each can have a larger margin per unit of his limited output to meet the higher unit overhead costs brought about by unnecessary multiplication of facilities. It is never quite clear why this situation does not lead to a bidding-up of producer prices in the struggle of each competitor to get closer to his optimum rate or scale of operations, and why this should not lead to the elimination of redundant intermediaries. The events which usually give rise to demands for limiting the number of competitors tend to support the view that excessive competition serves to raise producer prices rather than to depress them.

Generally, if there were significant economies in the operation of fewer and larger buying posts, abbatoirs, or ginneries, the interests of traders or processors and the responsiveness of producers to higher prices would promote a market structure comprising a few large establishments, without the need for compulsion. If there were substantial economies, intermediaries operating on a larger scale would be able to offer higher prices for produce. In practice, they would be prepared to pay the higher prices to attract supplies, even though their current rate of purchases might temporarily be below the rate at which the full economies are secured. They would be prepared to absorb the costs of growth if substantial economies were in prospect. The action of intermediaries and producers would tend to bring about a situation in which the possibilities of economies were appropriately balanced with producers' valuations of marketing convenience and the availability of a range of alternatives. Those who advocate the compulsory restriction of numbers in the interests of the economies of operations at optimum rates of output seem to betray some lack of confidence in the premises on which their schemes are based.

Where competition is said to work against the interests of producers for the reasons discussed in the previous section, it is usually proposed that the number of competitors be reduced by state action. The extreme version of such proposals requires the establishment of zonal monopolies. Here the limiting case of the zonal monopoly will be considered. The discussion will throw into clear relief the issues which are involved. The reader will readily be able to allow for the necessary qualifications when numbers are restricted less rigorously.

The establishment and supervision of zonal monopolies raise problems of their own, which tend to negate the superficial attractiveness which such systems present because of their ostensible tidiness, orderliness, and amenability to official control. The absence of the spur of actual or potential competition and the costs and inconvenience of supervision are likely to offset any small economies that may result from compulsory zoning. In underdeveloped areas a system of zonal purchasing monopolies might also imply either local monopolies in the sale of imported merchandise or a duplication of facilities for the purchase of produce and for the sale of merchandise which would offset or more than offset any theoretical saving in resources. In any event, there would be a reduction in the facilities available to the population which would impose inconvenience and possible hardship on some producers (and consumers). It would also greatly increase the feeling of dependence of customers on whichever firm was selected as the zonal monopolist.

As a corollary of the existence of a zoning system, it would be necessary for government to attempt to control the level of costs and profits of the zonal monopolists. The necessary calculations would pose difficult problems of assessment and allocation of costs and revenue; for the calculations would often have to refer to a small part of the interrelated activities of large organizations engaged in a great variety of geographically dispersed and dissimilar lines of business. In practice, even in the simplest cases, official allowances for costs and profits tend to be wide of the mark. The allowances are prone to be generous; indeed, they cannot err on the side of underestimate, since the desired services would then not be provided. Under competitive conditions over-generous allowances are competed away in favour of the parties between whom the intermediaries operate. With zonal monopolies, however, any additional profits or savings in cost would not be passed on to customers but would be retained by the monopolist.

If the interests of producers are to be protected, it is necessary for the type of service provided by the zonal monopolist to be defined and its supply supervised. If this is not done, the monopolist may reduce his services to cut his costs, even though this action would put a disproportionate burden of costs on the producers. This is more likely to happen if government officials are prone to judge efficiency in terms of the level of costs. Alternatively, the monopolist may tend to improve or elaborate his services and so raise his costs and the price paid by producers for the services. This is likely to happen if there is confusion of technical or operational excellence with economic efficiency. However, if the type of service to be supplied is laid down and controlled, marketing arrangements may become inflexible and fail to be adapted to changing needs – particularly as producers with new or changing needs will not be able to express their requirements in the most effective manner, that is, by removing their custom to more amenable suppliers. The dangers of a rigid system of marketing facilities and the costs of inflexibility are likely to be very high in territories in which the area of cultivation is still extending and in which large areas have not yet been integrated (or fully integrated) with a developing market economy. The cost, in terms of frustrated development, may not be visible or measurable. Finally, detailed official control of the type of service rendered by the zonal monopolists cannot take account of the (possibly widely) divergent needs and requirements of different producers. Those who prefer more (or less) service at higher (or lower) prices are precluded from satisfying their needs or preferences. At best, producers are likely to be offered no more than a few alternatives.

In addition to these effects of the establishment of zonal monopolies on the position of producers as customers of the zonal monopolist, there are also the effects of any organized (especially statutory) restriction of numbers on the potential competitors of the monopolist or protected group of firms. The establishment of a monopoly may bear harshly on those whose entry into the monopolized activity is barred and who may have to content themselves with less preferred alternatives. In many countries, and more particularly in the so-called 'underdeveloped' world, the potential entrants into trade are frequently agriculturalists. The produce buyer or village trader is quite often the farmer who thinks it worth while to collect and market his neighbours' produce or to cater to their simple requirements. Trading intermediaries are often members or former members of the agricultural community. Thus the establishment of zonal monopolies is likely to harm some producers not only in their capacity as customers of the monopolist but also as potential entrants into trade.

It may seem that many of the difficulties enumerated in the preceding section could be avoided if producer-controlled organizations were appointed as the zonal monopolists. When a producers' agency is the zonal monopolist, it may appear as if the interests of the producers and those of the trader or processor (monopolist) are identical. However, in practice, the harmony of interests is not easily established or maintained. Those managing the monopoly are likely to regard their organization as an end in itself, with interests possibly different from, and even antagonistic to, those of the majority of producers. The divergence of interests and outlook between the administrators of large organizations and their unorganized and often uninformed constituents is a conspicuous feature of modern economic organization in large-scale public and private enterprise and labour organization. It is all the more remarkable that this dichotomy of interests is so frequently disregarded in proposals for marketing reform in immature societies, in which the detailed and democratic control by producers over the administrators of the organizations supposed to be acting on their behalf is likely to be feeble and in which a check on the formulation of policy and economy of operations is likely to be absent or weak. Where such organizations are endowed with statutory monopoly powers, a further check is removed, in that dissatisfied constituents cannot transfer their custom elsewhere. At the same time, new entrants cannot expose the possible inefficiency of the organization or the feasibility of other more attractive price and marketing policies.

Moreover, it is not correct to assume or to imply that the interests of all producers of a particular product are necessarily the same. The producer-constituents of a marketing organization are likely to have conflicting views on many matters. These include the determination of prices for different grades of the same product; the question of whether each producer should bear his own transport costs or whether these costs should be averaged; the distribution of representation on the governing body; the timing of payments to producers; and the types of marketing service to be provided. The specific policies adopted by an organization are certain to affect some of its constituents less favourably than others; those who feel that their interests are adversely affected relative to those of their fellow-producers are unable to seek better treatment elsewhere.

These considerations suggest that the objections to the establishment of zonal monopolies listed in the previous section largely apply also to producer-controlled monopolies. These organizations are open to a further objection when they control a large share of the market supply of the product. Here the monopolist-buyer is also a monopolist-seller, and there is an obvious possibility that consumers may be exploited. This danger is often overlooked because the divergence of interests between different sections of the population is not recognized. The danger that the monopoly powers will be used to the detriment of consumers is obvious where the actual producer-constituents are in effective control of their organization. But it is likely to be present even where the policy of the organization is effectively directed by its permanent administrators. The latter may wish to justify the organization to their producer-constituents by selling at high prices; moreover, of costs are inflated as the result of the operation of the monopoly, the administrator will have a direct incentive to take advantage of the selling monopoly.

Proposals for the establishment of zonal monopolies, especially of producer-controlled monopolies, are sometimes advanced to redress alleged inequalities of bargaining power. It is implied that producers, being more numerous and individually less wealthy than merchants, suffer from an inherent selling weakness which causes them to dispose of their produce at lower prices than are warranted by market conditions. It is therefore advocated that their bargaining power should be increased by collective action or, alternatively, that the right to buy should be confined to a single organization specifically charged with the duty of safeguarding producer interests or to one which is officially supervised and controlled for that purpose.

Such proposals overlook several important considerations. Where several buyers act independently, their competition will ensure that producers receive prices in line with market conditions, even though producers are far more numerous than traders and operate on a far smaller individual scale. Under West African conditions even a small number of buyers appears to be sufficient to ensure that producers receive such prices. Even if there are a few buyers, they may still have to pay prices not much out of line with commercial values (as governed by the prices they themselves secure) as long as entry of new buyers is easy; in such conditions any attempt to depress producer prices would be upset by the competing offers of new buyers. In other words, bargaining power – that is, the ability of sellers to secure the full market value for their produce – depends primarily on access to independent alternatives and not on differences in wealth between sellers and buyers. Moreover, even if there are only a few buyers and the possibility of new entry is limited, it does not follow that the position of the actual producers would be improved by the establishment of producer-controlled or other zonal monopolies. As has already been shown, the disadvantages inherent in monopoly buying are not automatically or necessarily removed by changing the organization of the monopoly or by providing for its official supervision and control.

Criticisms of so-called 'excessive competition' among middlemen are often supported by the view that such competition forces middlemen to cheat producers more scandalously. Abuses, such as the use of false weights, feature prominently in discussions of marketing reform, particularly in the underdeveloped areas; official supervision of transactions or the establishment of producer-controlled statutory marketing monopolies is recommended to protect the producer against such practices. Here both the diagnosis of the effects of abuses and the wisdom of the remedy are questioned. The discussion relates to marketing in underdeveloped areas.

Where there is competition among middlemen and where new competitors are able to enter the market without difficulty, their gross-profit margins and their earnings are closely determined by the level of rewards obtainable by them in other available occupations, such as small-scale retailing. Earnings, whether from fair trading or improper practices, cannot remain long above this level; competition among the existing traders or from new entrants soon reduces margins. Competition forces buyers to pass on the equivalent of illicit gains to the producers; for example, if debased weights are used, under competitive conditions buyers would be forced to offer correspondingly higher prices per debased unit in their search for business. From the producers' point of view the result is the same as if he had received payment for the full weight at a lower price per unit.

In general, it does not much matter whether or not producers themselves are familiar with standard weights and measures. The individual producer is concerned with sales of specific lots of produce, and he will endeavour to obtain price offers for such lots from itinerant traders and/or from traders at one or more trading centres. Even though ignorant of weights, the producer is able to judge which is the most favourable offer. The conclusion is not substantially affected in circumstances in which the producer does not receive several offers simultaneously for the same lot of produce; his knowledge of his trading opportunities is widened by his contact over time with different traders and the treatment and terms offered by each. The ability to make such comparisons and the competition of buyers ensure that producers tend to receive competitive prices for their produce, irrespective of apparent abuses. The ultimate incidence of the alleged abuses tends to be neither on the producers nor on the ultimate buyers of the produce; paradoxically, the illicit gains to middlemen from the use of false weights (for example) are more apparent than real.

On the other hand, the producer is liable to exploitation where he has no choice of buyers and is forced to deal with a sole buyer or a concerted group of buyers, whose monopoly position is protected from competition by barriers in the way of new entry. Here again the question of abuses is largely irrelevant; for, by one means or another, the monopolist will continue to pay an effective price just sufficient to elicit the supply he desires. It is of little consequence whether this effective price is expressed in terms of standard or of debased weights. In both cases the returns to the producers will be less than the competitive returns.

The discussion suggests that statutory measures to prevent specific abuses are unnecessary under competition and likely to be ineffective under monopoly. In produce buying in underdeveloped areas the situation is usually a mixture of competitive and monopolistic elements. Accordingly, the statutory measures for the elimination of abuses are partly unnecessary and partly ineffective.

Measures to eliminate specific abuses are likely to be costly and may themselves contain strong possibilities of abuse to the detriment of producers. Most of these measures postulate the establishment and maintenance of expensive control, supervisory, and inspecting staff. The administrative costs can be reduced by requiring all transactions to take place at a limited number of centres, at which the necessary supervisory services are provided. But here the costs are merely transferred to the producer or the smaller trader who has to travel longer distances to market. Moreover, the inspecting staff generally consists of large numbers of petty officials with extensive powers over the activities of traders and producers. They are therefore in a position to abuse their powers for their own profit. Such abuses, however, differ greatly from abuses in competitive markets, in that their fruits will not be competed away in favour of producers. And the backwardness of the producers tends to deprive them of the opportunity of obtaining redress from higher authority.

The foregoing discussion is not meant to deny that it is generally desirable that traders should be honest and that standard weights and measures should be used. It shows that competition among traders safeguards the interests of producers even if traders resort to sharp practices and that in some circumstances the elimination of specific abuses may be ineffective as well as disproportionately costly.

In preceding sections the case for a variety of marketing reforms has been questioned on the ground, *inter alia*, that competition among intermediaries protects the interests of producers. It is necessary to look at this general proposition more closely and to deal with a specific omission.

At any moment of time the activities, offers, and transactions of buyers in a competitive trade may be conceived of as forming part of a process of experimentation and movement leading toward the equilibrium price which is 'justified' by existing supply and demand conditions. Until this equilibrium is reached, competing traders may be offering different prices for the same goods in the same circumstances; or the differences in the prices offered for differing goods in the same or different circumstances may not be the same as the price differentials justified by the given supply and demand conditions. Some price offers may be 'too high', others 'too low'. Discrepancies will tend to be reduced as equilibrium is approached. But as the details of supply and demand change all the time, the scope for the emergence of discrepancies is continuous.

Generally, knowledge of alternatives and luck determine the particular point on the path to equilibrium at which a particular seller disposes of his produce. The producer cannot offer his goods to all buyers at the same time; he cannot know full details of the changing offers of all accessible buyers; and, having taken his produce to one buyer or place, he cannot, without cost, transfer it to some otherwise more advantageous alternative. Thus some producers receive overpayments and others underpayments, as compared with the (theoretically) justified or equilibrium price. The spontaneity of price adjustments in theoretical models and the device of recontracting are absent in workaday agricultural markets. While competition, in the long run, safeguards the average returns to producers as a whole, the returns to individual producers are affected by skill in marketing or by good or bad fortune, even if buying is competitive.

The significance of the discrepancies between competitive price offers and the (theoretical) equilibrium price is likely to be greatest where short-period fluctuations in supply and demand are frequent, where producers are widely dispersed, and where transport and communications facilities are poor. The discrepancies are therefore likely to be prevalent in agricultural markets generally; in such markets in underdeveloped areas they are likely to be particularly common.

It may be argued that the overpayments and underpayments (as defined here) of competitive markets constitute a defect and that the workings of such markets are the cause of an additional class of risk. This line of reasoning is likely to appeal to reformers with strong egalitarian leanings; such reformers may feel that all producers in the same situation should be treated alike and that rewards and penalties should not be left to chance or depend upon the possession of particular marketing skills and knowledge. It is not improbable that such sentiments may be shared by many producers as well.

The practical question is, how much weight should be given to the particular point raised in this section? This is a matter of judgement, of weighing advantages against disadvantages. In our view the indicated disadvantage of competition is not significant when compared with the general benefits of competitive trading and with the drawbacks and dangers which attach to alternative systems of controlled or regulated marketing or controlled monopoly buying. This judgement applies with special force to conditions in the underdeveloped areas. Here the benefit of competitive trading and the drawbacks of regulated marketing are likely to be marked. Thus in conditions in which the price discrepancies are greatest, their significance is, in our opinion, overshadowed by these other more important considerations.

The underpayments and overpayments arising out of the price discrepancies effect a redistribution of a total return among the individual producers. The safeguarding or raising of the total seems to be more important than the achievement of what may be thought, by some, to be a more equitable distribution of the total. Even where the price discrepancies are greatest, their effect is cancelled out if the community of agricultural producers is considered as a whole. Moreover, the absolute magnitude of the discrepancies and their effects on individual producers are never likely to be great. To the extent that the incidence of price discrepancies is distributed by chance factors, the returns to the individual producer are little affected over a period of time. To the extent that luck does not enter into the results, the superior marketing knowledge, skill, and effort of some producers give them higher returns than are enjoyed by other producers. There does not seem to be any good reason why such skill, etc., should not be rewarded; in underdeveloped areas it is one way of encouraging the development of some of the qualities which are necessary if members of the rural population are to become fitted to take an active part in economic advancement. Moreover, the superior ability and effort of the successful producers indirectly help their fellows; for they improve the effectiveness of competition among traders and tend to reduce the magnitude and duration of the discrepancies.

We now turn to a group of measures of intervention in agricultural marketing, the advocacy of which is independent of these alleged shortcomings in competitive markets. In effect, the advocates of this group of measures refuse to accept market prices either as a guide for the decisions of producers or as a basis for their remuneration.

A good example is the imposition of a system of control whereby all producers in a zone receive the same price for their produce, irrespective of the distance of the point of first sale from the consuming market(s). It involves a transfer of income from favourably to unfavourably situated producers. Uniform price systems are a common feature of agricultural price and marketing control schemes.

Uniform price systems are superficially attractive because they appear to treat producers equally and equitably. Advocates of such systems claim that it is unfair that certain producers should receive lower prices merely because they happen to be further away from the markets, and they suggest that the 'burden of transportation' should be borne equally by all producers near or far. It is also contended that uniform prices encourage production by favouring marginal producers. Moreover, uniform price systems sometimes appear to be preferable on administrative grounds; there are fewer prices for the administration to worry about. But these arguments are not convincing.

Production in distant areas requires more transport (i.e. the use of additional scarce resources) to move the crops to the ports or to the markets; it does not seem inequitable, and it certainly promotes the husbanding of scarce resources, to require that the additional costs should be borne by the producers who cause the additional resources to be used and who benefit directly from their use. The suggestion that uniform prices stimulate marginal production overlooks the distinction between the intensive and the extensive margins. Producers near the markets or centres of communication can expand their output of the crops by more intensive effort, by more costly methods of production, or by diversion of effort and resources from other productive activities. Uniform price systems reduce the returns to nearby producers (or prevent the returns from being raised) and so discourage these types of marginal production. Such systems may encourage the extension and cultivation in distant areas, but only by making expansion of cultivation less attractive in more favourable locations.

The enthusiasm for uniform price systems may be less pronounced if it is realized that they result in the use of greater amounts of scarce resources in transportation to yield a given total output of the crop and that they penalize some producers for the benefit of others. These considerations are particularly important in underdeveloped areas which are poor in capital equipment, especially in transport, and where producers, even those favourably placed near the main markets, are generally poor. If it is intended to develop outlying areas, this can be done much more efficiently by direct subsidy or grant than, covertly and wastefully, by uniform price systems.

The administrative convenience of uniform price is also more apparent than real. If the transportation of the crop to the principal markets is undertaken in the first instance by intermediaries, they naturally have to be reimbursed for transport charges incurred. The marketing authority therefore has to deal with countless claims and to institute burdensome systems of checking, filing, and repayment. These costs of administration are additional to the extra demands on transport resulting from uniform price systems. In the final analysis they amount to reductions in the incomes of producers as a whole. Alternatively, predetermined transport allowances may be given to intermediaries; and, for the reasons given earlier, these are likely to err on the side of generosity.

In several countries the government or marketing organizations with statutory powers determine that produce which falls short of stipulated minimum standards or grades may not be exported. Such measures indicate that the authorities concerned do not accept world market valuations as being relevant for production and export decisions; for the minimum exportable qualities which are prescribed are always higher than the lowest qualities that are acceptable and have a price on world markets. Such policies may have serious consequences.

The prohibition of the export of inferior, but commercially marketable, qualities of produce necessarily brings about one or more of three kinds of result which adversely affect the interests of producers and of the economy as a whole. Such measures either: (i) frustrate the export of substandard output already produced; (ii) induce the uneconomic expenditure of additional resources; or (iii) deflect production into less valuable activities. These individual effects, to which we now turn, and possible combinations of them exhaust the possible economic consequences of these measures when applied to most export products.

First, the efforts of producers may result in substandard output, that is, output which has a commercial value abroad but which falls short of the official minimum exportable quality. Whenever resources (including time and effort) yield a substandard output which is not marketed, there is economic loss, both to producers and to the economy, of a kind which could have been avoided if the export of the produce had not been banned. Second, producers may attempt to raise substandard output to reach the minimum exportable level. Whenever producers are induced to devote additional resources in an attempt to raise the quality of their output merely because lower qualities may not be exported, the cost of the additional resources must exceed the increase in the commercially realizable value of the output; that is, there is a net loss to the economy. Third, producers who are aware that substandard output may not be sold and that the improvement of substandard produce requires additional resources may be deterred from embarking on the cultivation of the product in question. Whenever producers transfer their resources to other activities because of the restrictions on exports, the value of their output in the other activities is less than the commercially realizable value obtainable from the production of the product in question.

The insistence on minimum standards and the obvious pride of some marketing authorities in the improvement of quality of export produce exemplify the confusion of technical and economic efficiency. If economic and technical efficiency were synonymous, it would be in the interests of the motorcar industry and of the British economy to restrict the output of the industry to Rolls Royces or similar high grades of motorcars. Inferior products would disappear, and the quality of output and exports would be greatly raised. Yet nobody would argue that technically inferior but cheaper cars should not be exported when there is a very large and profitable demand for them in world markets. Similarly, there is no case for refusing to allow the export of inferior grades of colonial produce for which there is a demand at a lower price, simply because they do not conform to certain prescribed technical standards.

In some countries the marketing organizations prescribe different producer prices for different grades of the same commodity; often the grade-price differentials are not the same as the price differentials which prevail in the markets in which the organizations resell their produce. For example, in the 1950s the world cocoa market did not generally distinguish between Grades I and II; very slight premiums of the order of £1 a ton occasionally emerged. The producer prices prescribed by the Nigeria Cocoa Marketing Board for the two grades have differed by £5 a ton. The Nigerian Oil Palm Produce Marketing Board distinguished between several grades of palm oil in the producer prices it prescribed. Each grade covered a stipulated range of degrees of free-fatty-acid content. There were differences in the buying prices for different grades far in excess of differences ruling in world markets or stipulated in the contracts of the board with the Ministry of Food. Within each of the grades, however, no allowance was made in producer prices for varying degrees of free-fatty-acid content, though in market contracts (as well as in the contracts between the Ministry of Food and other suppliers) the price paid made allowance for small differences in free-fatty-acid content. Thus, within each grade, producer-price differentials (which are zero) were smaller than market-price differentials.

The prescription of grade-price differentials in payments to producers which are out of line with market-price differentials either encourages uneconomic expenditures and effort or discourages economic expenditures and effort. On the one hand, where the producer-price differentials exceed the market-value differentials, producers are induced to spend additional effort and resources to raise the grade of some of their output which would not have been profitable (to the producers) if only commercial differences were paid. This means that some producers incur expenditures to raise quality which are in excess of the additional proceeds received by the marketing organizations on the sale of higher grades of produce. If the marketing organizations are considered as the guardians of the collective interests of producers, the results are patently uneconomic; if, however, the organizations are considered by others or by their executives as something apart from the producers, then this policy might nevertheless be favoured, since the extra exertions and expenditures of the producers would not fall upon the marketing organizations, whereas the higher receipts in world markets would obviously raise their sales receipts.

On the other hand, where the producer-price differentials are smaller than market-price differentials, the effect is to inhibit improvement in the quality of output. Some producers are discouraged from improving quality whenever the additional costs are greater than the resulting increase in prescribed producer price, even though they are smaller than the resulting increase in market value. The improvement which is frustrated would be in the interests of producers and of the economy as a whole and also in the interests of the marketing organizations, if these interests require the maximization of net sales proceeds.

It is conceivable that arbitrarily wide producer-price differentials may result in some loss of production, somewhat analogous to part of the loss brought about by arbitrary minimum exportable standards. Some producers may find that the price paid for the lowest grade is not sufficient to induce them to produce at all and that the extra costs of qualifying for the higher grades (and therefore better producer prices) are beyond their reach for technical or economic reasons. This may affect the distribution of effort between different crops.

Grade differentials in producer prices which are not the same as those obtaining on world markets penalize some producers and favour others. The redistribution that takes place is not likely (save accidentally) to accord with predetermined canons of equity or equality. If redistribution of income according to a given pattern is desired, it is better to achieve it directly; apart from anything else, the uneconomic expenditure of resources and effort, which arbitrary price differentials induce, will be avoided.

The policies of imposing minimum standards for export produce and of arbitrarily wide premiums and discounts for different grades of produce are sometimes defended on the grounds that the consequent improvement in the quality of export produce strengthens the position of the exporting country, particularly in unfavourable world market conditions. This view is untenable, for the buyers on world markets judge the produce they buy according to its quality in the light of market requirements. Since the policies under discussion stimulate expenditure of resources and effort in the exporting countries which exceed the additional value in world markets, the result is uneconomic.

It has sometimes been argued that, without restrictions on exports, severe competition among traders forces them to accept all produce offered to them, irrespective of quality, and that, in consequence, competition leads to a constant undermining of the quality of export produce, to the long-run detriment of producers. The argument was used to support the prescription of minimum export standards in Nigeria. The reasoning is not acceptable. If competition is severe, it forces merchants to adjust the prices they pay for different qualities closely in accordance with market valuations. They buy only if they can sell abroad; and the more intense the competition and the finer the margins on which they work, the more closely the shippers have to adjust their buying prices to market requirements. Competition does not cause a constant deterioration in quality; rather, it leads to increasing vigilance in seeking out supplies acceptable in world markets and in linking buying prices with world market values, which is to the advantage of producers.

The prescription of minimum export produce standards and of producer-price differentials for different grades requires a system of compulsory inspection of produce to check whether the minimum standard is reached or to determine the particular grade of the produce. The cost of inspection and grading services is an additional disadvantage of the policies in question. The cost is likely to be onerous in poor countries, where administrative talent is scarce. In such countries it is likely that the authorized buying or inspection posts will tend to be curtailed in order to reduce the costs or to raise the technical efficiency of the service. This expedient may be inconvenient and costly to some producers, besides tending to make trading facilities less adaptable to changing and growing needs.

In underdeveloped areas, where many producers and small-scale traders are illiterate and unaware of their legal rights, a system of compulsory inspection and grading may place them at the mercy of the inspecting officials. Thus in many parts of Nigeria petty tyranny and corruption in the operation of produce inspection have been widespread and oppressive. On extended visits to Nigeria in 1949 and 1950 it was found (by one of the writers) that in many places there was a recognized scale of (illicit) fees payable to produce examiners to have the produce inspected and passed. Those who refused to pay were made to wait for such long periods that their produce deteriorated and fell below standard; or they were forced to sell their produce uninspected to other intermediaries at less than its commercial value; or they were liable to have their produce down-graded by the examiners. Wherever possible, the traders or clerks tried to placate the examiners by paying the recognized tariff; and, indeed, they were likely to complain only when the exaction exceeded the conventional level which was regarded as reasonable. It was universally recognized that complaints to officials would be useless, since formal proof would be difficult to secure. Moreover, a successful denunciation of one examiner followed by his conviction would be noted by his successor, possibly to the serious disadvantage of the protestant. It seems that only very compelling reasons would justify the conferment of such drastic powers under conditions in which they are so widely abused. The preceding analysis has failed to reveal any such reason.

The foregoing discussion does not imply that the grading of produce may not be a useful marketing device in appropriate circumstances. Accurate and acceptable grading facilitates commercial communication and reduces marketing costs. It is for this reason that systems of grading (and consequential arbitration of contract) have been developed in world commodity markets voluntarily by those in the markets. However, such systems of grading are voluntary; transactions in ungraded produce or in produce below the standard of the lowest recognized grades take place if it suits the parties concerned. From the point of view of producers and produce exporters, the market recognition of grade specifications may be very convenient, and they may voluntarily submit their produce for inspection and grading.

In export trades in which many shippers are small firms which are not well known or represented in world markets, difficulties arise when supplies on arrival are found to be below contract specification in quantity or in quality. As consignments are normally shipped under a letter of credit, the importer has difficulty in obtaining redress because the exporter has by then received most of the contract price. Moreover, as the exporter is not represented abroad, disputes cannot be referred to arbitration. In these circumstances the additional risks of the importer may seriously retard the development of the export trade. In such cases, however, it is in the interests of both importers and exporters to devise methods to provide safeguards to traders. They are thus likely to support firms of cargo supervisors who can inspect consignments before they are accepted. Where the establishment of such agencies is difficult or delayed, a voluntary inspection scheme by the government may provide the necessary service. Both the problem and the two types of solution are illustrated by developments in the Nigerian export trade in logs.

For many years African traders exported timber in logs. In the 1940s this was the only branch of the export trade in which African shippers were prominent. With the influx of many new shippers after the war, there was a considerable increase in consignments of logs which, on arrival, were seriously below specified weight or quality. Importers were increasingly reluctant to place contracts with the smaller exporters. Eventually, one of the smaller merchant firms undertook to act as cargo superintendent. The services it provided were not regarded as sufficiently extensive, and in 1949 the Nigerian authorities devised a simple and practical scheme which proved an immediate success. It provided for voluntary inspection; the exporter on application could have his consignment inspected during the loading of the timber onto the ship; an official certificate of quality and quantity was issued, which was presented by the exporter to the bank with which the letter of credit was arranged. The actual amount paid out to the shipper under the credit depended on the contents of the certificate, a procedure which effectively safeguarded the overseas importer against loss. The scheme was widely publicized in the principal overseas markets for Nigerian timber, and importers generally insisted on this inspection certificate before arranging a letter of credit.

The scheme was voluntary and served to protect the interests of the trade without investing the inspectors with compulsory powers and without introducing arbitrary grading or quality standards. Yet it fulfilled all the essential requirements, since, unless the overseas importer neglects to take advantage of this simple facility, he is effectively safeguarded against loss and need have no reluctance in placing an order with a shipper unknown to him.

The results and implications of the measures discussed in this paper are so far-reaching and in important respects so contrary to their avowed objectives that the question naturally arises as to how they come to be proposed and adopted in practice.

The centuries-old belief in the unproductive nature of trading activities and of traders is still widespread. Even those who do not subscribe to crude notions about the alleged unproductivity of trade are frequently not able clearly to analyse the nature of the services performed by traders. Hence legislators, administrators, and others may start off from the idea that marketing arrangements offer a fruitful yield for reform. Moreover, administrators and investigators usually are familiar with marketing arrangements in advanced economies; when they come to examine marketing arrangements in underdeveloped areas, they may fail to allow for the differences in conditions and in relative factor prices and so make misleading comparisons. They may fail to see the economic rationale of the existing arrangements.

This particular influence may be reinforced by another factor. It may be that when various commissions or committees investigate the marketing problems (or, for that matter, other economic problems) in underdeveloped territories, they feel that their mission is not justified unless they submit recommendations for extensive changes in existing arrangements. They feel that they are expected to recommend policy rather than to describe and analyse an existing situation. And they see, perhaps subconsciously, a justification for their appointment in a list of recommendations.

Some of the measures examined in this paper stem from a confusion of technical efficiency and economic efficiency. This is clearly present in the prescription of arbitrary price differentials for different grades of produce or the banning of the export of produce below prescribed technical standards. It may also be a factor in the advocacy of limitations on the number of traders or processors so that each may have a larger 'plant'. The technician may be aware of the greater technical efficiency of larger units but less conscious of the additional costs and inconvenience which consumers and/or producers have to bear if there are fewer trading or processing centres. The technician is also often unaware of the varied and important advantages to the population flowing from the presence of independent alternatives and of the stimulating effects on economic growth of the activities of traders. Moreover, the rationale of the price mechanism and the economic functions of prices are not always clearly understood.

2

MARKET CHANNEL COORDINATION AND ECONOMIC DEVELOPMENT

C.J. Slater

The political art of development consists of balancing capacity-expanding, income-concentrating activities with demand-expanding, income-redistributing or 'unconcentrating' activities. The Keynesians in the crowd have undoubtedly moved ahead in the argument and have begun to question the efficiency with which savings get translated into income changes through the financial markets in less-developed communities. It is precisely because the financial markets in less-developed communities are often inefficient in translating private savings into effective income-expanding investment that improvement in the efficiencies with which marketing channels operate may be crucial in improving the reinvestment process.

Improvement in the vertical coordination of marketing channels can induce the entrepreneurs operating in these distribution channels to become more efficient by increasing their output through utilizing more efficiently their present capacity with relatively small increments in capital. The argument also suggests that horizontal improvement in the coordination and efficiency with which market channels operate can be important in the demand-expanding side of the equation of development. To the extent that market channel coordination fosters fuller utilization of the productive capacity of the community, it contributes to increased output.

MARKET PROCESS COORDINATION

If one postulates certain behavioural assumptions regarding consumers' propensities to consume, save, and shop as well as producers' attitudes

---

Source: Condensed from the chapter in L.P. Bucklin (ed.) *Vertical Marketing Systems.* Scott, Foreman & Co., Glenview, Illinois, 1970.

toward risk and uncertainty in relation to their output levels, a theory can be suggested:

1. Income redistribution as a goal of development can be achieved by lowering the relative prices of necessities such as food, since these necessities absorb a very large share of low-income family purchases.

2. Prices of necessities in large, rapidly growing urban centres can be lowered by inducing two changes in retail marketing:

(a) increases in scale of markets from individuals each selling a single line of products to a supermarket selling over 1000 items;

(b) variations in pricing so that selected, high-volume necessities are sold at lower margins while less frequently purchased items usually purchased by higher-income families are sold at higher margins.

3. In order to secure the larger quantities of the commodities sold at lower prices, coordination vertically between retailer and suppliers becomes necessary. Reduction in market risk associated with contract purchase by retailers, and increases in scale of assembly, transport, and wholesale functions can contribute to efficiency and, thus, to lowered costs.

4. Selected producers, given the greater assurances of marketing arrangements through the discount supermarket, will in some cases elect to expand their production by adopting changes in technology.

5. Given the atomistic market structure at retail and wholesale levels for foods in the large, rapidly growing cities of some low-income areas, the reduction in retail prices for selected commodities consumed by low-income families will induce limited price reductions by small atomistic retailers. Because the margins of these small retailers are so small, they must achieve price reductions by improving their buying capability. Retailer cooperative-buying programmes can achieve some of the risk reduction and channel simplification that the large retailer can enjoy.

Thus, the twin goals of development – increased level of real income and redistribution of income – can be fostered by marketing reforms. Income will increase due to producers responding to the more certain market opportunities of contract production and income redistribution through consumer price reductions for necessities.

Inducing the process requires a systems approach, for there are many public facilitative actions needed as well as supervised credit and technical assistance. It is appropriate to examine the evidence from large-scale studies and the results of a systems analysis to identify and quantify the changes induced by market channel coordination.

Before examining the community studies it is useful to point out that a critically different assumption is used in these arguments from that found in traditional economic analysis of these problems.

Postulated behaviour

The average propensity to consume food in the less-developed community is as high as 0.7 for lower-quartile income consumers and is often below 0.3 for the upper quartile. It is further postulated that the upper quartile has a propensity to save 0.05 or greater. Given these propensities, reductions in food prices will redistribute real income, thereby expanding the effective demand. Similarly and unfortunately, a reduction in the proportion of income that can be available for savings results from reductions in the price of necessities.

It can be further postulated that producers take a low-risk, low-output option, given the risks and uncertainties of the market and technologies of production available to them. When a reduction in the risks of markets occurs such that the quantity demanded for products that are practical to produce can be assured at dependable prices, producers will tend to move toward technologies that incur somewhat greater risks due to fuller use of available resources including producer credit to marginally expand the presently available resources employed. Critical to this notion is to induce producers to expand output by low-cost, fuller utilization strategies rather than to induce the creation of new institutions of production to expand capacity.

It is further postulated that capacity may not be fully utilized uniformly by producers and distributors, and small changes in capital often can use more fully the available capacity of production in the less-developed community. Labour is similarly in surplus so that wages are usually not driven up by output changes. Instead, usually more people are employed for a greater amount of time; thus, increases in real wages for the community occur even though the wage rate may not be sharply affected by changes in output.

These postulated conditions are at some variance with the traditional stagnation thesis offered in Schulz (1964). The stagnation equilibrium implies that capacities are presently utilized so that increases in demand would increase the prices of the presently scarce resources. Even though many areas are tilled, the poor and limited uses of non-labour inputs imply a low rate of use. Given the low rate of capacity utilization characteristic of many less-developed communities and the high rate of unemployment, the postulated conditions of the stagnation equilibrium seem inappropriate.

The market integration thesis

With the above postulated conditions characterizing the less-developed community, it is suggested that horizontal market coordination is a critical first step in the market integration process. Horizontal coordination is defined as a consolidation of distribution from the atomism that is the hallmark of merchants in less-developed communities the world over. Horizontal coordination implies larger-scale retail outlets selling several products – including necessities to low-income people – and following 'modern' practices of charging low margins for high-turnover staples and higher margins of low-turnover luxury goods. The scale economies due to increases in the sales per worker cut costs of operation of such horizontally integrated retailing institutions somewhat, but the key reductions in costs result from the opportunities created to vertically coordinate the marketing system and to cut the cost of acquisition, particularly for low-priced consumption goods that are the 'leader' items.

The vertical coordination of the marketing system reduces the intermediaries' risks associated with transactions. The capital available to the intermediary is expanded because of the dependability of the order placed by the horizontally integrated larger-scale retailer. The vertical coordination of the higher levels of the marketing system with horizontally integrated retail operations also creates a set of producer expectations that can result in fuller utilization of presently available productive capacity. The increments in investment that are necessary to utilize more fully the productive–distributive system can often be paid for by reduced transaction costs of the better capitalized system.

When markets have these conditions of horizontal and vertical coordination, it is often more attractive to introduce additional capital because of the more-dependable operating institutions available to use the capital.

A necessary but not sufficient condition to induce the process of horizontal and vertical coordination is public facilitative changes to improve market environment, i.e.

1. Better producer market information through government communication services concerning price, product, and terms of sale.
2. Improved transportation facilities including penetration roads to reach more producing areas rather than reduction in the cost of importation of goods and services by installing trunk roads or import harbour facilities.
3. Improvement of operational, locally understood product specifications and grading.
4. Extension services to provide for application of improved technologies.
5. Improved credit facilities and technical assistance through supervised credit.

These public facilitative changes are themselves not a sufficient condition to induce change unless market coordination is a practical option for a selected few progressive operators in the system. The thesis outlined above suggests a sequence of priorities for internal market process improvement as an aspect of the overall task of development.

1. *Describe the marketing system*, (i) in order to know the average propensities to save, the shopping mobility and potential shopping mobility of all income groups; (ii) to appreciate more fully the expansion opportunities and problems for horizontal market coordinating institutions; (iii) to learn the special problems of vertical market coordination; (iv) to learn the needs for facilitative changes that may best be carried forward by government action.

2. *Induce or encourage horizontal market coordination in the private sector* by encouraging innovative operators with modern ideas and basic faith in the expansibility of income to create or expand low-price, multi-product outlets.

3. *Vertical market coordination can be fostered at the retail level initially*, once horizontal market coordination is underway and discount supermarkets are selling necessities to low-income people. It is important to note that suboptimal maximization is often associated with producer vertical market coordination. Dairy marketing in the United States has been a prime example of producers increasing their income at the expense of the ultimate consumer. Frequently, however, forward vertical market coordination is associated with export commodity promotion schemes and marketing boards useful to foreign-exchange expansion. The objective is vastly different where a nation seeks to increase producer income such as coffee and sugar marketing for export.

4. *A selective development of facilitative reforms can foster the improvement of horizontal and vertical coordination.* Communication, transportation, commodity exchange, and public storage regulation are examples. The facilitative reforms are, however, no substitute for market reforms internal to the system.

5. A concomitant reform needed when market integration is fostered is to *redeploy distribution labour as improvements in efficiency cut distribution labour costs.* A careful appraisal is needed of the benefits through increases in real income and redistribution of income to lower-income groups versus the offsetting costs associated with displaced distribution labour. A general systems simulation model appears to be an efficient mechanism for assessing the benefits and costs associated with an ongoing programme of internal market reforms. Later, the results of this systems analysis will be reviewed briefly.

MARKET PROCESSES IN DEVELOPING COMMUNITIES

With this thesis or theory of internal market process reform in mind, let us turn to some of the limited evidence generated from three large-scale studies of the marketing systems of San Juan, Puerto Rico; Recife, Brazil; and La Paz, Bolivia.

These studies were sponsored principally by USAID with local host government and Ford Foundation support and were conducted by the Latin American Market Planning Center, Michigan State University, from 1965 through 1968. The purpose was first to evaluate the role of food marketing in economic development and second to diagnose and analyse the marketing changes appropriate to fostering improvements in the internal market.

There was strong evidence, contrary to traditional wisdom, that horizontally integrated retailers charge lower prices for necessities than do the small, single-product or narrow-line retailers. Supermarkets charged lower prices on five out of nine items. In one very large discount centre, rice, an important staple, was 12.8% lower in price. This was found to be due to vertical coordination of acquisition procedures.

Some reduction in rice marketing costs was being realized through a new vertically coordinated arrangement which linked a Recife supermarket (Bom Preco) with a large rice mill in the São Francisco area. Through an aggressive retail pricing policy plus a lower cost procurement system this system was underselling the more traditional rice marketing system by 10–15% at the retail store level.

Reduction in spoilage and waste due to the reduction in the number of transactions appeared to be an important gain from vertical market coordination. Staple spoilage (rice, beans, and manoic) was about 4.6% of the value of sales on a weighted basis for atomistic retailers and 0.5% for self-service retailers.

The efficiency with which marketing systems function is reflected in the interest rates charged for supplying the needed capital for operations. The annual cost of using credit experienced by wholesalers of perishables (bananas and tomatoes) in Recife was 290–562% and the cost of credit to wholesalers of dry staple products was 24–59%. These are weighted and annualized figures that reflect the almost loan-shark practices used to finance the necessities in the atomistic markets of Recife. To the extent vertical coordination reduces risk, these interest penalties can be reduced as well.

Summary of operating evidence

The three communities study provides some partial evidence to support the thesis of market process coordination. The postulated behavioural conditions of consumer consumption propensity and shopping propensity as well as limited producer response were to some extent supported. The horizontal coordination of retail food markets seems to achieve price reductions. Vertical market coordination thrives under appropriate facilitative conditions, and lack of appropriate price expectations limits supply response.

The thesis may be generally conceded, but unless the benefits outweigh the costs, other more traditional development strategies will continue to win the scarce dollars of support. Thus, we need to assess the benefits and costs by simulation.

GENERAL SYSTEMS SIMULATION

If the full benefits of internal market coordination are to be achieved, a marketing development programme must take its place alongside the other major development priorities in areas such as the north-east of Brazil. While much of the funding involves local currencies, the changes in business regulations and the employment-reducing effects of distribution labour displacements are serious political issues, as is the fundamental position of income redistributing effects of consumer price reductions.

A marketing development programme must then compete with such programmes as industrial expansion, agriculture productivity improvement, public health, education, and infra-structure buildup. Pay-offs must be more carefully assessed, since marketing is a new contender for scarce development energies.

The costs of inducing reforms and the impact of these reforms upon society need to be assessed. It is useful to summarize the reforms and impacts in order to appreciate better the benefits and costs of reforms (Table 2.1).

The five changes and the five impacts are, of course, operating in a dynamic setting, and political reaction from the vested interests of present institutions is likely to affect the quality of actions. Improvement in internal market coordination will likely have the following sequence of effects:

1. A reduction in retail prices of necessities.
2. A reduction in wage costs, interest costs, and spoilage or shrink in supply for the large discount retailers.
3. An initial increase in supply due to spoilage reduction, later an increase in supply due to vertical coordination of markets.
4. Because the average propensity to consume food, while high, is less than one, the reduction in food prices will expand the demand for non-food consumption goods.
5. The increases in production cause wages and non-wage income to expand. Given the overall condition of unemployment and underemployment characteristics of the less-developed communities, the effect will no doubt be employment expanding rather than wage-rate increasing in net effect.
6. Offsetting all these gains are the distribution worker displacements as more food sales go through the large outlets with high sales per worker.

Table 2.1. Summary of reforms and impacts

Reforms and institutional charges	Probable impact on the community
1. The creation of a large-scale discount supermarket featuring necessities at low prices	1. Reduction in volume of sale through small, atomistic, single commodity retailers
2. Vertical coordination by large-scale retailers seeking dependable suppliers often by direct purchasing	2. Reduction in revenue for small wholesalers whose function it has been to supply the small-scale retailers
3. A technical assistance and retailer cooperative organization to assist the small retailers reacting to the competitive pressure of the large outlets	3. A further displacement of middlemen and wholesalers as the small retailers rationalize their supply channels
4. A farm input, technical improvement and promotion programme directed at the suppliers responding to the large retailers' vertical coordination activities	4. A displacement of farm assemblers and farm-input suppliers as the large retailer programmes reach back to affect, selectively, the responsive farmers
5. Improvement in the infra-structure for farm market-production feeder roads, market news programme, extension service expansion, rural credit expansion	5. A displacement of older institutions and sources fo rural revenues such as local turnpike taxes on produce moving to the city

These six forces need to be assessed in balance. Is the income gain greater than the distribution wage loss? A balanced internal market development programme can treat several points in the system at the same time and stimulate output as demand is fostered.

SUMMARY AND CONCLUSIONS

The essential conclusions from the limited data available suggest that the market process integration thesis merits attention from development planners. Market coordination fosters development when the conditions of government-directed facilitative actions are appropriate.

The twin goals of development – increasing real income and redistributing income to foster a broader base of effective demand – are seemingly both well served by marketing channel coordination. The approach sketched here is one that depends upon a systems approach to the problem. It does not fit into the sector development approach now widely popular – ruralists fostering agricultural programmes and urban and industrial specialists following an equally independent course. Marketing channel coordination cuts across this sectoral approach and thus finds political support difficult to obtain from the traditional parties interested in development investment.

Two researchable areas of potential contribution to economic growth through market process integration may merit consideration: (i) the impact on investment of the income-redistributing effects of market process reforms; and (ii) the potential impact of market integration on farm income and, thus, upon the urban migration problems of developing communities.

The primary driver of the two general systems models thus far developed has been the effects of price changes upon demand and, hence, supply. This implies a supply response to demand changes, when the demand changes have originated from shifts from savings to consumption. In the short run this seems appropriate, but evidence needs to be collected concerning the capital markets of developing communities. Theorizing from the classical competitive model is no more reliable or satisfying here than in the case of consumption propensities.

The second problem concerns the migration from rural areas to urban communities. Efforts to improve the lot of urban people are confounded by the universally high rate of urban migration in developing societies. Productivity changes in agriculture have often meant reduction in the man-to-land ratio, thus pouring migrants into the cities. Special benefits thus accrue when agriculture-output increases occur without forcing labour off the farm. Yet, much effort to improve agriculture fosters technical, intensive-output improvements.

REFERENCE

Schulz, T.W. (1964) *Transforming Traditional Agriculture.* Yale University Press, New Haven.

3

THERE IS METHOD IN MY MADNESS: OR IS IT VICE VERSA? MEASURING AGRICULTURAL MARKET PERFORMANCE

B. Harriss

The mid and late 1960s were an era of pioneering research in the economics of agricultural marketing in underdeveloped countries. In Africa teams from several American and African universities studied the systems of Sierra Leone (Illinois and Njala), Nigeria (Michigan State, Stanford, Stanford Research Institute, Ife, and Nigeria), Kenya (West Virginia and Nairobi), and Ethiopia (Cornell and Stanford Research Institute) resulting in, for example, studies by E. Gilbert (1969), C. Ilori (1968), A. Thodey (1968), and A. Whitney (1968). See also Q.B.O. Anthonio (1968).

The application of a practical methodology for analysing market performance took place simultaneously in India (see the studies by R. Cummings, Jr., 1967; U. Lele, 1967, 1971; A.S. Holmes, 1969; Z.V. Jasdanwalla, 1966; and R.C. Gupta, 1973) and Bangladesh (see Muhammad O. Farruk, 1970). Essentially the same methodology is being used a decade later as identified by my studies of rice marketing in Southern India, H.M. Hays, Jr.'s studies of cereals marketing (1975, 1976, 1978), N.O.O. Ejiga's study of cowpea marketing in Northern Nigeria (1977 from Cornell), and the studies for Niger and Upper Volta of the CILSS/Club du Sahel (1977) 'étude diagnostique' from the University of Michigan (see D. Kohlers, 1977; Elliott Berg, 1977).

STRUCTURE, CONDUCT, PERFORMANCE METHODOLOGY

The methodology used is an adaptation of 'structure, conduct, performance' analysis. This is an attempt to compromise between formal structures of economic theory and empirical observations of organizational experi-

---

Source: Condensed from *Food Research Institute Studies*, Stanford 17(2), 1979. Reproduced with permission.

ence in imperfect markets. It is a standard tool for market analysis in the United States and the United Kingdom (Bain, 1959; Bateman, 1976).

Market structure consists of 'characteristics of the organization of a market which seem to influence strategically the nature of competition and pricing within the market' (Bain, 1959, p. 7). In particular, these are the degree of seller and buyer concentration, entry conditions, and the extent of agent and product differentiation. R.L. Clodius and W.F. Mueller add the distribution of market information and its adequacy in sharpening price and quality comparisons and in reducing risk (1967, pp. 345–50).

Market conduct is the 'pattern of behaviour which enterprises follow in adapting or adjusting to the markets in which they sell (or buy)' (Bain, 1959, p. 9), in particular methods employed to determine price, sales promotion, and coordination policies and the extent of predatory or exclusionary tactics directed against established rivals or potential entrants.

Market performance represents the economic results of structure and conduct (Bain, 1959, pp. 10–12), in particular the relationship between distributive margins and the costs of production of marketing services. In particular, time series price data are used to throw light on the degree of competition in marketing systems:

1. through intermarket price correlation to indicate the degree of market integration;
2. through the relationships between transport costs and intermarket price differences (via graphical plots, regression analysis, and the analysis of average margins) to indicate the competitiveness of interrelational trade; and
3. through the relationships between seasonal price fluctuations and storage costs to indicate market competitiveness through time, as well as the calculations of annual and longer term moving averages to investigate longer period cyclical changes in the price level.

About this methodology, which has achieved the status of orthodoxy, W.O. Jones, whose responsibility it was to organize and coordinate the pioneering African research and to synthesize the results, asserts: 'Primary emphasis in evaluating efficiency was placed on the determinants of price'. The investigations were formulated in terms of commodities. 'The desirability of pursuing a commodity approach became increasingly clear as price analysis and the field studies progressed' (Jones, 1968, p. 96), and 'In some ways the measurement of market performance as manifested by the behavior of prices was more satisfactory than that based on identifying imperfections' (Jones, 1974, p. 17). In 1974 in a review of the studies of the 1960s, Jones writes, 'I have never published a formal critique or evaluation of the way in which those studies were conceived and executed. The seriously interested student could extract and reconstruct all of this from the final report, although there I was not primarily interested in reviewing defects and deficiencies in our concepts or performance' (1974, p. 3).

[Harriss then first looks at the methodology for examining market performance and the quality of the data used. She then examines the relationship between the data and the conclusions drawn from it. She questions the use made of the price correlation coefficient as an index of market competitiveness and integration.]

CONCLUSIONS

Five general points are worth making at this juncture:

1. The authors of the majority of economic analyses of agricultural marketing for West Africa and South Asia using structure, conduct, performance methodology (or some personalized variation of it) examined here display a serious lack of logical relationship between the data presented and the conclusions derived. In no other branch of economics does it seem possible to elevate so many value judgments to the status of scientific conclusions. There is a serious lack even of simple comparison of the results. To date and to my knowledge there has been no comparative critique.

2. The conclusions to be drawn from the research are confusing, and attempts to synthesize by unifying consensus seem to be guilty of over-simplification. For one example the reader can refer back to the conclusions of Jones for tropical Africa, mentioned at the beginning of this paper. For another, Berg's attempt to summarize studies of the Sahelian states results in 'there is little empirical evidence for monopsonistic grain markets' (CILSS/Club du Sahel, 1977, Vol. 1, p. 11), and he concludes that Hays' and Gilbert's research 'indicate reasonably competitive rural grain markets, storage behavior in line with what one would expect from a prudent farmer and no severe rural indebtedness' (a subject actually under-researched by both writers) (CILSS, 1977, Vol. 2, p. 25). More research along structure, conduct, performance lines will only resolve this confusion if it is consistent. The past record suggests the probability of this happening as unlikely indeed.

3. The polar assumption that commodity markets are either perfectly (or effectively) competitive or monopolistic (an assumption reflected in any content analysis of vocabulary) allied to the equilibrium assumption that markets can be judged to be in a state of relative competitiveness for all plannable time, is clearly false. Jones writes skeptically, 'the concept of the conditions for a perfectly competitive market is useful in determining how a market is inefficient, but it is not very helpful in determining how inefficient a market is' (1974, p. 16). One would like to add from experience, 'where, when, and why'. Indeed, evidence is being interpreted on the assumption that the theory is right: evidence must be squared with theory or explained away.

Transactions even using the simplest classification of formalist economic theory may take one of nine forms as shown by Wiles (1961) and then by Gross (1966, p. 63):

	Sellers		
Buyers	One	Few	Many
One	o,o	o,f	o,m
Few	f,o	f,f	f,m
Many	m,o	m,f	m,m

Agricultural marketing economics has been obsessed with the issue of whether markets in both an economic sense and in geographical aggregates of various sizes belong to the top left or bottom right corners of this diagram. Our unwillingness to locate rigorously an analysis of market behaviour anywhere than in the boxing ring corners has been nothing short of cowardly. Also, given any one of these nine configurations of buyers and sellers and any distribution of concentration of trades and businesses, any distribution of profitability following from it, the use of aggregate (annual) average returns on money invested in trade tell us very little about the form of the process of resource extraction from agriculture and of its accumulation in trade. Further, the discipline has also very largely assumed the geographical linearity of trade. If the unidirectional assumption about trade is relaxed it follows that evidence on competition based on rates of return to trade, based in turn on the relationships between annual average price differences between markets and average transport costs, is likely grossly to underestimate the profits made by two-way, or multidirectional, trade, profits which result from the differences which will be cancelled out in annual or in seasonal averages.

4. A competitive market may be necessary, but it is clearly not sufficient for the maximization of productivity. To concentrate attention on the concepts of competition diverts attention from the structural interrelations between production, exchange, and consumption. To concentrate attention on the behaviour of the commodity market (because of the relative ease of access of data on price, however poor) diverts attention from the interrelations between several commodity markets and between the circulation of commodities and that of money. These are essential to an understanding of the role that agricultural markets play in economic development including technological change in agricultural production.

5. The fetishism of competition, however, is not entirely devoid of purpose, but its raison d'être is ideological: related to a laissez-faire aversion to the type of state intervention which replaces rather than regulates private commodity markets. The tenor of the policy recommendations following the conclusions in this school of theses is strongly anti-interventionist and pro-infrastructural. The recommendations stress the value of state intervention in sectors of the economy such as transport and communications, physical market sites, and such aspects of marketing as information, grading, standardization, processing, and packaging (see Thodey, 1968, pp. 64-76; Gilbert, 1969, pp. 276-77; Ejiga, 1977, p. 26; Hays, 1975; Jones, 1968, p. 98; Olatanbosun, 1975, pp. 111-20; and Helleiner, 1974, p. 69). More controversial reformist proposals include the creation of storage facilities and subsidized credit to larger traders or entrepreneurs for them to expand operations (Whitney, 1968; Anthonio, 1968).

These conclusions can only follow logically from a verdict on the operation of the commodity markets as essentially competitive. Even so, they fail to face questions of the means and the nature of policy implementation and of the form of ownership of the proposed infrastructure. Jones writes, 'it probably could be demonstrated that we have got into trouble when we overlooked some of the assumptions underlying the models we were using. But in many instances more precise examination of the extent to which basic assumptions were satisfied would not have helped because, theory frequently does not predict the consequences of lifting assumptions' (1974, p. 23). When the assumption about market competitiveness is lifted, it does not follow that the infrastructural improvements will diminish the antisocial behaviour that is elsewhere denied to exist.

For the present, therefore, a question mark must be placed not simply beside the methodology of conventional agricultural marketing economics in the structure, conduct, performance tradition, but also beside the history of the interpretation of the results.

REFERENCES

Anthonio, Q.B.O. (1968) The marketing of staple foodstuffs in Nigeria: a study in pricing efficiency. Ph.D. dissertation, University of London, London.

Bain, J.S. (1959) *Industrial Organization*. Wiley, New York.

Balasubramaniam, A. (1978) Bulletin of market intelligence, Coimbatore District. *Annual Bulletin of Market Intelligence*, Department of Statistics, Madras.

Bateman, D.I. (1976) Agricultural marketing: a review of the literature of marketing theory and of selected applications. *Journal of Agricultural Economics* **27**(2), 171-226.

Berg, E. (1977) Upper Volta. In: CILSS/Club du Sahel, *Marketing, Food Policy and Storage of Food Grains in the Sahel: A Survey. Vol. 2: Country Studies.*

Center for Research on Economic Development, University of Michigan/U.S. Agency for International Development, Ann Arbor, Michigan.

CILSS/Club du Sahel (1977) *Marketing, Food Policy and Storage of Food Grains in the Sahel: A Survey, Vol. 1: Synthesis with Statistical Compilation and Annotated Bibliography* and *Vol. 2: Country Studies.* Center for Research on Economic Development, University of Michigan/U.S. Agency for International Development, Ann Arbor, Michigan.

Clodius, R.L. and Mueller, W.F. (1967) Market structure analysis as an orientation for research in agricultural economies. *Journal of Farm Economics* 43(3), 515–533.

Cummings, R.W., Jr. (1967) *Pricing Efficiency in the Indian Wheat Market.* Impex, New Delhi, India.

Ejiga, N.O.O. (1977) Economic analysis of storage, distribution and consumption of cowpeas in Northern Nigeria. Ph.D. dissertation, Cornell University, Ithaca, New York.

Farruk, M.O. (1970) *The Structure and Performance of the Rice Marketing System in East Pakistan.* Department of Agricultural Economics Occasional Paper No. 31, Cornell University, Ithaca, New York.

Gilbert, E.H. (1969) Marketing of staple foods in Northern Nigeria: A study of staple food marketing systems serving Kano city. Ph.D. dissertation, Stanford University, Stanford, California.

Gross, B.M. (1966) *The State of the Nation.* Barnes and Noble. New York.

Gupta, R.C. (1973) *Agricultural Prices in a Backward Economy.* National, Delhi, India.

Hays, H.M., Jr. (1975) *The Marketing and Storage of Food Grains in Northern Nigeria.* Samaru Miscellaneous Papers No. 50, Institute of Agricultural Research, Samaru, Zaria.

Hays, H.M., Jr. (1976) Agricultural Marketing in Northern Nigeria. *Savanna* 5(2): 139–148.

Hays, H.M., Jr. and McCoy, J.H. (1978) Food grain marketing in Northern Nigeria: spatial and temporal performance. *Journal of Development Studies* 14(2), 182–192.

Helleiner, G.K. (1974) The marketing board system and alternative arrangements for commodity marketing in Nigeria. In: Onitiri, H.M.A. and Olatunbosun, D. (eds), *The Marketing Board System.* NISER, Ibadan, Nigeria.

Holmes, A.S. (1969) Marketing structure and conduct and foodgrains pricing efficiency in a North Indian Tahsil. PhD dissertation, University of Maryland.

Ilori, C. (1968) Economic study of production and distribution of staple foodcrops in Western Nigeria. PhD dissertation, Stanford University, Stanford, California.

Jasdanwalla, Z.Y. (1966) *Marketing Efficiency in Indian Agriculture.* Allied Publishers Private Ltd., Bombay, India.

Jones, W.O. (1968) The structure of staple food marketing in Nigeria as revealed by price analysis. *Food Research Institute Studies* 8(2), 95–123.

Jones, W.O. (1974) Regional analysis and agricultural marketing research in Tropical Africa: concepts and experience. *Food Research Institute Studies* 13(1), 3–28.

Kohlers, D. (1977) Niger. In: CILSS/Club du Sahel, *Marketing, Food Policy and Storage of Food Grains in the Sahel: A Survey. Vol. 2: Country Studies.* Center for Research on Economic Development, University of Michigan/U.S. Agency for International Development, Ann Arbor, Michigan.

Lele, U.J. (1967) Market integration: a study of sorghum prices in Western India. *Journal of Farm Economics* **49**, 47–59.

Lele, U.J. (1971) *Food Grain Marketing in India. Private Performance and Public Policy.* Cornell University Press, Ithaca, New York.

Olatunbosun, D. (1975) *Nigeria's Neglected Rural Majority.* Nigerian Institute of Social and Economic Research and Oxford University Press, Ibadan, Nigeria.

Thodey, A.R. (1968) *Marketing of Staple Foods in Western Nigeria. Vol. I: Summary and Conclusions.* Stanford Research Institute, Menlo Park, California.

Whitney, A. (1968) Marketing of Staple Foods in Eastern Nigeria. Report No. 114, Department of Agricultural Economics, Michigan State University, East Lansing, Michigan.

Wiles, P. (1961) *Price Cost and Output.* Basil Blackwell, Oxford.

4

MARKETS AND STATES IN TROPICAL AFRICA: THE POLITICAL BASES OF AGRICULTURAL POLICIES

R.H. Bates

The states of Africa differ in important ways in their policy choices, but behind the differences lie basic similarities in their emergent political economies.

Fledgling industries locate in the urban areas. Workers and owners, while struggling with each other for their share of industrial profits, possess a common interest in perpetuating policies that increase these profits. They therefore demand policies that shelter and protect these industries. They also demand policies that promise low-cost food.

Because of the public purposes they espouse, African states seek to advance the interests of industry. To secure revenues to promote industry, they therefore seek taxes from agriculture. By maintaining a sheltered industrial order, they generate economic benefits for elites, as well as resources for winning the political backing of influential groups in the urban centres. To safeguard their urban-industrial base, they seek low-cost food. This aim therefore leads them to intervene in markets and to attempt to depress the level of farm prices.

Governments recruit partners in the countryside. Their rural confederates include tenants and managers on state production schemes. They also include elite-level farmers, as well as the more widespread group of progressive farmers who have become dependent on state-sponsored programmes of subsidized inputs. These are the rural allies of African regimes – groups that find it privately advantageous to support the governments in power even though they impose disadvantageous agricultural prices.

The bureaucracy is another key element in this emergent social order. It spans the markets which governments manipulate. In accordance with public policy, it sets prices within these markets and thereby creates

---

Source: Passages from the book published by University of California Press, Berkeley, 1981. Copyright the Regents of the University of California.

noncompetitive rents. Some of these rents the bureaucrats surrender to the governments in the form of taxes, some they consume themselves, and the remainder they use to build up cadres supportive of governments in power. Through the public management of economic resources, the bureaucracies help to institutionalize a structure of relative advantage – a structure within which they themselves occupy positions of privilege and power.

Scattered around these charter members of the emergent social order in Africa lie the mass of rural producers. They suffer from government attempts, to implement an adverse structure of farm prices, and they fail to benefit from the compensatory payments conferred through subsidy programmes. Because of official repression, they lack political organizations with which to defend their interests. Furthermore, governments separate the interests of potential rural leaders – the larger farmers – from those of the mass of rural dwellers. As a consequence, instead of pursuing their collective interests, villagers pursue their private interests. They do so by fighting for political favours and by exploiting alternatives left open to them in the private market.

Owners and workers in industrial firms, economic and political elites, privileged farmers and the managers of public bureaucracies – these constitute the development coalition in contemporary Africa. It is they who reap the benefits of the policy choices made in formulating development programmes. The costs of these choices are distributed widely, but fall especially hard on the unorganized masses of the farming population.

Throughout this book we have noted a series of factors that influence the making of agricultural policy. These factors furnish important sources of variation. They affect the extent and form of government intervention, and they generate differences in the relations between farmers and the state.

HISTORICAL FACTORS

Many African governments inherited marketing boards from their colonial predecessors. While sharing colonial roots, the boards nonetheless differ in their historical origins. Those in East Africa had often been founded by the producers themselves, whereas many of those in West Africa were formed by governments in alliance with trading interests. Though both later became agencies of the independent governments, they differed in their sensitivity to the interests of producers. The East African producers tend to receive a higher proportion of the world market price for their crops than do their West African counterparts.

A second factor is the social and economic base of the coalition that captured power at the time of self-government. In many instances, the coalition was dominated by urban interests; this was notably the case in

Ghana and Zambia, where rural producers formed the backbone of a defeated opposition and thus failed to seize power when independence came. In other instances, rural producers dominated the nationalist movements which succeeded the colonial governments; this was true in the Ivory Coast and, in a more complicated fashion, in Kenya as well. Generally, where urban interests came to power, they have adopted policies more hostile to the interests of farmers; governments dominated by rural producers have intervened less forcibly in the market for outputs and have also provided more favourable subsidies in the markets for inputs.

THE CLAIMANTS

Other factors that affect the ways in which governments intervene in agricultural markets have to do with the claimants for resources from agriculture. Among the most important claimants are the governments themselves, and, as we have seen, one of the most important factors influencing their behaviour is the nature of the 'revenue imperative'.

The demand for revenue from agriculture, and thus the manner in which governments set prices against it, varies with two factors: the level of governmental commitments to spending programmes, and the availability of nonagricultural sources of funds. These help to account for important differences in government pricing policies, and influence the ways in which government treats different segments of agricultural industry.

The importance of the first factor is evident in the changes in policy that took place in many states soon after independence. Self-government brought radically increased commitments to programmes of public spending, and as the demand for revenues rose, so did the level of taxation on export agriculture. The importance of the second factor may be seen in the fact that governments with access to nonagricultural sources of funds often impose lighter levels of taxation on export crops. For example, when the oil revenues received by the government of Nigeria increased threefold between 1970 and 1973, one consequence was an upward adjustment of the prices that the marketing boards offered to producers of export crops.

The way in which a government intervenes in the markets for food crops is strongly influenced by the magnitude of the fiscal resources at its command. Its ability to increase food supplies, either through imports or through the creation of production projects, is limited by the amount of funds it has to spend. Thus the decline of copper revenues in Zambia in the late 1970s led to cutbacks in government programmes to support subsidized urban food prices; and conversely, the major programmes of farm subsidies so characteristic of production in Nigeria began with the rapid expansion of revenues from oil.

The *types* of fiscal resources available, however, produce different effects. The greater a government's *nonagricultural* revenues, the greater its efforts to increase urban food supplies and subsidize costs to consumers; the higher its subsidies on farm inputs; and the higher the prices it offers to producers of export crops. Greater *agricultural* revenues, on the other hand, promote the same interventions on behalf of the consumers of food crops and in support of farm inputs; but they lead a government to offer lower prices to export crop producers, because the costs of these programmes must be covered by taxes on farm products.

Variations in the strength and structure of the revenue imperative thus help explain variations in the way governments intervene in agricultural markets.

Governments are not the sole claimants of resources from agriculture, however. Processing industries seek raw materials, and industrial workers and owners of firms seek low-cost food. Variations in the strength and behaviour of these groups also influence the ways in which governments seek to regulate agricultural markets.

Evidence of the importance of local processing interests is contained in the changes in the prices for export crops offered by the government of Nigeria. As we have seen, beginning in 1973, domestic prices rose as the government secured increasing amounts of its revenues from oil. But for cotton and groundnuts, domestic prices never reached parity with export prices, and an important reason was pressure from domestic processing industries – textile manufacturers in the case of cotton, and producers of vegetable oils in the case of groundnuts. The larger the fraction of output consumed by domestic processing firms, the greater the pressures on governments to keep down the prices offered to farmers.

Whereas the processors of raw materials look for low-priced cash crops, workers and industrialists look for low-priced food. When urban consumers are poor, expenditures on food consume a higher percentage of their incomes; the lower their incomes, the more they benefit from reductions in food prices. Nations with lower per-capita incomes are thus more likely to adopt policics in support of low-priced food. Average incomes in Africa are so uniformly low as to prevent observation of the importance of this factor; but if it is of obvious importance in explaining the strength of the pressures upon African governments to appease their urban constituents by providing low-cost food. Moreover, this reasoning has been applied to account for variations in the agricultural policies of different nations of the world in the contemporary era, and for variations in the policies of particular nations as they pass through the development process.

Among the basic determinants of the demand for low-priced food by owners of firms are the proportion of their costs represented by wages and the degree to which they can pass higher costs on to consumers. It should be noted that for governments, wages represent a high fraction of their costs, and that few of these costs can be transferred to the consumers of government services. Little wonder, then, that among the industries of Africa, it is the governments themselves which are among the most vocal in calling for low-priced food.

Also important is the degree to which increased food costs are translated into wage claims by workers. In conjunction with what has been argued thus far, we can therefore see why nations with socialist governments adopt agricultural policies that differ little from those of their capitalist neighbours. The stress which socialist regimes place on upgrading the incomes of the rural poor meets a powerful counterforce in the making of farm policy: the higher costs they face from an increase in the price of food. Because they provide a more abundant level of services, they have a greater number of employees; because they own more firms, they are more directly affected by wage claims; and because of their ideological commitments, they are often more responsive to the demands of their workers. An increase in the price of food therefore imposes higher costs on socialist governments, leading them to adopt food-price policies that resemble those of other nations to a greater degree than their official policy positions would lead one to expect.

CHARACTERISTICS OF PRODUCTION

In addition to factors arising on the side of the claimants of resources, important variables operate on the side of the production and marketing of agricultural commodities. These, too, shape the policy choices of governments.

One factor is the nature of the crop. We have noted, for example, that in the case of export crops, governments directly intervene in product markets in attempts to depress prices. For food crops, intervention often takes less direct forms: attempts to change crop prices by altering the availability of supplies or the price of farm inputs. The nature of the crop influences the form of intervention. Export crops often grow in specialized regions. Not all consumers can buy them; they must have access to foreign markets or to expensive processing equipment. And these crops must pass through specific locations – ports and harbours, for example. By contrast, most food crops are grown by most farmers. They may be sold to anyone, for they are often easily processed and directly consumed; and they need not pass through highly specialized marketing channels. It is therefore simply much easier to control the marketing of export crops. As a consequence, in efforts to alter produce prices, governments can more directly intervene in the market for export crops; almost literally, they can seize control of the market. But for food crops, government manipulation must be more indirect; it must take the form of altering supplies and costs of production.

A second major factor is the structure of production. Large farmers often possess close social and political ties with governing elites; and this is of increasing importance in Africa. One consequence is that crops whose production is dominated by large farmers tend to be less heavily taxed. Rice, which is grown by elite farmers in Ghana, is subsidized; Ghanaian cocoa, which is grown largely by smallholders, is heavily taxed. The producer price of coffee grown on estates in Kenya lies at over 90% of the world price; for coffee grown by smallholders, the producer price stands at less than 66% of the world price.

The size distribution of production is also important because it affects the incentives and capacity of farmers to organize in defense of their interests. When the interests of farmers and governments do conflict, large farmers can more readily organize efforts to alter government policy. Groundnuts, for example, are produced in both Nigeria and Senegal. In Senegal, the Marabouts produce one quarter of the crop; in Nigeria, no comparable group dominates groundnut production. Although in both countries the crop is heavily taxed, political protest led by the Marabouts in Senegal resulted in an upward revision in the producer price of groundnuts; in Nigeria there were no organized protests, and the export of groundnuts was banned in an effort to lower the prices paid by local industries. A multitude of small farmers is far more vulnerable to adverse actions by government.

Another factor is the degree of relative advantage that producers hold in the production and marketing of a crop. Ironically, the stronger their relative advantage in production, the weaker their political position. For the stronger their relative advantage, the longer they will persist in growing a crop under conditions of falling prices – the more thoroughly they can be 'squeezed', in short, by adverse pricing policies. Where producers hold an advantage in the market for a crop, however, then the obverse is true. If consumers have few alternatives, producers can demand higher prices for their products, and they can more vigorously defend their position in the formulation of pricing policies.

An example of special environmental and ecological conditions that give economic advantages to producers would be the forests of West Africa, which create highly favourable conditions for the production of cocoa. Taking advantage of the inferiority of the producers' second-best alternatives, West African governments have long subjected cocoa to severe levels of taxation. And now that the locational advantage of the producers has been severely eroded by the additional costs imposed by government policies, the advantage of the governments is weakening; as we have seen, producers are increasingly evading the policies of governments by entering the production of alternative crops. And there is increasing evidence that government policies may change.

The effect of consumer alternatives on the pricing policies of governments can perhaps best be illustrated by the different treatment given to export crops and food crops. Export crops are sold in international markets where purchasers can choose between sources of supply. Food crops are sold on domestic markets were alternate sources of supply must come in the form of imports. Because most often imports must come from distant food surplus regions, they represent a costly alternative. All else being equal, the producers of food crops are therefore in a stronger position to secure higher prices than are the producers of export crops – which is another reason for their more favorable treatment by governments.

Differences in historical background, in the characteristics of claimants to resources from agriculture, and in factors associated with the production and marketing of agricultural commodities are all differences that cause variation in the choices made by African governments. Their importance can be illustrated by citing divergent cases.

Palm oil in Southern Nigeria in the 1960s was produced in a nation where marketing boards had been set up by the government in association with merchant interests. Government revenues derived from export agriculture, and popular demands for government services were strong; local processors consumed a growing share of the industry's output; farmers had few alternative cash crops; and production was in the hands of small-scale, village-level farmers. The industry was subject to a high level of taxation. Only when farmers began to abandon the production of palm oil for other crops, and when the government found different sources of revenue, did the government relent and offer higher prices for the crop.

The production of wheat in Kenya offers a striking contrast. Historically, the marketing board for wheat had been set up by the producers themselves and prosperous indigenous farmers had played a major role in the nationalist movement which seized power in the post-independence period. The government derived a relatively small portion of its revenues from agriculture; farmers had attractive alternatives to the production of wheat; consumers had a strong preference for wheat products, and alternative sources of supply lay in distant foreign markets. Wheat production was dominated by a relatively small number of very large farmers; and elite-level figures had direct financial interests in wheat farming. The result was a set of policies providing favourable prices for wheat products and extensive subsidies for farm inputs.

A major objective of this study has been to identify factors that help account for the ways in which governments intervene in agricultural markets. We have located three types of factors. And, as we have seen, they help to account for differences in public policies and for variations in the relations between the farmer and the state in Africa.

ALTERNATIVE FUTURES

The agricultural policies of the nations of Africa confer benefits on highly concentrated and organized groupings. They spread costs over the masses of the unorganized. They have helped to evoke the self-interested assent of powerful interests to the formation of a new political order, and have provoked little organized resistance. In this way, they have helped to generate a political equilibrium. But, in the longer run, the costs inflicted by these policies are being passed on to members of the policy-making coalition, and the configurations that were once in equilibrium are now becoming politically unstable.

Among those excluded from the immediate rewards of the new political order are the mass of farmers. For the benefit of others, they are subjected to policies that violate their interests. But the effects of these policies are increasingly harmful to everyone. Reducing the incentives to grow food leads to reduced food production; the result is higher food prices and waves of discontent in the urban centres. The coups and counter-coups that have swept West Africa owe their origin, as we have noted, in part to discontent over higher food prices. And they show how policies that have been designed to serve the interests of powerful groups impose costs which in the long run affect everyone, thereby undercutting the positions of advantage they have helped to create and disrupting the political order.

Reducing the incentives to produce export crops is also proving politically costly. The result of adverse incentives has in some cases been a measurable decline in the production of exports, with a resultant loss of public revenues and foreign exchange. Pressed for revenues, governments have to cut back on politically popular programmes – food subsidies to appease the urban consumer, input subsidies to tame the rural elite, or schools, roads, and clinics to reward communities that have kept political good faith.

As governments throughout the developing world have learned, financial retrenchment generates political crisis. In the face of fewer revenues, governments do have an alternative: they can spend beyond their means. But this choice, too, fuels political disaffection by strengthening the forces of inflation. Shortages of foreign exchange have meant decreased imports, and shortages of imports provoke cutbacks in production by industry and in services by governments. The result, once again, is growing political discontent. Too often, then, policies adopted to extract revenues from export agriculture have led to an increasing scarcity of the resources used to underpin the political order.

The costs that were inflicted on farmers are thus increasingly being transferred to those who initially benefited from the selection of agricultural policies. The basis of the equilibrium erodes. Incentives are thereby created for altering policy choices. Here the question arises: Are there politically workable terms under which the farmers could be admitted to the governing coalition, and thereby receive support for policies more favourable to their interests?

The following scenario might work for food producers. In nations with a major extractive industry, a coalition of urban industry and food producers could unite against the extractive industry. Food producers could use the tax revenues generated by the extractive industry to subsidize the cost of inputs. Government-owned railways could charge high prices for carrying the freight of the extractive industry in order to subsidize the rail charges for agricultural products, for example. Food producers could also use public revenues to support high producer prices; they can use them, for example, to finance storage and exports, thereby withdrawing excess supplies from the domestic market and protecting a high level of domestic prices. Industry, for its part, could seek an overvalued currency and tariff protection. An overvalued currency would facilitate imports of capital, paid for with foreign exchange earned by the extractive industry; tariff protection would shield local markets from foreign competition and also offset the increase in wages resulting from higher-priced food. Alternatively, industry could secure the use of government revenues to subsidize urban food prices, thereby offsetting the threat of higher wage demands.

Coalitions such as these have in fact formed in Africa, as in the settler territories of southern Africa. They may now begin to form in areas where discoveries of oil or uranium may bring radical increases in the wealth of certain black African nations. And they become more likely as greater numbers of politically influential persons begin to engage in domestic food production.

In the case of export crops, the scenario differs. The most obvious tradeoff that would bring the producers of export crops into the dominant coalition would be one in which measures to reduce the overvaluation of the national currency are exchanged for measures to reduce the cost of industrial labour. Other things being equal this would greatly increase the incomes of the producers of export crops. But it would do so at the expense of others, such as industry, who must import from abroad. To secure a coalition between export agriculture and industry, compensation must be offered for this loss of advantage. In part, such compensation can take the form of higher levels of prosperity among rural producers, for this expands the size of the markets for manufactured products. But their prosperity also poses a threat, for it leads to competitive bids for labour by the more prosperous countryside. The most direct form of compensation that can be offered, therefore, will consist of measures designed to cheapen the costs of labour: repressive labour laws which limit collective action by workers, and laws which give foreign workers relatively open access to the domestic labour market. Export-oriented policies on the one hand, and policies in support of cheap labour on the other, thus form the basis for a coalition that incorporates the interests of export agriculture.

Such a bargain has in fact been struck on one nation – the Ivory Coast. As students of that nation have long appreciated, an important basis for its policies is that members of the elite derive large portions of their incomes from the production of export crops. This fact assumes particular importance in the context of our argument. For just as overvaluation cheapens the costs of imported equipment for firms, it also cheapens the costs of imported goods for consumers. The elite, being rich and having 'modern' tastes, consume a disproportionate share of imports, and therefore can be expected to resist devaluation of the local currency. Insofar as the elite draws a major portion of its income from exports, however, the benefits of the change in commercial policy compensate for its loss in foreign purchasing power. As less sacrifice is required from the politically influential, it is politically easier to secure this shift in commercial policy. Here again we can see the importance of the rural origins of the political elite for the making of public policy.

These scenarios portray politically workable grounds for incorporating the interests of the two kinds of producers into the policy-making coalitions of Africa. In both instances, exceptional conditions give political impetus to the favoring of their interests. In most cases these conditions will be absent. As a consequence powerful actors – revenue-starved governments, price-conscious consumers, profit-seeking industries, and dependent farmers – will persist in seeking their individual, short-run, best interests, and they will continue to adhere to policy choices that are harmful to farmers and collectively deleterious as well. Producers will prosper, then, insofar as they successfully evade the prescriptions of their governments.

Alternatively, in response to the erosion of advantages engendered by shortfalls in production, the dominant interests may be persuaded to forsake the pursuit of unilateral short-run advantage, and instead to employ strategies that evoke cooperation by sharing joint gains. In the face of mounting evidence of the failure of present policies, people may come to believe that short-run price increases for farmers may in the longer run lead to more abundant supplies and less costly food; or that decreases in tax rates may lead to greater revenues as a result of increased production; or that positive incentives for greater production may lead to greater production and lower prices and leave only the most efficient farms in production, thereby accelerating a shift of resources from agriculture to industry. The growth of an awareness that present measures offer few incentives for farmers to play a positive role in the great transformation may thus provide a foundation for attempts to reform the agricultural policies of the nations of Africa.

5

FOOD SYSTEM ORGANIZATION PROBLEMS IN DEVELOPING COUNTRIES

H.M. Riley and J.M. Staatz

Recently, agricultural economists' traditional approaches to analysing marketing problems and designing public sector interventions have been subjected to severe criticisms. One of the most important of these critiques argues that the static, perfect competition model, though still useful, must be supplemented by conceptual and analytic procedures that give more attention to the dynamics of the development process and to the roles of institutions, policies, and rules in achieving market performance goals.

In all food systems there are multiple performance goals, and many of these goals can be specified only in the context of a particular country and culture. But all economies also share certain performance goals – notably, equity, progressiveness, and the traditional goal of economic efficiency. We need a conceptual and analytic approach that will encompass these commonly sought goals.

In addition, there is a growing concern among agricultural economists that 'a food system perspective' be adopted as a framework for assessing the workability of alternative market interventions. Such a perspective broadens the scope of marketing research to include not only the product movements and transformations that occur after farm-level production but also farm input marketing and policies on prices, trade, and institutional reforms.

The first four major sections of this report deal, respectively, with traditional approaches to marketing policy and the changes that are occurring in these approaches; the body of research on marketing, which includes studies by geographers and anthropologists as well as economists; some issues in future marketing research; and a number of institutional issues, among them the need to strengthen professional capabilities in developing countries. The final section of the report outlines follow-up actions recommended.

Source: Agricultural Development Council, New York. Report no. 23, December 1981. (Condensed.)

MARKETING POLICIES: GOVERNMENT AND DONOR APPROACHES

Governments of low-income countries and donors have increasingly recognized that the agricultural marketing system plays a crucial role in economic development, not only by physically distributing increased production but by influencing production incentives and distributing the benefits of growth. As a result, governments and donors have tried many approaches to marketing improvement, but with varying degrees of success. For one thing, these efforts have often been colored by stereotypical views of marketing as an essentially unorganized, exploitive, and nonproductive activity that is not really amenable to scientific analysis. In addition, past policies have often assumed that a clear dichotomy existed between production and marketing and that the major task was to combat the monopoly position of private-sector merchants. Finally, many past efforts have underestimated not only the technical and management expertise required in marketing but the large number of the very poor who are marginally employed in the marketing system.

Creation of public marketing agencies

Many countries have established parastatals, marketing boards, and cooperatives – often with statutory monopolies – to handle the marketing of certain agricultural products. These government-backed agencies, however, have been less successful in handling food crops for domestic consumption than in handling export crops. There are three primary factors in this lower success rate: the greater complexity of domestic food marketing systems, which involve thousands of assembly and distribution points; the generally lower level of value-added in the processing of domestic food crops as compared with export crops; and the existence of a well-established private trade. The performance of public marketing agencies has often been hampered by high overhead costs; few incentives for efficiency, particularly when deficits can be covered by a public treasury; and a lack of marketing expertise, especially during initial years of operation.

One reason why governments and special interest groups have continued to support public marketing agencies, despite their poor performance, is the ability of these agencies to grant preferential market access to favored groups; for example, farmers in remote areas and politically powerful urban consumer groups. Another, more positive reason for continued support of public agencies is the fact that, where they have not been granted statutory monopolies, they may have improved private trade performance by increasing competition.

Facilitation of private trade

Governments and donors have also attempted to strengthen marketing generally by providing physical infrastructure, information services, and credit and extension programmes aimed at improving private trade performance. Improving infrastructure has often led to better market performance, but this method has had two shortcomings. First, the type of infrastructure provided has frequently been more appropriate to relative factor prices in Europe or North America (where the infrastructure plans are usually formulated) than to those in low-income countries. As a result, expensive imported capital has often displaced cheap domestic labour in technically complex, new processing and marketing facilities. Second, when they have stressed physical infrastructure, governments and donors have sometimes neglected to provide other important public goods – such as improved information systems, uniform weight and measures, and changes in laws and regulations – that would encourage innovations. Alone, an individual market participant might not find it profitable to adopt such innovations; if they were adopted by all participants, however, these innovations would greatly enhance food system productivity. It may be that such public goods have been underemphasized because it is difficult to estimate their costs and benefits *ex ante.*

Training of professionals and market participants

Graduate, undergraduate, and technical training in agricultural marketing has been provided by governments and donors for some time. However, professional training has been hindered by several factors: the failure of many programmes to include the practical, business-oriented training necessary for successful management of private or parastatal entities (e.g. basic budgeting and accounting); the lack of teaching materials specifically relevant to marketing in low-income countries; and the lack of field experience in how marketing systems in low-income countries actually work. Because of these weaknesses in training programmes, students often find it difficult to apply the general principles they have learned to the practical marketing problems they face in their own countries.

Training programmes have, to a great extent, neglected short-term, in-service training for market participants. The extension programmes of most low-income countries are oriented almost entirely toward farm production and give little attention to marketing. Providing small merchants and shippers, for example, with basic instruction in accounting, inventory management, and packaging techniques might contribute substantially to market efficiency.

Recent changes in approaches to marketing

Experience over the past 30 years has demonstrated that without appropriate government action there is no guarantee that a marketing system adequate to the needs of a rapidly growing country will evolve spontaneously. At the same time, it has become clear that good marketing performance requires both technical expertise and incentives, and that simply replacing private traders with state agencies in no way assures such performance. As a result, governments and donors have begun not only to give more attention to agricultural marketing but to attempt to differentiate between marketing system functions that are best handled by centralized means – for example, through the government – and those that are better handled by relatively decentralized means – for example, through private trade. Many countries have given greater emphasis to the role of private trade in an effort to use both the human capital already in the system and the strong incentives for marketing efficiency that often exist in the private sector. Other nations continue to see a strong role for public marketing organizations but have given considerable attention to the design of incentives for good performance by such organizations.

Many donors and governments have come to see the old dichotomy between marketing and production as arbitrary and misleading. Like farming, marketing involves production processes that use inputs and create value; it includes the off-farm elements of the food production system. Marketing also includes the system by which production is allocated among consumers. Increasingly, therefore, marketing discussions are taking place in a *food systems* context and include broader issues such as food subsidies, trade policies, dual pricing schemes, and the impact of the market structure on nutrition and on production incentives.

MARKETING RESEARCH: STUDIES TO DATE

Work by economists

Many of the studies by economists have been largely descriptive, whereas others have been diagnostic and prescriptive. The conceptual approaches to these studies have in general taken two forms: approaches that emphasize the perfectly competitive market as a norm and those that use a broader, food systems framework of analysis.

Research using perfect competition norms

These studies have often evaluated market efficiency by comparing price differentials through time and space with the costs of spatial and temporal arbitrage; by calculating net margins for various marketing functions; and by evaluating the degree of intermarket relatedness, using correlations of price movements across markets (Jones, 1972; Center for Research on Economic Development, 1977; Southworth *et al.*, 1979).

Research using perfect competition norms has made several important contributions. First, it has provided a good empirical description of how several important food marketing systems work, information that is essential to intelligent policy making. Second, it has challenged many of the prevailing stereotypical notions of indigenous marketing systems and of the market behaviour of farmers and merchants in low-income countries. It has demonstrated that these systems are often not as exploitive as supposed and that, given the institutional and infrastructural setting in which they operate, they tend to be fairly efficient. Furthermore, by showing that 'traditional' market participants are indeed 'economic men and women' who respond to market incentives in predictable ways, this research has revealed that standard economic policies can be used to influence market behaviour. By focusing on market efficiency, the research has also drawn attention to unexploited economic opportunities within existing marketing systems and has outlined ways in which infrastructural and policy constraints have hindered such efficiency.

The perfect competition approach also has several limitations. Consistent with the model, market performance has been defined almost solely in terms of static economic efficiency. Relatively less attention has been given to other dimensions of performance such as stability of product flows and prices, product suitability, and equity. Furthermore, following a structuralist view of the industrial organization framework, this approach has tended to focus much more on issues of horizontal concentration within a subsector than on vertical coordination issues (such as the effect of wholesaling arrangements for a particular commodity on farmers' incentives to produce). The perfect competition approach also does not address some of the dynamic aspects of market development, such as the effects of economies of size in marketing and processing. And the approach has been criticized for sometimes drawing unwarranted conclusions from correlation analyses based on unreliable secondary data (Harriss, 1979).

A final limitation of the perfect competition approach is that it has sometimes exaggerated the importance of improving the physical infrastructure of marketing systems relative to the importance of changing institutions, standard operating procedures of firms and government agencies, and market rules – all of which might contribute substantially to improved market performance. This tendency probably reflects the fact that it is easier to assess the costs and benefits of infrastructural changes than those of institutional changes. At an extreme, the policy recommendations of studies using perfect competition norms might be characterized (some would say, caricatured) by the Schultzian 'efficient but poor' hypothesis as applied to marketing, arguing that nothing but physical infrastructure need be improved.

Systems-oriented research

Another group of economists has attempted to look at food production and distribution more as a unified system and has stressed the interdependence of activities at different levels in that system. These researchers have argued that small increases in productivity in one part of the system (for example, improved inventory management at the wholesale level) may greatly improve the potential of the whole system. Similarly, they argue that failure at any level may cause stagnation in the entire system.

This systems approach shifts the focus of research from the farmer or merchant acting as individual to all market participants acting as a coordinated group. A major goal of this research has been the discovery of ways to facilitate better coordination among participants at different levels of the food system – such as developing new operating methods, rules, and institutions – in order to increase the productivity of the system as a whole. This approach contrasts rather sharply with the perfect competition approach, which emphasizes increasing the efficiency of individual market participants within a given institutional framework.

According to the systems approach, food system performance includes many dimensions not stressed by the static economic efficiency approach. Thus, for example, systems-oriented researchers have emphasized the influence of the form of market organization on economic growth and equity. The systems approach views markets as the means by which linkages between different sectors of the economy are activated, and it stresses the importance of laws and regulations in shaping the behaviour of the market participants.

By stressing the unity of production and marketing and the multidimensional nature of market performance, the systems approach has provided a broad framework for the analysis of marketing problems. The emphasis of this approach on dynamic issues, such as the possibilities for capturing external, system-wide economies through the introduction of new technologies and institutions, has helped to place marketing policies firmly in a developmental framework.

Like the narrower, perfect competition approach, the systems approach has its drawbacks. Its performance norms, based in part on the concept of *workable competition*, are much less clearly defined than are those of the former approach. In addition, despite its emphasis on institutions and their effects on market behaviour, the systems approach lacks a well-developed methodology for *ex ante* evaluation of the performance consequences of alternative institutional arrangements. Used indiscriminately, this approach would be completely unwieldy: everything in the food system would affect everything else in the economy and vice versa. Used with caution, however, the systems approach helps place marketing problems in their long-term developmental context. As outlined in a later section, a major challenge to economists is to make the systems approach more operational by developing methodologies to evaluate the performance consequences of alternative interventions in the food system.

Work by geographers and anthropologists

Geographers and anthropologists have stressed spatial considerations in the organization of markets and the relations between economic organization and other aspects of culture. Geographers, developing concepts such as *central place theory*, have built on Von Thunen's concepts of the spatial organization of markets and production, concepts economists have sometimes disregarded.

Marketing research by anthropologists falls into three main categories:

1. *Regional analysis* addresses many of the issues investigated by geographers, examines the spatial organization of markets and the causes, consequences, and correlates geographical patterns of market organization (see, e.g. Smith, 1976).
2. *Microbehavioral studies* investigate decision making by individual market participants, comparing the outcomes with those predicted by microeconomic theory (see, e.g. Gladwin and Gladwin, 1971).
3. *Organizational analyses of social interaction in the marketplace* study the interaction between social relations and economic processes, examining, for example, how kinship networks influence the flow of information and the structure of retailing in a given market and how these factors, in turn, reinforce certain kinship obligations.

Clearly, geographers and anthropologists have viewed marketing from a different perspective than that chosen by agricultural economists. However, partly because these researchers have traditionally been excluded from programme design and implementation, their work has often been purely descriptive; they have rarely diagnosed specific problems or prescribed specific methods of improving market performance. When these researchers have addressed the economic performance of markets, they have frequently used perfect competition norms.

Integrating the work of anthropologists and geographers with that of economists is often difficult because each discipline has its own conceptual framework and asks a different set of questions. As a result, the information collected by one discipline may not directly address questions that are of central interest to others. Merging the conceptual approaches of the three disciplines would endanger the unique contribution of each. Greater coordination of the three can probably be achieved, however, if members of the three disciplines work together in specific problem-solving situations.

ISSUES IN FUTURE MARKETING RESEARCH

The urgency of marketing problems in many countries dictates the need for improved policy-oriented research. Increased population, urbanization, commercialization, and the other dislocations associated with the structural transformation of an economy are straining existing food marketing systems. Policymakers need to understand these systems better if they are to design effective marketing policies, and there is a critical need for marketing research to be more closely integrated with action programmes aimed at improving food system performance. Such integrated programmes should include components for project monitoring, evaluation, and redesign. And we can increase the cost-effectiveness of these components by involving graduate students, particularly students from the developing countries, in this type of project-related research.

Immediate research tasks

Many policymakers still need basic descriptions of how food systems operate in their countries. Descriptive studies can do much to demystify marketing. Such studies should include discussions of who the major market participants are, what these participants do, and how they make market decisions. (Anthropologists can make important contributions to this type of study.) These studies should also make preliminary assessments of major problems in food systems, based on discussions with market participants and local officials. It is also very desirable, in any descriptive study, to outline product flows, price surfaces, and transfer costs within the marketing system and to estimate the variability of these factors. Such information helps identify the sources and approximate magnitudes of inefficiencies within the existing system and serves as a guide in developing actions to reduce these inefficiencies.

The major conceptual and analytical approaches used to date in marketing studies have difficulties, as we have pointed out, in predicting the performance of alternative government interventions and in quantifying the trade-offs between efficiency and other dimensions of market performance. Faced with these difficulties, some argue that researchers should press ahead in the short term with 'old-fashioned empiricism', gathering information on how various food systems work in the hope that this information may suggest new conceptual models. There is a danger, however, of doing a large number of studies of individual food systems that are so location specific that the information obtained does not build up a generalizable body of knowledge. If research is to generate information useful to policymakers in areas outside a specific research site it must focus on underlying economic and social relations as well as on idiosyncratic characteristics of individual marketing systems. A balance between theory building and pure empiricism is needed.

Long-term research challenges

In the future, a major challenge to researchers will be to develop a conceptual approach to marketing that will be both broader than the perfect competition model and more operational than existing systems-oriented research. At the most fundamental level, marketing research should evaluate how well the food system of an area works relative to the goals defined for that system by the residents of that area. Thus 'appropriate' market organization and institutions will vary according to the social, political, and cultural situation of each country.

Measuring market performance

A major task in carrying out applied market research is discovering the local definitions of good market performance and developing workable norms against which to measure current performance. Since there are likely to be many aspects to performance, evaluating it requires a multidisciplinary approach.

Given the rapidly changing demands on most developing countries' food systems, it is not enough to simply describe and evaluate the performance of current marketing systems. As already indicated, researchers must develop methodologies capable of projecting the consequences of alternative market interventions (one of which should be the continuation of current policies). Possible methods include formal and informal simulation modeling and comparative institutional analysis. *Simulation modeling* would permit researchers to vary important policy parameters and trace their consequences over time. Formal modeling of an entire marketing channel, however, can be very demanding of data, trained personnel, and computer time.

Comparative institutional analysis involves evaluating the feasibility in one locale of market interventions that have worked in another, or trying several interventions during project implementation and redesigning them in the light of their comparative performance. In selecting alternative interventions, it is important to choose those that encompass both systems-wide economies (e.g. standardization of shipping containers, or regional specialization in production) and the institutional changes necessary to capture those economies.

Ensuring acceptable performance

A second major task for researchers is to investigate ways in which imperfectly competitive marketing systems can be made to perform better. Frequently markets in developing countries are too small to support more than a few modern processing plants or certain types of exporters or importers. Little research has addressed the question of how to ensure that such components of the marketing system, whether private or public, perform acceptably. In the past, policymakers seem often to have assumed that if the activities of processors, exporters, and the like were controlled by the private sector there was no way to induce good performance, but that if they were controlled by the public sector good performance was guaranteed. In reality, the performance of both public and private entities is conditioned by the incentives and sanctions provided to individuals in these organizations (e.g. tax policies, or rules governing the setting of wages). Nevertheless, little research has considered how to design incentives and sanctions that will elicit better performance from both public and private marketing organizations.

Developing cost-effective methodologies

Developing more cost-effective research methodologies could involve making use of the data base (some of it from previous marketing studies) that already exists in many countries. Cost-effectiveness could also be increased by developing quick, preliminary survey techniques, based on informal interviewing and inspection of market facilities, that would attempt rapid identification of the critical constraints in a marketing system.

Other crucial research issues

Determining the relative roles to be played by the public and private sectors in food system development represents, for every country, a fundamental political–economic decision. As countries develop and new demands are made on their food systems, the 'appropriate mix' undoubtedly changes. Researchers can play a significant role by helping to shape the discussions of this issue, by outlining some consequences of alternative actions, and by helping to design appropriate interventions once a political decision regarding the 'appropriate mix' has been reached. When some private trade is allowed to coexist with public marketing organizations, as is the case in most countries, a critical research issue is the design of policies to coordinate the behavior of private and public market participants so as to ensure price stability, food security, and other food-system goals.

In many countries, a critical research issue is the design of market arrangements that will give small-scale participants, both small farmers and traders, viable market access. Equity is an important aspect of market performance, and if food system development is not to be characterized by a 'trickle-down' approach, questions of market access for small-scale participants (e.g. through cooperatives) need to be carefully analysed.

Because food prices are both incentives for agricultural production and major determinants of the real income of the poor, many countries face a major political and economic dilemma when establishing their agricultural price policies. In most low-income countries, population and income growth make it critically important to boost agricultural production. But it is becoming increasingly difficult to raise prices in order to increase production because of the adverse effect higher prices would have on the growing number of landless and urban poor. Marketing researchers face one of their very greatest challenges in attempting to discover ways of insulating low-income consumers from higher food prices while still providing farmers with adequate incentives to increase production.

Strengthening professional capabilities in developing countries

In nearly all developing countries there is a serious lack of local professionals trained to carry out tasks that are essential to the development of a dynamic, efficient, and equitable food production and distribution system. Even in countries where the numbers of trained professionals are actually growing, talent is often drawn off into administrative positions or is underutilized because of political instabilities and ineffectively organized institutions. When governments and institutions rely on foreign technicians and advisors, provided by donor agencies, projects are often not well integrated into local institutional operations and programmes lack continuity because of the constant turnover of expatriate professionals. Thus there is a critical need to create and expand core groups of indigenous professionals who can assume major leadership and supportive roles in food system organization and management in both the public and the private sectors within each developing country.

Building in-country capabilities for training and research

It is generally recognized that the development of indigenous training and research institutions will take time. Such investments in human resources, however, can produce high rates of return in the long run.

Universities in the United States and other developed countries have played a major role in the establishment of educational and research institutions in a number of developing countries.

Since the early 1960s there has been a substantial flow of developing country students through United States university graduate programmes in agricultural economics. A recent assessment of this activity (Fienup and Riley, 1980) indicates that a high percentage of United States-trained professionals have returned either to their own countries or to the region of the world from which they came and that nearly all are employed in positions that make use of their training. Professionals surveyed gave relatively high ratings to the usefulness of their training. They also made suggestions for further strengthening United States training and for encouraging collaborative efforts to build in-country capabilities for training and applied research.

During the 1970s there has been a significant shift by USAID and other donor agencies away from support for graduate degree training and toward short-term, project-related training activities. Although they recognized the contribution such short-term training can make to project effectiveness, workshop participants expressed the firm belief that donor agencies and developing country governments should re-examine the long-term consequences of reducing their investments in graduate degree training, both in-country and abroad.

It was also the consensus of the group that the relevance to developing country conditions of United States university degree training of professionals preparing for overseas careers should be increased. This is particularly important in the subject matter areas useful in food system organization and management. High priority should be given to arranging thesis research in developing countries and, when possible, involving local professionals in field supervision. Efforts should also be made to include United States university PhD candidates on research teams financed by donor agencies and local governments.

The group recognized the importance of extending professional development beyond the completion of formal degree programmes and through collaborative research projects led by mature professionals and staffed by young, less experienced researchers. It was also agreed that participation in professional networks, short courses, and sabbatical programmes is important for the continued growth of young professionals. In developing countries, the latter are frequently handicapped by isolation; often they lack the stimulation and reinforcement that come from professional interchange. Finally, the participants noted an apparent need for short, in-service training programmes for mid-career administrators and young professionals who find themselves thrust into administrative positions for which they are unprepared.

Strengthening research on food system organization and operations

The weaknesses of past research, as outlined in the second major section of this report, suggest the need for institutional arrangements that will reduce the frequency of nonadditive, fragmented studies of limited long-term usefulness. Greater efforts should be exerted to develop a systematic, longer-term research strategy that would provide for a reasonable balance between specific problem-solving studies and the kind of subject-matter research that serves a variety of information needs in policy making, teaching, and the planning of further studies.

Specific suggestions on the organization of research activities are:

1. Broaden the scope of 'marketing research' to include topics such as the public distribution of food to low-income urban families; the performance of marketing boards, especially in comparison with alternative institutional arrangements; pricing policies for agricultural inputs and products; and the regulation of foreign trade.
2. Incorporate 'marketing' subprojects into larger, more comprehensive projects on agricultural production, post-harvest handling, nutrition, and integrated rural development.
3. Combine applied research, training, and extension activities with operational market intervention programmes.

REFERENCES

Center for Research on Economic Development (1977) *Marketing Price Policy and Storage of Food Grains in the Sahel: A Survey.* Center for Research on Economic Development, University of Michigan, Ann Arbor.

Fienup, D. and Riley, H.M. (1980) *Training Agricultural Economists for Work in International Development.* A study sponsored by the American Agricultural Economics Association available from the Agricultural Development Council, New York.

Gladwin, H. and Gladwin, C. (1971) Estimating market conditions and market expectations of fish sellers in Cape Coast, Ghana. In: Dalton, G. (ed.) *Studies in Economic Anthropology.* American Anthropological Association, Washington, D.C.

Harriss, B. (1979) There is a method in my madness: Or is it vice versa? Measuring agricultural market performance. *Food Research Institute Studies*, **17**(2), 197–218.

Jones, W.O. (1972) *Marketing Staple Food Crops in Tropical Africa.* Cornell University Press, Ithaca, N.Y.

Smith, C. (1976) *Regional Analysis. Vol. 1, Economic Systems.* Academic Press, New York.

Southworth, V.R., Jones, W.O. and Pearson, S.R. (1979) Food crop marketing in Atebubu District, Ghana. *Food Research Institute Studies* **17**(2), 157–195.

6

MARKETING, THE RURAL POOR AND SUSTAINABILITY

J.C. Abbott

Marketing is commonly linked with those who have surplus to sell. It also has a role that is vital for the poor, notably where governments are unable to undertake major transfers of resources and income raising activities must be self supporting. Consideration of this potential is the more timely when many governments of poor countries are curtailing their responsibilities under structural adjustment.

POVERTY, GOVERNMENTS AND EXTERNAL AID

While real incomes in the developing countries have risen by 50% since 1950, distribution between and within these countries has been uneven. Using 1985 figures the World Bank has put the number of people below its poverty line – an income of $370 per year – at 1.1 billion (World Bank, 1990). One-third of these are in India. By this definition half of the people in Africa below the Sahara are below the poverty line, so are most of those in north-east Brazil and the Latin American mountain zones with high American-Indian populations. Their numbers have since increased, reflecting:

1. population growth;
2. economic recession with lower prices for exports, reduced capital inflows, high indebtedness;
3. structural adjustment programmes resulting in reduced government expenditures and higher prices for food that have impinged negatively on those who were not producers;
4. drought and civil war in a number of countries.

Populations continue to grow – in some countries at an alarming pace. Growth rates of 4% per year prevail in North African countries making them increasingly dependent on food imports. Kenya also with a 4%

annual increase in population will have increasing difficulty in providing lasting food security. Fundamentalist pressures in North Africa and the fragility of some other African governments have blocked effective promotion of birth control where it was needed. In China and India there were determined efforts to limit population growth; in recent years these have been relaxed.

Historically the way out for the poor in a disaster prone area, or one overpopulated in relation to its resources, has been to emigrate. Many workers on cocoa farms in Ghana and Ivory Coast came from the Sahel. People from there also went to Sudan when irrigated cotton growing in the Gezirch set a new demand for labour. Such movements continue. They are facilitated by transport and information arrangements developed for marketing. It has been said that if research on crops for dryland agriculture does not make a break-through soon most people will have left it.

Only the governments of the countries concerned can make the shifts in resources and services needed to alleviate large-scale poverty. Their scope for action depends on the resources available to them and the capacity of their administration.

There are some Third World countries with small populations and substantial income from petroleum resources. Here governments have been able to ease the life of the poor, providing services and cash allowances. Others are aware of a considerable disparity of individual incomes, but see the tax base as too small and the poverty sector too large for significant income transfers to be practicable. Even so they could extend basic services – primary education, low cost health support. Some of the services they do provide – distribution of low-cost food grains, for example – are poorly targetted. They go to government personnel and urban middle-class leaving aside the rural areas where 80% of the poor live. That more is not done for the rural poor, including a shift in the relative burden of taxes from agriculture to urban activities, has to be attributed to political preference for the status quo.

While the provision of food grain rations at low prices in India has benefited mainly urban residents, food for work programmes have been organized there for the rural poor. Men and women without work during the slack season have been engaged in building up public infrastructure, as under the Maharastra Employment Guarantee Scheme (Lipton, 1989).

Lasting solutions to rural poverty require, however, that the people be enabled to earn more, or support themselves better, on a self-sustaining basis. This was the conclusion of the 1979 FAO World Conference on Agrarian Reform and Rural Development. Broadly it recommended:

1. Improved access to land, and irrigation to make it more productive.
2. Better access to markets, inputs and related services.

3. Provision of advice and training on improved agricultural production methods.
4. Development of non-farm rural activities.

The conference also stressed people's participation and the role of women. Here we shall concentrate on the second heading, access to expanding markets, and the inputs needed for increased production. The activities involved in this – the transport, processing and distribution of farm output, provision and maintenance of the supplies and equipment needed and meeting the associated demand for production inputs, and handicrafts using local materials – are also likely to provide the more immediate off-farm work opportunities in poor rural areas.

External aid in initiating projects to alleviate rural poverty on a self-sustaining basis will be illustrated through the experience of the International Fund for Agricultural Development, Rome (IFAD). This was set up in 1978 as a penance, it has been said, by the petroleum exporters for the damage they did to the developing countries by forcing up its price. It is small by World Bank standards. Its loans totalled $2.4 billion over the first 10 years. But it focuses specifically on the poverty issue.

IFAD seeks out rural sectors with incomes averaging $120–140 per year. First it concentrated on raising their production capacity turning away from what in some development circles was deemed commercial approach. It found that the continuing success of its projects hinged on a parallel attention to marketing. The late 1980s saw a conscious attempt to strengthen marketing systems for small farmers, fishermen and livestock raisers through better services and credits to improve roads, storage, processing, and stimulate competition in marketing channels.

ROLE OF MARKETING IN POVERTY ALLEVIATION

The dynamics of marketing in development are well known. They apply equally to the poor. Even the subsistence small-holder must sell some produce if he is to have the cash to pay for inputs and services that will raise his output and his level of living.

The tasks and responsibilities of marketing may be summarized as:

- finding a buyer and transferring ownership;
- assembling, transporting and storing;
- sorting, packing and processing;
- providing the finance for marketing and risk-taking;
- assorting and presenting to consumers.

If marketing is to fulfil its role of stimulating and extending development, specific enterprises must be responsible for finding foreign or domestic buyers for various types and qualities of produce. They must be able to arrange assembly from farms; packing and presentation in appropriate containers; storing according to buyers' requirements; transport to buyers' depots or markets which they attend; storage to extend the availability of seasonal commodities and processing to extend the time and range of sales outlets. The enterprises must provide the necessary investment capital for fixed facilities, and the working capital to carry purchases from farmers until resale proceeds are received. Implicitly, these enterprises must possess the financial resources, the qualified managerial, sales and technical personnel, together with the initiative and willingness to accept business risks, which are necessary to perform these tasks efficiently. In export marketing, or in substitution for imports on domestic markets, they must be able to match the competence of rival enterprises in other countries.

Marketing enables a person with some land to move from semi-subsistence to growing produce regularly for sale. Correspondingly, it allows an increasing proportion of a country's population to live in cities and buy their food nearby. Marketing also provides an incentive to farmers to grow produce for export. This increases farmers' income so that farmers form a growing market for domestic industry as well as earning foreign exchange to pay for essential imports.

An efficient marketing sector does not merely link sellers and buyers and react to the current situation of supply and demand. It also has a dynamic role to play in stimulating output and consumption, the essentials of economic development. On the one hand, it creates and activates new demands by improving and transforming farm products and by seeking and stimulating new customers and new needs. On the other hand, it guides farmers towards new production opportunities and encourages innovation and improvement in response to demand and prices. Its dynamic functions are thus of primary importance in promoting economic activity, creating employment. For this reason an efficient marketing sector has been described as the most important multiplier of economic development (Drucker, 1958).

As indicators of the benefits to small farmers and fishermen, and landless workers that accrue from effective marketing initiatives, it may be noted that the tea processing and marketing system operated by the Kenya Tea Development Authority provides a profitable outlet for 138 000 small growers. Their incomes are well above the average for their area, likewise those of 28 000 growers and associated workers for the Mumias Sugar Co. For a number of crops smallholder production is advantageous because the incentive for care in harvesting is direct. This is the case with tea in Kenya and marigolds in Ecuador, and also with tobacco in various countries. Smallholder production of sugar is now preferred because of the risk that organized labour on plantations would disrupt harvesting at the time for optimum yield. People with no land of their own have been brought into profitable poultry raising systems. The Charoen Pokphan enterprise in Thailand saw a profitable market in Japan for boned-out chicken. It furnished young chicks together with the feed, drugs and veterinarian services required to raise them. Those who had no chicken houses could build them with bank loans it negotiated. Large numbers of women were employed in dressing and packing plants.

Cattle owners in thousands have benefited from small private and larger cooperative milk marketing developments in India and from livestock marketing initiatives that have expanded outlets for traditional suppliers in Botswana, Chad and Swaziland. Continuing access to the favourable market provided by the Botswana Meat Commission has stimulated a livestock off-take much higher than on similar range grazing conditions in other parts of Africa. It has been of broad and immediate benefit to the low income livestock-dependent rural population.

Up to 1960 shrimps caught by Indian fishermen were discarded, or spread around coconut trees as fertilizer. In 1980, exported frozen, they earned $45 million from US markets alone (Abbott, 1988).

For rural people without access to land, employment is the core factor in alleviating poverty. In addition to stimulating production that gives employment, agricultural and fish marketing enterprises also employ large numbers of people directly. Reliable statistics are scarce. Processing grain, cowpeas and groundnuts, and fish has been a major source of gainful employment for women in Africa (Simmons, 1975). The Asian Productivity Organisation estimated the annual rate of growth of employment in agricultural and marine products processing during 1970–1975 at 7.9%. Very large numbers are also put to work in the production and marketing of fruit and vegetables needing careful handling to obtain the best prices. New transport technology and marketing initiative developed snow pea production in Guatemala. Income gains were highest on the smallest farms and among landless labourers who found increased employment (Von Braun *et al.*, 1989).

MARKETING STRUCTURES FROM THE VIEW OF THE SMALLHOLDER

The complexities and significance of the market process have often been underrated, at considerable cost to development. Smallholders tend to incur higher transaction costs than larger production units, because the quantities of inputs they need and of output they sell, are smaller. They are often less well informed and have less bargaining power. Important for them are the consideration and assistance they receive from the marketing enterprise. Will they be treated fairly at the local buying stage? Will they be advised what to grow and when, and on how to raise its market value? Will

Table 6.1. Relative advantages for small farmers of alternative marketing structures

Marketing structures	Sales position of small farmers vis-à-vis larger farmers	Sales position of small farmer vis-à-vis buyer	Technical assistance	Seed/planting materials provision	Fertilizer supply	Other credit	Government support required
Independent private firms	Bargaining weight can be used against small farmer	Advantage of access to alternative outlets	Advice based on local experience	May supply on credit	May supply on credit	Consumption credit often available in addition	Provision of market infrastructure and information services; maintenance of competition, some price stabilization
Marketing and processing transnational	Equitable prices if can make a contract	Dependent but secure if can meet product quality requirements	Can be direct and intensive	Direct supply on credit	Direct supply on credit	No	Should negotiate participation for small farmers and prices

Cooperatives	Equal if operates successfully	Favourable provided cooperative operates efficiently	Collaborate with government service	May arrange a supply	Direct supply on credit	May supply where supported by cooperative bank	Continuing financial support, supervision and protection generally needed
Marketing board/ state trading agency	Equal price if can reach official buying station	Depends on access to buying station; may be subject to illicit charges	Usually left to government service	Assistance rare	Assistance rare	No	Insistence on measures to help small farmers at rural buying points
Development company/ authority	Equal prices	Protected provided can meet product quality requirements	Can be direct and intensive	Direct supply on credit	Direct supply on credit	No	Major financial input or privileges usually required

they be helped with credit, and with access to the inputs they need? Table 6.1, based on farm level observation and group discussions, compares the main types of marketing enterprises in these terms (Abbott, 1987).

The enterprises that constitute the driving force in marketing may be independent individuals and partnerships, joint stock companies based locally or in another country, cooperatives and marketing boards, corporations or authorities set up by government.

Independent private marketing enterprises

These have shown themselves able to start up and go a long way with very little capital. They are great builders of capital assets. Their operators tend to be economical – even parsimonious – in their personal expenditure, very careful in their business outlays, stringent in their requirements of performance from paid staff. They are able to operate at very low cost. Only those staff who make a positive contribution to the enterprise are employed. Full use is made of family labour. Outlays on equipment and other capital expenditure are kept to the minimum.

Because of their local knowledge and use of family labour they can combine economically the purchase of farm output, sale of farm inputs and of consumer goods on a scale adapted to many rural areas with scattered population and low purchasing power.

Because of flexibility in decision making, private marketing enterprises have shown themselves especially well suited to handle perishable produce, livestock and meat and eggs and to see new opportunities for marketing initiatives. The costs incurred by a firm exporting fresh green beans from Senegal to France are shown in Table 6.2. The exporter pays the farmer 70 cents per kg. He finances the whole operation and carries the risk. Any price at Paris less than $2.50 per kg would mean a loss.

A structure of private marketing enterprises has the important advantage of continuity without drain on public resources. The critical factor for the small farmer is competition. Very large numbers of small farmers are helped with advice, inputs and cash advances by private traders seeking their business. The small farmer tied to a merchant money lender by accumulated debts has been a folklore figure. Systematic studies in Thailand, for example, recognise that there is a cost in transferring from one buyer/credit supplier to another, but show that farmers could carry it. They did move their business to another trading partner when it was advantageous to do so (Siamwalla, 1978).

The main role of government with a private marketing structure is to promote competition by assuring free entry of new enterprise, a legal basis for transactions and ease of access to price and other market information.

Table 6.2. Marketing costs of exporting Senegal green beans to Paris

	$ per kg
Price to grower	0.70
Assembly, grading, packing, loading	0.54
Air freight to Paris	0.81
Customs duty, handling charges	0.10
Wholesale market commission (10%)	0.24
Total	2.39

Source: Harper M. & Kavura R. *The Private Marketing Entrepreneur and Rural Development.* FAO, Rome, 1982.

Governments will also be called upon to moderate extreme price fluctuations where practicable.

Transnational marketing enterprises

These have helped substantially the small farmer able to participate in intensive production/marketing contract systems. He is assured a market outlet. He receives a full set of services on credit. The extension provided is better than that normally available because it is tailored to the needs of the market outlet served and the processes used. Initiated for export crops, this approach has been extended to sugar, tomatoes and other crops for domestic markets. Even so, only a small proportion of farmers can expect to obtain contracts with such enterprises.

The disadvantage is captive status. Mediation by a government department, as in Taiwan and Thailand, may be needed for equitable pricing and quality determination.

Farmers' associations or cooperatives

These can enable small farmers to economize on transport to distant outlets, undertake initial processing themselves and increase their bargaining power. Success depends on their managerial capacity, and their cohesion in face of opportunities for individuals to benefit by going outside.

In principle, cooperatives constitute a very favourable instrument for improving small farmers' bargaining power on the market and channeling to him new inputs and technologies. In practice, in the developing countries they have done best in assembling not very perishable crops such as coffee and cotton. In India and Kenya they have managed the regular routine of the dairy.

Conditions favouring cooperative or other group marketing are:

1. Specialized producing areas distant from their major markets.
2. Concentration upon and homogeneity of farm production for market.
3. Groups of farmers dependent on one or a few crops for their total income.
4. Availability of local leadership and management.
5. An educated membership.
6. Members with strong kinship or religious ties.

Intrinsic handicaps of the formal cooperative for very small farmers are lack of capital, if each member is to contribute an equal share; difficulty in finding competent motivated managers; and politicization. With their bureaucratic procedures, many cooperatives have difficulty in competing with private traders in prices and services to small farmers. In most countries they have needed special protection and substantial government financial and administrative assistance.

Development authorities

Typically these provide market, input supply, credit and technical assistance services to the farmers in a designated area associated with land settlement or reform. Sale of produce through the authority facilitates recovery of land charges and credit. Small farmers are cared for under such an authority if it is managed well. Replication of the models financed by external aid in the 1970s did not go very far because of the recurrent costs of such bodies and declining government revenues during the 1980s.

Parastatal marketing organizations

Through the 1960s marketing boards and other autonomous public bodies assigned an export monopoly over certain products by government, and demonstrated ability to carry out major assembly and sales operations. They standardized quality and packaging and achieved benefits of scale in transport, processing and sales. Farmers could count on receiving a pre-announced minimum price if they delivered their produce to the board's buying station or agent. Usually this price was cushioned against sharp fluctuations on international markets by a reserve fund. The defect of this system was that the prices paid to producers tended to decline as a proportion of the f.o.b. price in real terms. Exchange rates were maintained at levels adverse to exports; the export board was a convenient mechanism for tax collection; more was paid into reserve funds than was paid out to growers. In Ghana, for example, this led to massive smuggling of cocoa into neighbouring countries and to declining production.

Relieved of these adverse features a parastatal can still be helpful to small producers. As an alternative to highly fragmented export marketing channels, the sponsors of the Kenya Tea Development Authority insisted that growers face current world prices. As low-cost producers they have continued to prosper. The parastatal Interbras exploited the advantages of scale in Brazil without needing a monopoly. Lobster fishermen with only one or two boats sold to Interbras directly. It sent the lobsters frozen to the USA. Controlling the bulk of the supply from Brazil, it used cold storage to manage the market there and obtain better prices. For highly perishable export crops, such as green beans, stabilization is more difficult. Management of seed allocations on the basis of market intelligence and diversion of surpluses to other outlets may be the best course.

Parastatals to stabilize supplies and prices of food grains on domestic markets followed the export model. They bought grains and other products at pre-announced prices when offered at their buying depots or to their agents. Maintaining a minimum price in an otherwise free market they have been a valuable protection to farmers. However, where operated as a monopoly to enable them to cover the costs of pan-territorial pricing and of holding reserve stocks their prices have often been a disincentive.

A question general to all marketing boards is whether an organization set up primarily to concentrate export bargaining power or to stabilize prices will concern itself very much with services to small farmers. Frequently their buying unit staff require tips before they will give farmers attention. Women farmers have been discouraged by marketing boards' refusal to buy produce in small quantities. Several have declined to enter into the provision of fertilizers or the handling of credit repayments. Many have little concern for what happens before produce arrives at the buying station, leaving the small producer to pay for transport or depend on an intermediary.

Parastatal marketing organizations designed specifically to serve smallholders and the landless have been tried in various countries. They have the best chance of viability where:

- the small farmers are concentrated in a particular area;
- the farmers offer produce in which they have a cost advantage.

This is the logic behind the establishment of a 'cooperative corporation' in the Andhra Pradesh tribal area of India. It bought other produce and sold inputs and consumer goods, but the handling of forest products brought in by the tribals was its lifeline. While requiring government capitalization and subsidies, it was near to breaking even on current operations in 1989. Its prices to tribal suppliers were claimed to be 30% higher than those offered by private traders in areas where it had no buying point.

Where smallholders live amongst and grow the same crops as larger farmers assistance through a special enterprise is more difficult. The Ghana Food Distribution Company was supposed to buy directly from small farmers, so it bought only small-sized lots. In practice, it has been buying a small part of various farmers' marketable surplus.

ROLE AND REACTIONS OF RURAL MARKET PARTICIPANTS

We can now look at the various participants in low-income marketing systems and consider how they will act in relation to the alternatives open to them and the likely impact on their welfare.

A *smallholder* can sell:

1. to a private trader who comes to his house or the village which may have supplied credit;
2. to a private trader at a local assembly market if there is one within reach – here he can compare prices and purchase fertilizers and other supplies;
3. through a cooperative or development authority – if available – where he also obtains his inputs;
4. to a buying depot of a parastatal if there is one within reach;
5. to a processor under a contract if there is one in the area and his land is suitable for the crop sought.

He can also try to sell the produce fresh or with some processing retail, on a door-to-door basis, by the roadside, or at a rural market (where he could dispose of produce still in hand to a wholesaler).

Sharecroppers commonly hand over part of their crop to a landlord who will sell it along with produce from other tenants as a large-scale market participant; a progressive landlord will also buy inputs for his tenant. The balance of the cropper's marketable surplus can be sold as a smallholder, or can be sold through the landlord. The sharecropper's choice of crops may be limited by the landlord for convenience in sharing and handling.

The *landless* can:

1. raise pigs, chickens, rabbits, etc. around their house, and market their output like a smallholder;
2. earn wages in cash and kind working for farmers whose operations have become more intensive because of improved market opportunities;
3. undertake small-scale assembly, processing and retailing provided they can acquire some capital. In West Africa, the Caribbean and elsewhere this is common recourse for women traditionally handicapped in obtaining access to land.

Labourers can:

1. seek wages in cash and kind working for farmers whose enterprises are becoming more intensive, and for marketing, transport and processing enterprises;
2. attempt small-scale assembly and retailing like the landless, if they can acquire some capital.

Herders:

1. are mainly concerned about forage and water for their animals, but must sell some each year to buy food, clothes and other consumer goods and to pay for medicaments, etc.;
2. may sell locally to a private trader with a traditional social linkage, accepting delay in payment;
3. can drive stock, or have them transported, to a specialized market or abattoir;
4. can drive other herders' stock to markets, acquire capital and transport, possibly moving on to wholesaling livestock and/or retailing meat.

Fishermen can:

1. deliver fish to family members who sell it retail locally, drying, smoking or cooking supplies that cannot be sold fresh for sale later to a wholesaler;
2. sell to a wholesaler who has financed a boat, nets or other equipment and fuel;
3. sell at an organized market, through a cooperative or parastatal, or to a specialized processor if there is one within reach.

In most societies movement between working for a wage, producing on one's own account and engaging in trade is open for those with initiative and ability. Most families trading in agricultural produce, livestock and fish began in farming, herding and fishing. Successful traders who began as landless pedlars have moved into processing and then into commercial farming, livestock raising and fishing. However, such movements are restricted by traditional and social attitudes to varying degrees. Many old established communities look down on commerce and leave it to aliens whom later they resent. Sub-groups in a particular society may concentrate on processing and marketing because they are denied access to land.

Physical constituents of smallholder marketing

Essential if marketing initiative is to have an impact on a poor under-developed area are roads, transport vehicles, storage and distribution facilities, processing technology and materials, means of communication and information.

Many places are cut off from sources of fertilizer and markets for their produce during certain seasons. There are many more where the cost of *transport* over bad roads, and the time it takes, inhibit marketing of the more valuable perishable products. A study in Ethiopia showed truck charges over earth roads with deep gullies and wet spots five times those over tarmac. In Zimbabwe, with a relatively good road system, the limit to lower cost group purchasing of fertilizers is set by accessibility by a loaded vehicle. Tarring the main road into a project area and opening up, or rehabilitating, feeder roads has led to reductions in marketing costs of up to 60% and the use of fertilizer on holdings that were inaccessible before.

For much of Africa it has been said that during the 1960s and 1970s the infrastructure for marketing was good, but government policies were adverse. In the 1980s policies improved, but achieved only modest results because the infrastructure had become defective. Port handling was delayed, electric light and power uncertain, spare parts and fuel for transport and processing were lacking. Not only do such uncertainties and difficulties add to the costs of marketing produce and obtaining supplies; they can reach the point of discouraging production. The small farmer who sees his produce deteriorating for lack of transport to a market does not easily lay out funds for production the following year. Those growing perishables at a distance from their market are especially vulnerable. The Kigezi Vegetable Growers Cooperative Federation with over 4000 small farmer members collapsed when regular transport to Kampala could no longer be assured.

Because of transport and access uncertainties, land-locked countries such as Malawi incur double and triple their normal costs on imported fertilizers. In addition, there are costs of holding large stocks to be sure that supplies are available when needed. Some of these extra costs are beyond the responsibility of national governments. Unfortunately the impact on farmers and fishermen of these delays and difficulties does not always receive the recognition in central government circles that it merits. Priority in the allocation of available foreign exchange does not go to the provision of fuel, transport vehicles and spare parts to keep them running. Nor, in many countries, is adequate attention given to equipping local repair services with the resources and materials needed for an adequate maintenance.

Storing grain in their own home, or in a special structure nearby, is most convenient for smallholders. It can be supervised directly. Participation in group *storage* arrangements involves transport to and from the store, and the contentious issue of assuring to each participant that the grain he takes out is equivalent in quality to what he puts in. Use of insect resistant varieties of maize, rice and cowpeas reduces storage losses; family and other consumers in the production area may also prefer them. Sachets of insecticide powder convenient for smallholders are available, but involve

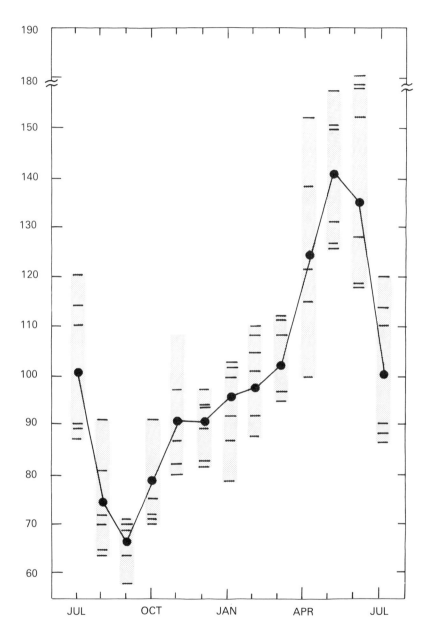

Fig. 6.1. Seasonal price movement for maize, Atebubu, Ghana 1965–1974. Source: Ghana, Ministry of Agriculture, *Monthly Food Situation Report.* Horizontal bars show prices in individual years, large dots show monthly averages for the period.

a cash outlay that may not always seem justified.

The main purpose of storage, in addition to conserving supplies for family use later in the year, is to avoid having to sell at the low prices usually prevailing just after harvest. Illustrative of seasonal price movements are those shown for maize in Ghana in Fig. 6.1. Wholesale prices of paddy in the Madras area of India rise seasonally by about 30%. A post-harvest price increase of 5–17% is needed to cover the cost of grain storage, depending on its duration and the interest rate on finance.

Price ranges between the peak maturing and off seasons for perishables can be very large. However, storage via refrigeration is expensive. The figures in Table 6.3 are derived from FAO experience in the Near East and show that storage costs can absorb much of a wide seasonal range in price. They refer to a price difference of $60 per tonne between the time of picking apples and 6 months later. The profit margin in storing was only $4.70 per tonne. Often it is better to look for changes in production and marketing that would reduce the need for storage, e.g. extend the harvesting season by planting early and late varieties, use of plastic to promote early growth or late maturity.

The use of suitable *processing* technologies can extend greatly the market for farm and fishery products. It can reduce waste, improve palatability, prevent spoilage and simplify handling and transport. It can adapt products to the needs and preferences of consumers in distant places.

Well suited to small-scale operations are the husking of paddy and grinding of wheat and maize for local consumption, the squeezing of oil from palm fruit, the processing of cassava to eliminate the bitter element and offer it for sale ready to eat as gari, the drying of tomatoes, drying and grinding of pepper, the drying and smoking of fish, the preparation of

Table 6.3. Storage costs and seasonal range in price of apples

	$ per tonne	
Estimate increase in price over 6 months		60.00
Storage costs		
Rent (or overhead cost if self owned)	21.00	
Storage losses (weight and quality)	23.80	
Rent of crates	4.30	
Interest (10%) on initial value of stocks	6.20	
Total costs		55.30
Net margin		4.70

livestock into various forms of meat, hides and skins and of milk into a range of dairy products. Many of these have special appeal that goes beyond purely local customers. Hygienically presented and packaged, they can be marketed successfully to higher income consumers. Use of an attractive traditional package will add to the appeal. The preparation and marketing of appealing processed products may be the fastest way of increasing smallholder incomes.

Caution is needed in proposing the adoption of equipment developed for different conditions. It was recognized early in a Nigerian fisheries project that the standard smoking kilns available might not be suitable for the people on the project. The first step would be to test them out with small groups of women, and modify them according to the comments received.

Small-scale *assembly, wholesaling and retailing* of farm and fishery produce can begin from the operator's house. It involves no additional cost, minimizes time spent going to and from work, and facilitates part-time use of family labour.

Sale at an organized local market has the advantage of access to alternative buyers and to information on current prices through enquiry of a number of sources. Table 6.4 presents information obtained by an FAO assisted survey of the smallest farmers in Kenya. It shows that a large proportion of them sold their produce at rural markets – though for many products guaranteed prices could be obtained from a marketing board. Less transport was involved. For some varieties, at certain seasons, and for produce sold retail, the price at the market would also be higher.

Such markets should be within reach by foot or bullock cart, i.e. within 12 km distance. The outlay on fixed facilities can be low, using local materials. Important is ease of access under all weather conditions, protection of market users against theft and harassment, and the provision of relevant information on prices. Extension of rural market networks is specifically helpful to the smaller growers. The larger farmer can more easily justify transport to an outlet that is further away.

Assembly markets attended by wholesalers with access to transport expedite sales to more distant outlets. Location on a trunk road and good communications, preferably by telephone, are the essentials. At Asseswa in Ghana large quantities of maize are assembled on the ground. Two market centres serve the mountain producers of European type vegetables in Java. The only physical facility is a blackboard showing prices reported by telephone from Djakarta and other cities. Growers meet wholesalers there in the morning. When a sale is agreed they go together with a truck to a place where the produce is stacked ready by the roadside.

Wholesale markets for fishermen are located at convenient landing points with a good road to consuming centres. Paved areas where fish can be displayed, shaded against the sun, with running water for washing them down are the basic facilities. Sales proceed by negotiation, or by auction where the quantities offered and the number of buyers justify it.

Central wholesale markets develop primarily to serve urban retailers of perishable produce needing fresh supplies regularly and should be located within easy reach by the transport vehicles they use. Ease of access from the main producing areas is another important consideration. Small growers at a distance usually send produce individually or in groups to commission agents. Where space is available they may also be allotted areas from which to sell directly.

The environs of such markets, and of rail and bus stations, are also favoured by migrants from rural areas selling produce retail. As mobile pedlars evading market charges, they hope to make a margin on a small capital outlay or credit. Local government and market administrators are often opposed to such informal traders. They are denied sales opportunities, and are moved on by the police. Beginning as petty retailers they earn some income when otherwise they would be unemployed. Some of them may build up the capital and 'know how' to undertake extended operations.

Essential for effective marketing, also at the smallholder level, are convenient means of *communication* and access to reliable *information*. The substantial differences in prices for similar produce within countries such as Nigeria reflect delays in the arrival of reliable information that could be the basis of arbitrage transaction. Use of standard weights, measures and quality specifications that are generally understood helps make possible sales by telephone or letter without the need for a personal visit to examine each lot. Uncertainty over the amount and quality of produce offered impedes sales over long distances and leads buyers to discount their offers. That regular traders obtain information by word of mouth or over the telephone while irregular market participants – especially the smaller farmers and consumers – have no such channel, can load price bargaining against them. Insufficient allowance is sometimes made for the effect of competition in generalizing the benefits from information received by private communication. Even so, market transparency can be improved greatly with the issuance of crop forecast data, setting up blackboards in assembly markets showing the prices of the previous day and those at other relevant markets, and regular reporting on the radio.

Many of the most conspicuous constraints on marketing by small farmers and fishermen flow from infrastructural inadequacies. The timely movement of crops or arrival of inputs can be blocked by roads that become impassable or a truck that cannot move. Produce can not be protected against humidity, materials needed for processing may not be available, information on imports or other events that will affect prices on smallholders' markets may not be passed on.

Table 6.4. Smallholder sales (%) by place: Kenya

	On holding	Rural market	Roadside	Processing plant, milk collection centre, etc.
Maize	26	45	8	21
Beans	22	57	11	10
Finger millet and sorghum	6	90	4	—
Sugarcane	39	55	3	3
Potatoes	47	51	2	—
Cabbages	44	48	7	1
Tomatoes	5	86	8	1
Bananas	32	64	4	—
Chickens	37	48	5	10
Eggs	14	71	10	5
Milk	21	7	28	44
Sheep	49	50	—	1
Goats	39	61	—	—
Cattle	46	43	3	8

Source: FAO/Government of Kenya. Integrated Rural Survey, 1975.

Institutions and policy

The elements of an economic environment favourable to development are well known: a realistic exchange rate; fiscal and price incentives for both exports and the production of domestic food supplies; availability of foreign exchange for transport and processing equipment, supplies and spare parts, and incentive consumer goods for rural populations; and confidence that a full range of marketing enterprises can continue to operate without abrupt changes in policy, arbitrary price controls and harassment by local officials. A price policy that makes imports expensive benefits small producers, processors and traders. It makes their output more valuable. The impact of such policies was evidenced by the expansion of agricultural production for export and to meet domestic food requirements in Ghana in the late 1980s. This was stimulated by high black market prices with only limited supplies on sale at the low official prices. To placate politically powerful urban consumers many African governments kept food prices artificially low. In compensation they offered subsidized fertilizers and credit, but supplies were rationed and tended to go to the larger farmers and the politically influential. Scarcity of foreign exchange and other priorities for its use left shops in the rural areas bare of imported items. Renewed availability of consumer goods helped revive agricultural production in Tanzania.

Finance is needed in rural areas for marketing as well as for the purchase of production inputs. Farmers and traders need credit to hold stocks to meet consumption needs and wait for favourable prices. This has long been recognized in India where public warehouses were established. They could issue certificates on produce in stores that were negotiable for credit at a bank. The produce could be taken out of store when the loan was repaid. This procedure works well enough with large standard lots that can be sold in store by description. For small farmers with small lots it is cumbersome. Most convenient for farmers is extension after harvest of a loan taken to pay for inputs. This need not present much difficulty where land titles can be offered in security. Where, however, allocation of the next year's loan for inputs is conditional on full repayment of the previous year's, it can disrupt timely arrival of the next year's inputs. Group management of credit for holding stocks on the farm or in joint storage should accommodate differing views on its duration.

The scope for raising the value of small farmers' market output through practical advice and the application of research is great indeed. Farmers need to plan their sales to the best advantage. Many lag in matching their output to market requirements. They tend to offer small lots surplus to family needs, irregular in variety, quality and maturity. Such suppliers cannot be counted upon from one season to the next, so they do not get priority from buyers serving an established consumer clientele.

Processing methods for farm and fishery products, the best form of packaging (taking into account constraints on materials) and use of market information to time sales to the best advantage are all matters for marketing extension.

It is now recognized that much less should be expected of government than in the past and more should be done by local initiative. Even so, a professional *marketing policy and support unit* is needed to:

1. guide policy on marketing enterprises and facilities, monitor their performance and ensure that credit, transport and essential supplies are available when needed;
2. coordinate continuing support services including market information, marketing extension, standardization of weights, measures and quality specifications, export quality controls, etc.;
3. maintain a continuing focus on the use of labour intensive procedures in marketing and processing, and on measures to assist the smaller production and marketing enterprises.

Some countries have such a marketing department, but it is still oriented to price fixing and parastatal intervention.

In most parts of the developing world *women* have a major role in agriculture as workers, and in some as farm operators. Because the land, however, is in the name of their husbands they have difficulty in obtaining credit, and sometimes also in obtaining use of transport for marketing. Because of long-standing male predominance in mixed group activities, they have little to say in decision making by cooperatives. It is in protest against this male predominance that womens' group savings and input purchase activities proliferated in Southern Africa, outside the cooperative system.

In the long run it would be preferable to integrate women into the main line of rural institutional development. This may not be practicable for them until they have full control of their assets and the product of their work. It was in reaction against discrimination over access to land that women in coastal West Africa, the Caribbean and the mountain regions in Latin America became so active in domestic marketing. Special programmes have been instituted to facilitate their operations – provision of child care and rest places at markets, access to information and credit. In Gambia, women can obtain credit if they are registered leaseholders of land (Jiggins, 1989). The critical requirement for women in agriculture and its marketing is that they be relieved of discriminatory handicaps, at least in access to official services. The full weight of government is needed to combat local traditions and promote market, input supply and credit systems where women participate on equal terms with men.

Structural adjustment and privatization

Features of the liberalization of agricultural marketing through the 1980s have been the relaxation of parastatal monopolies, and of controls on prices and movements of produce. Exchange rates and subsidy policies were oriented to provide more incentive for exports and for individual initiative. Private enterprises became fashionable because they were self-supporting and not in continuing need of state assistance. These changes are expected to promote the movement of resources into activities with a natural comparative advantage. Will this mean that, while some producers and traders go ahead more rapidly, others are left behind and the poor become still poorer?

How well has the private sector been able to fill the gap left by the retreat of government? In most situations of direct government intervention, a viable private trading sector remained. It took several forms: as buying or distribution agents for the intervention agency; as trade channels for quality grades and preferred varieties to serve market segments not covered by the intervention agency; as standby speculators stepping in whenever an agency ran out of purchasing funds or storage space; or as organized smuggler systems where official crop procurement prices eroded by inflation provoked the establishment of direct uncontrolled links with more remunerative cross-border markets. Classic examples of smuggler systems operating during the late 1970s and early 1980s were Ghana/Ivory Coast (cocoa), Gambia/Senegal (groundnut) and Burma/Thailand (cattle) (Reusse, 1987).

Few, if any, intervention agencies dealing with local produce for domestic consumption handled more than 50% of the quantity marketed. In some countries they operated in parallel with free market enterprises. In others where they had an official monopoly of wholesale transactions there was substantial evasion. So in most countries farmers and small traders had considerable operational experience of domestic marketing. The inability of governments to guarantee supplies at official prices that camouflaged inflation gave them an opportunity to build up capital.

In export and import marketing, by contrast, the private sector was often replaced completely, apart from cross-border smuggling. Here, however, in many countries old established export/import houses from pre-intervention times maintained a legal and operational frame with a latent ability to return to action. Their proprietary links with foreign merchant houses permit most of them to draw on external resources for the financing of import or export transaction, a reason for governments under severe budget and balance-of-payments constraint to welcome their re-entry on the market, as in Nigeria when the export boards were abolished, and in Zaire.

The main constraint on small private marketing enterprise in expanding in line with the opportunities open to it is usually finance. The abolition of ONCAD in Senegal left forthwith a much larger role to indigenous private enterprises. Initial assembly from small growers was undertaken by traders whose limited capital base (often US$30 or less) permitted them to buy only 3–15 kg at a time. On completing a sack-full they had to sell it before they could start buying for another. Finance for such operations came from personal funds, relatives and larger wholesalers; only through the wholesaler would some bank credit reach the small assembler level. The average cost of this informal credit was around 7% per month in contrast to official bank charges of 1.1% monthly (Newman *et al.*, 1985). In Tanzania and Zambia the banks have been slow to lend to private marketing because they had so much credit still outstanding to the parastatals. Access to vehicles has also been an obstacle. In Tanzania, small-scale traders handle 60% or more of internal produce marketing yet can only obtain vehicles through unofficial channels.

A clear lead from government is needed to develop confidence. Investment in storage and other permanent facilities is discouraged where continuity of operation is uncertain, as when governments concede it grudgingly under external pressure. With no assurance of continuity, private enterprise is likely to concentrate on the most immediately profitable activities. It will be unable to cope with an exceptional harvest calling for greatly increased financial outlays and risk acceptance. It will be reluctant to serve outlying production areas with poor transport connections.

Risk of exploitation

Small farmers are exploited when, for lack of information, because of indebtedness or the absence of an alternative outlet, they are paid a lower price than necessary to cover the costs of the marketing enterprise concerned, including a reasonable return for risk and management. They are also exploited when obliged to sell to a parastatal monopoly which offers only low prices because of excessive costs or discriminatory government policy.

Under the old private marketing system in Ethiopia many farmers were paid low prices because of high transport costs, wide margins and some cheating over weights. Overall, however, grain markets there were said to be well integrated and prices reflected costs. Under the state system of the 1980s those farmers who had access to private buyers received much better prices than through the official channel. The Siamwalla (1978) review of marketing in Thailand noted that farmers there were able to break credit ties with private traders. Margins were in line with costs, except for some crops where transnationals had invested in new facilities and were using a semi-monopoly position to recoup their capital outlay.

The inference is that the small farmer is less likely to be 'exploited' where he has a choice of buyers, each with easy access to finance and transport, than when he can only sell to a parastatal – unless that parastatal is subsidized from non-agricultural resources. Access to a subsidized reserve outlet would, of course, strengthen his position. Liberalization and financial constraints in many countries will not preclude retention of some government mechanism for purchasing to prevent market failures for small farmers. Arranging transport to such a buying point would be a logical recourse for a group of farmers dissatisfied with the other outlets open to them.

Steady pursuit of a policy of lowering the barriers to market entry is likely to be the most economical and enduring way to prevent exploitation. Legal monopolies can be abolished. Under donor pressure President Marcos of the Philippines finally freed coconut oil exports from the monopoly assigned to the mills of his associate Cojuangco. Unimpeded transport access is a vital condition. News of supplies awaiting buyers should then attract new enterprises. Provision of market information and making available transport vehicles and finance for purchasing and storage are the most direct measures. The entry of participants with a different profile of interests helps sharpen competition, e.g. a buyer with his own retail outlet, one bypassing an intermediary market to serve directly a more distant consumer centre, one buying for processing, farmers who take up the marketing of their own and others' produce, and farmers' associations.

Limited sales space at assembly and central wholesale markets inhibits new competition, particularly if local authorities require all produce sold wholesale to pass through such markets. Pressure can be brought for the enlargement, relocation, supplementation or bypassing of such markets. Reservation of space at assembly and central markets for irregular and seasonal operators – producers and occasional traders – on payment of a small fee, eases their way into marketing operations.

A path-breaking initiative by IFAD has been the provision of finance for small-scale producers, traders and transporters to enable them to go further in marketing and so intensify competition. It was stimulated by the de-regulation privatization process.

In some project areas previously benefiting from pan-territorial pricing there was no buyer for the main cereals at harvest time. Vehicles were provided; farmers were allowed credit to hold stocks to take advantage of higher prices after the postharvest low point. The accumulation of stocks of processed produce to await a favourable sales opportunity was another basis for credit.

Access to inputs

Commercial planting materials, fertilizers and pesticides have been used in the developing countries mainly for export crops and by larger farmers. They became important for the main body of smaller farmers in Africa, particularly, with development of the higher yielding hybrid maize. They must obtain new seed each year and apply fertilizers to realize its potential.

Requirements for efficient provision of these inputs are:

1. timely access to supplies on the part of distributors;
2. pricing procedures that cover essential costs and provide an incentive for performance;
3. access to finance at each stage in the channel through to the user;
4. suitable transport access and availability of vehicles and servicing.

All these requirements present problems under the financial limitations of sub-Saharan Africa. In most countries distribution has been through an official body or a government cooperative system. Prices to farmers have been equalized over the whole country by subsidy, but supplies have not always been available when needed. The policy stance under structural adjustment is to shift to competing private enterprise channels and reduce the burden of subsidy.

Fertilizers and other agrochemicals merit import priority and internal subsidy up to the point where they cease to pay off in foreign exchange earned by production for export and savings on food imports averted. For users within these criteria, supply services should receive high priority in timely allocation of foreign exchange, freedom of individuals and enterprises to take new initiatives, use of port handling facilities and storage, and access to transport means. For users left aside under such a programme, research and extension should be strengthened to develop alternative ways of maintaining soil fertility, crop yields and farm incomes.

Where foreign exchange, domestic subsidy funds and transport means are all constraints, such inputs as can be imported or produced domestically should go to the crops and land affording the greatest return – net of input and crop-to-market transport costs. In many places the use of fertilizer on traditional varieties of maize is uneconomic without a subsidy. Competitive distribution with free pricing is likely to promote an economic pattern of use.

The main roles for government are then to:

1. promote competition by removing obstacles to enterprise in obtaining bank finance, access to transport and spare parts, and ensuring that these are available on equal terms;
2. intensify local research and extension on cropping and other husbandry procedures to maintain soil fertility and crop yields for areas where mineral fertilizers are uneconomic.

The selection, development and distribution of advantageous seeds will continue to be important for both high- and low-input agriculture. Generally, competition with imported lines distributed through domestic wholesalers and retailers is needed to ensure an adequate and timely service, and likewise for specialized agrochemicals.

The overall drift of such a set of actions is to put input supply, for the main body of farmers, along with crop and livestock marketing, onto a self-sustaining basis. Governments of poor countries will then be in a much better position to collaborate with external aid in devising enduring ways of helping those with the lowest incomes.

Exports and the rural poor

There can be no doubt about the advantages of producing for export where the conditions are favourable. While earning the foreign exchange needed to pay for essential imports, the original settlers in the Gezira became the aristocrats of agriculture in Sudan. Now their children pay immigrant workers to pick the long staple cotton. The returns from coffee production in Kenya in 1989 were about 30 cents per work hour once the trees were established, as against 6 cents from maize for local consumption. Farmers in Guatemala growing snow peas for the North American market grossed 14 times more per hectare in 1989 than from maize. Net of costs, their incomes have doubled. Income gains were highest on the small farms that provided their own labour and among landless households that found increased employment (Von Braun et al., 1989).

Production for export markets and for domestic food intake were proposed as contrasting alternatives in socially oriented writings of the 1970s. There were situations where concentration on well-paid export crops left the local population short of food supplies. Food failed to arrive from elsewhere in response to the new market demand because of lags in marketing initiative and drought. This led tea growers in Rwanda, for instance, to neglect their tea and plant food crops between the rows. Most programmes that produce sugar, tea, tobacco, etc., for export processing now foresee this issue. Food crops are incorporated into the rotation. Alternatively, land is set aside for subsistence cropping.

Generally, export cropping enhances food crop production because farmers learn the advantages of applying fertilizers and other inputs. This is well demonstrated in cotton production in countries such as Burkina Faso. A food crop following cotton benefits from fertilizer residues left in the soil. Use of purchased inputs on food crops is also favoured where the proceeds of export crop sales are available to repay credit.

Increased incomes for those engaged in production and marketing for export have sometimes been seen as detrimental to the poor in the area concerned because they force up the cost of essentials for living. The Dolefil pineapple operation in the Philippines has been criticized for this (Tavis, 1982). The immediate beneficiaries of an enclave-type processing operation may indeed be 200–400 families, plus the direct and indirect employees of the processor. In fact, the benefits of intensive pineapple and sugar production, broiler raising, shrimp farming, etc. extend much further in that they support new demands for housing materials, consumer goods, feed grain and a range of services. Satisfying market demands resulting from increased income in these related sectors extends further the indirect benefits.

Sociologists have stated that cooperative dairy projects in India for improving middle-class diets in distant cities operate at the cost of lower nutrition in the production area. George (1986) maintained that the mobilization of feed resources for small-scale commercial milk production in India absorbed grass and crop residues formerly available free to lower segments of the rural population. Using statistically significant sampling techniques, however, Mergos and Slade (1987) concluded that participants in the Madhya Pradesh dairy project benefited from it; to non-participants it was neutral.

Phases of relative disadvantage for particular sectoral groups are inherent in the development process. Those who feel left behind at one stage should benefit eventually from the new economic environment that has been created and the opportunities for employment and enterprise that it offers.

Participation and sustainability

A continuing thread through the preceding discussion has been how to maintain marketing improvements and services when governments have not the means. An undertaking to maintain facilities, institutions and services after aid funding ceases is of little value if the government has no money. They must, therefore, generate their own income or be conditional on voluntary contributions of labour by prospective users and the establishment of procedures for maintenance by them subsequently. The management responsibility must be hired off to local organizations, traders' associations and private enterprises that will seek payment from users (Mittendorf, 1989).

Recent IFAD projects have stressed user participation in planning and labour contributions for access road improvement. The Madagascar Highlands project has trained 80 roadmen to guide maintenance by villagers. Potential users of a new cattle trekking route from Nyala to Omdurman in Sudan are being invited to enter a private company that will manage the wells and collect charges.

IFAD has moved away from the provision of standard processing and storage equipment. For a Nigerian fisheries project it was recognized that the usual smoking kilns might not be suitable. The first step would be to test them out with small groups of women, and modify them according to the comments received. It has accepted the smallholders view that credit for storage construction is often used most economically through adapting part of his house. Of the metal bins envisaged for a project in Honduras only half were put to the use intended. Storage away from the smallholder's home means recurrent journeys for the user and scepticism over its management. In Bangladesh, IFAD noted that village stores were not a success and opted for assistance in storage at the farm.

High performing local marketing extension staff can be rewarded from aid budgets; but how can their interest be maintained when they fall back on local salaries? Some arrangements whereby they participate in the benefits from the intelligence and advice they provide need to be tried out. They might begin with small payments in kind.

In the 1970s external aid encouraged the governments of low-income countries to help the poorest of the poor. The intensive development area approach placed a heavy burden on local administration. There could be revenue gains in the long run; in the meantime there were recurrent costs.

The strategy for poverty alleviation implicit in this paper combines it with growth. Improvements in the poverty sectors strengthen the overall economy if they are self-supporting. Our intention is to promote initiatives for progress through incentives to individuals. The main requirement of government is then to assure the necessary physical infrastructure, provide strategic support services, and maintain consistent policies for employment creation and initiative. The prices of imported capital equipment and of energy can be raised to promote the use of labour-intensive alternatives. Tax and other incentives can be provided for enterprise in putting people to work. Barriers to new entrants into marketing and processing can be removed, thereby increasing competition as well as opening up ways to additional income and employment raising activities.

REFERENCES

Abbott, J.C. (1987) *Agricultural Marketing Enterprises for the Developing World.* Cambridge University Press, Cambridge.
Abbott, J.C. (1988) *Agricultural Processing for Development.* Gower Publishing Co, Aldershot.
Drucker, P.J. (1958) Marketing and economic development. *Journal of Marketing*, January.
George, S. (1986) *Operation Flood.* Oxford University Press, New Delhi.
Jiggins, J. (1989) How poor women earn income in Sub-Saharan Africa and what works against them. *World Development* **17**(7), 953–963.
Lipton, M. (1989) New strategies and successful examples for sustainable development in the Third World. Report 170 IFPRI, Washington.
Mergos, G. and Slade, R. (1987) *Dairy Development and Milk Cooperatives: The Effects of a Dairy Project in India.* World Bank, Washington.
Mittendorf, H.J. (1989) Improving agricultural physical marketing infrastructure in Africa through more self-help. *Journal of International Food and Agribusiness Marketing* **1**(1), 9–27.
Newman, M., Sow, P.A. and Mdyoe, O. (1985) Regulatory uncertainty government objectives and grain market organization and performance: the Senegalese case. Paper presented at the XIX Conference of the International Association of Agricultural Economists, Malaga.
Reusse, E. (1987) Liberalization and agricultural marketing. *Food Policy* **12**, 299–317.
Siamwalla, A. (1978) *Farmers and Middlemen: Aspects of Agricultural Marketing in Thailand.* UN Asian Development Institute, Bangkok.
Simmons, E. (1975) The small scale rural food processing industry in Northern Nigeria. *Food Research Institute Studies* **14**(2), 147–162.
Tavis, L.A. (ed.) (1982) *Multinational Managers and Poverty in the Third World.* University of Notre Dame Press, Notre Dame, Indiana.
Von Braun, J., Hotekkiss, D. and Immink, M. (1989) *Non-traditional Export Crops in Guatemala: Effects on Production, Income and Nutrition.* IFPRI, Washington.
World Bank (1990) *World Development Report 1990.* World Bank, Washington.

II
MARKETING ENTERPRISES

Private marketing enterprise was out of favour through the 1960s and well into the 1970s. This was the era of development planning. An array of small family businesses was thought difficult to manage. Van der Laan's book on the Lebanese traders in Sierra Leone is notable for a measured treatment of a group often thought highly exploitative. For a popular view see *Mr Khoury* (J. Bingley, Constable, London, 1952). Case studies of individual contributions to development through marketing initiatives became available in the books of H. de Farcy, e.g. *Commerce Agricole et Développement* (Spes, Paris, 1966), the FAO meeting report *The Role of Entrepreneurs in Agricultural Marketing* (Rome, 1972), and *The Private Marketing Entrepreneur and Rural Development* edited by M. Harper and R. Kavura (FAO, Rome, 1982). Published in Bangkok, the significance of Siamwalla's paper summarized here was recognized later. Reviewing various studies he noted that for most crops there was little evidence of exploitation of Thai farmers by the prevailing structure of private marketing enterprises. While many farmers took credit from them they were still able to shift from one trader to another for better terms. See also G.J. Scott's *Markets, Myths and Middlemen: A Study of Potato Marketing in Central Peru* (International Potato Center, Lima, 1986).

Much detailed information on small-scale marketing enterprises has been assembled by sociologists. Examples are A.G. Dewey's *Peasant Marketing in Java* (Free Press of Glencoe, New York, 1962), papers by S.B. Mintz on the Caribbean and more recently *Markets and Marketing*, edited by S. Plattner (University Press of America, 1991).

In the 1970s the aid agencies came under criticism for neglecting sociological considerations in their development projects, so they included sociologists in their integrated rural development teams. However, it was not easy to obtain constructive proposals. T. Scarlett Epstein (*Urban Food Marketing and Third World Development*, Croom Helm, London, 1981) was one of the first to recognize this. She helped found the periodical *Rural Development in Practice*.

In the 1950s and 1960s the sociologists observed small-scale trading by women along with men. They saw women operating as highly rational economic beings. In coastal West Africa they handled 80–90% of the food trade. They wielded considerable power. When Sekou Touré's government in Guinea tried to force them into a communist mould they raided the police station and tore up the files. In Ghana they were said to be behind a coup that brought down a government. With the feminism of the 1970s, attention shifted to the difficulties they faced. The passages included in this chapter are adapted from Janice Jiggins, a leader in this field.

Transnational marketing enterprises have grown steadily in spite of expropriation in some countries. Their size and power sparked a reaction in the 1970s. A United Nations

Centre on Transnational Corporations was established followed by a spate of books on their negative aspects, e.g. Susan George's *How the Other Half Dies* (Allenheld Osman, Montclair N.Y., 1977), B. Dunham & C. Hines' *Agribusiness in Africa* (Earth Resources, London, 1983). D. Morgan's *Merchants of Grain* (Viking Press, New York, 1979) provides the lively detail of the investigative journalist.

By the end of the decade the transnational issue had gone off the boil (P. Streeten: *Multinationals Revisited*, Finance and Development, June, 1979). New forms of transnational marketing enterprise were appearing, as examined in Chapter 10. Of the authors of this paper, R. Saint Louis is professor in agricultural economics at Laval University, Quebec. His co-authors combine studies there, at Michigan State and at Montpellier.

Cooperative studies figured prominently in the marketing literature of the 1960s. Most were descriptive, and sympathetic when things went wrong. It was 1970 when R. Spinks pointed out that governments often turned their backs on the realities of the policies they applied – 'Attitudes towards agricultural marketing in Asia and the Far East' (*Monthly Bulletin of Agricultural Economics and Statistics* **19**(1), 1970. FAO, Rome). Uma Lelé's 'Cooperatives and the poor: a comparative perspective' (*World Development* **9**, 55–72, 1981) reflected recognition that cooperatives were being captured by the dominant elements in rural society. The first influential warning, however, that trusting to government-installed cooperative systems could be directly prejudicial to development came from Guy Hunter of the Overseas Development Institute, London, and G. Hyden of the University of Dar es Salaam (*No Shortcuts to Progress: African Development Management in Perspective*, Heinemann, London, 1983). Hunter's conclusions drawn from a seminar at Uppsala are presented here along with passages from an appraisal of cooperative marketing achievement in developing countries prepared by J.Feinberg for COPAC, the joint agency agricultural cooperative promotion unit in Rome.

Experience with marketing parastatals up to the 1960s was brought together in the FAO Marketing Guide No. 5 *Agricultural Marketing Boards: their Establishment and Operation* by J.C. Abbott and H. Creupelandt, 1966. Papers presented at international meetings on marketing boards in Africa are reproduced in *The Marketing Board System* (eds H.M.A. Onitri and D. Olatunbosua, Nigerian Institute of Social and Economic Research, 1974) and *Marketing Boards in Tropical Africa* (eds K. Arhin, P. Hesp and H. L. van der Laan, Kegan Paul, London, 1985). W. Jones' paper presented here was prepared for the Leiden meeting, 1983. Van der Laan and his colleagues at the African Studies Centre have followed up with constructive papers, as also have G. Antony and E. Fleming with 'Statutory marketing authorities in the third world: recent changes and conclusions' (*Journal of International Food and Agribusiness Marketing* **3** (2 and 3), 1991).

Performance analyses of particular parastatals are few. Notable for its basis on inside experience is that of the Food Corporation of India by V.K. Garg *State in Foodgrain Trade in India* (Birla Institute of Scientific Research, Delhi, 1981).

Case studies of marketing enterprises by types and an assessment of their relative suitability for differing tasks and conditions are presented in Abbott's *Marketing Enterprises for the Developing World*, (Cambridge University Press, 1987).

7

THE LEBANESE TRADERS IN SIERRA LEONE

H.L. van der Laan

THE PRODUCE TRADE

In 1970 produce receiving (or buying) had become a routine matter, both for farmers and traders. Farmers seemed well informed about prices and qualities and knew the approximate weight of their parcels, presumably because they had measured the volume before bringing it to a trader. The agents and sub-agents of the Marketing Board could be recognized by their bigger stores and the weighing machines at the doors. There were further smaller traders who might be called unofficial buyers, because they were not recognized by the Board. Their stores were smaller and they still used hanging scales or table scales for weighing. Several Lebanese operated as unofficial buyers. Usually they were poor and dependent on a relative living nearby, who was an agent or sub-agent, to whom they sold their produce. It seems that the agent or sub-agent paid somewhat more than the producers' price to the unofficial buyer to keep him in business, so the relation contained an element of subsidy or charity. On the other hand, it was convenient for the agent or sub-agent that he now had the opportunity to send some farmers on to the unofficial buyer when he himself was very busy. Usually, farmers with small parcels, diffident sellers, and the women with a few pounds of palm kernels were sent on.

I estimate that the Lebanese, as agents, sub-agents, and unofficial buyers, received some 70% of the scheduled produce in 1970. This high percentage could hardly have been predicted in 1949, when the future of the Lebanese in the produce trade seemed bleak. One can attribute this unexpected development partly to the resilience of the Lebanese, but there is another factor as well. Many Lebanese have an unshaken faith that the produce trade justified their presence in the country in the past, and that it

---

Source: Selected passages from *The Lebanese Traders in Sierra Leone*. Copyright Mouton & Co., The Hague, 1975.

still does so today. It is in the produce trade that the traditional character of Lebanese enterprise reveals itself most strongly. Participation in it, even as poor unofficial buyers, is dictated by the past, often by the expressed – or assumed – wishes of the trader's father. Second- and third-generation immigrants dominate in the produce trade; new immigrants are rare. There are more conservative men in the produce trade than in any other activities in which the Lebanese are engaged. Fortunately, there are still opportunities for a conservative trader, as many farmers remain faithful to a particular one. Villagers who travel to a town may take their produce to the trader who formerly lived in their village, and they can enjoy talking about common acquaintances. Often the son inherits these customers, or, if he grew up in the village, he is patronized by the friends of his youth. We must, however, expect a slow decline in the Lebanese share in the produce trade as personal links between farmers and Lebanese traders fade away.

The European trading companies CFAO, PZ, and SCOA still played a role in 1970 in financing both their own stocks and the operations of their sub-agents. They maintained stores in Freetown, in which stocks were sometimes kept. Stores belonging to the Companies or to bigger Lebanese agents provide a temporary solution when the Board's stores cannot take delivery of all produce and a long queue of lorries is waiting at the quay to be unloaded. Bags of produce which are rejected by the Board's storekeepers are taken to the private stores of the agents to be cleaned, dried, and rebagged. As some products are easily affected by rain during transport, e.g. coffee, the need for private produce stores in Freetown will continue.

There was also unscheduled produce in which the Lebanese could trade without needing the approval of the Board. These products were piassava, ginger, and coffee, and our interest is centred on the trading situation in the years that these products were not on the schedule. Normally the Board has been unwilling to put piassava on the schedule. The market is small, and buyers are few and have special requirements. Moreover, the export trade is concentrated in Bonthe and therefore difficult for the Board to control and direct. It seemed for some time during the 1950s that piassava would be re-routed from Bonthe and Sulima to Freetown, but many producers and traders preferred Bonthe and continued to deliver it to the exporters there. When the Board put piassava on the schedule in July 1965 the producing areas were not in favour of it. They probably feared that it would draw trade away to Freetown. However, the Board descheduled piassava in 1968 and the product has continued to be exported via both Freetown and Bonthe ever since. The Lebanese share of exports was 41% in 1954 and 61% in 1955. The European share was 41 and 26% for those two years. Since then the Lebanese share has continued to increase at the expense of the European Companies.

In ginger the Lebanese have been particularly strong. They accounted for 85% of exports in 1954 and for 93% in 1955. After ginger was scheduled in 1961 the Lebanese Buying Agents accounted for 94% in 1961/2 and for 98% in 1962/3. The repeated de- and re-scheduling of ginger during the 1960s caused great problems to Lebanese traders and seemed unfair to them.

Coffee was on the schedule of the Board from 1949 to March 1953. Very little coffee was sold to the agents and the Board was convinced that most of the crop was smuggled out of the country. In the end the Board decided to deschedule. This had the desired effect: legal exports of coffee increased quickly. For eight years coffee was free. There are no figures to show which percentages were handled by the companies and by the Lebanese, but after coffee was scheduled again in 1961, the Lebanese buying agents accounted for 39% of the Board's purchases in 1961/2 and for 29% in 1962/3.

THE KOLA TRADE

Everywhere in West Africa kola nuts are transported northwards from the forest zone to the savannah zone, and the natural destination of kola nuts from Sierra Leone is Guinea, Guinea-Bissau, the Gambia, and Senegal. Long ago African traders had realized that the kola could be transported to these countries by ship and they organized the trade by sea. After the construction of the railway the interior began to yield kola, too. The people harvested more from the existing trees and planted additional ones. Having set aside some nuts for their own use, they prepared ('washed') the remainder for sale. The Lebanese were prepared to buy kola, for instance in the railway towns beyond Blama. They sent their consignments to their partners in Freetown for sale or for export. Exports rose quickly after 1905. The additional quantities available for export came from the interior, while the supplies from the coastal belt remained constant. The rail-borne supplies soon equalled and then exceeded the water-borne supplies of the coastal belt, and the strength of Lebanese kola traders in Freetown grew proportionately. They may have had another advantage as well, for it is very likely that the prices which the Lebanese paid in the interior were lower than those paid by African traders in the coastal belt. This would mean that Lebanese traders in Freetown obtained cheaper supplies than their African rivals. They did not fail to exploit the new opportunities and began to export. The fact that they spoke Arabic may have been an excellent introduction to the Muslim importers along the coast. By 1914 they dominated the kola export to Dakar. By 1920 the trade was almost entirely in the hands of the Lebanese.

The kola trade was the most promising line for the Lebanese in Freetown in the 1920s. Of the 21 Lebanese that were listed as traders in the Handbook of Sierra Leone in 1925, eighteen were described as trading in kola. It was fortunate for them that they faced little competition from the companies. Their success should not be belittled, though. Kola exporters needed a great deal of capital. This was because stocks of kola deteriorated more quickly in Bathurst and Dakar than in Freetown because of the lower humidity in the north. Therefore the importers in these towns kept their stocks low and frequently – but irregularly – ordered further supplies from Freetown. The burden of holding and financing stocks was shifted as much as possible to the exporters in Freetown. Although it required capital there is no doubt that this was profitable for them – indeed, any Lebanese who exported kola in the 1920s could be assumed to be well-to-do. One Lebanese exporter, Michael Abdallah Blell, formed a partnership with Sarkiss Mahdi and asked him to settle in Bathurst to sell kola from Sierra Leone. The Freetown partner regularly shipped kola to his partner in Bathurst and profits on these consignments were shared. This is an interesting parallel with the partnerships which some Lebanese had formed to make the best use of the railway. For many years kola was the second export earner of Sierra Leone, surpassed only by palm kernels.

In 1954 the Lebanese still held 70% of the kola export market, but in 1970 the number of Lebanese kola exporters had fallen to three. Those who gave up switched over to the general merchandise trade, and Mandingo, Temne, Susu, and Hausa traders have taken over. This has happened without any assistance from the Government and is an interesting example of African commercial advance.

THE RICE TRADE

Rice has been the staple food of Sierra Leone for many centuries but trade was limited until the emergence of large concentrations of 'non-farmers' in the towns and in the mining areas. The 20th century has witnessed a continuous, at times stormy, expansion of the demand for rice, occasionally necessitating imports. As trade expanded more working capital was needed to finance the increasing tonnage. At first there was sufficient African capital, but at a certain stage – a few years before World War I – Lebanese capital was drawn in. Soon a few Lebanese dominated the wholesale trade in rice. The urban consumers often complained that they hoarded rice to raise the price. A serious shortage of rice in 1919 was followed by looting of Lebanese shops and stores. Since then the rice market has always been on the Government's agenda.

The rice trade in the villages also had some undesirable features. Many farmers did not estimate their own requirements properly and sold too much of their crop in October. When they discovered in May that they had not enough seed rice for their farms they turned to the traders to borrow rice. Most Lebanese traders seem to have advanced seed rice on the following terms: the farmer had to repay two bushels of rice at harvest time for each bushel which he borrowed at sowing time. These terms were not unreasonable because the price of rice was much higher in May than in October. We note that the trader gave credit for about four months. In some areas the farmers pledged or 'trusted' their rice farms to the lending trader so that he obtained a formal claim to the harvest. Many farmers were grateful to the trader who advanced seed rice to them because without it they would have prepared their land in vain. They saw the arrangement as a joint investment of the trader and themselves. But even if we accept this view, the dependence of the farmer on the trader was beyond dispute and the danger of increasing dependence obvious.

The year 1954 was a turning point in the rice market. Before that time rice was imported only occasionally and the decision to do so was usually taken after long deliberation. Since 1954 rice has been imported every year and the orders were placed in the early months of each year, the only point of debate being the amount that was required.

As the Government, through its Rice Department, assumed overall control of the rice market in 1955, the position of the rice wholesalers in Freetown changed completely. It seems that the Lebanese rice wholesalers were not unhappy with the change. Working with large consignments from abroad was easier than accumulating stocks piecemeal from various areas and buyers. They were interested in obtaining important licences and in buying large quantities from the Rice Department. Since they possessed capital and storage space and had contacts with the market women, they were accepted by the Rice Department as an essential link in the distribution of imported rice. However, African traders were also interested in this part of the rice trade and the Tender Board decided to give preference to African rice traders, allocating most if not all rice quotas to them. This seemed the end of Lebanese trading in imported rice, but several African traders who had neither the experience nor the storage space to handle their quota approached the Lebanese and proposed some form of co-operation with regard to their quota. Other African traders found it easier to sell their quota to a Lebanese wholesaler. In this way some Lebanese continued to participate in the trade, in spite of the efforts of the Government to squeeze them out. One of the wholesalers, the firm of J. Milhem & Sons, occupied a special position because it succeeded in importing 6000 tons of rice at short notice in August 1959 when a sudden shortage caught the Rice Department unawares. When private importation was allowed again in 1960, the firm of Milhem was the only non-African wholesaler to receive a licence to import rice – a reward for its quick action the previous year.

Pressure to bar the Lebanese completely from the trade in rice was mounting further in those years and culminated in a parliamentary motion to bar non-Africans from the rice trade, which was adopted in December 1962.

THE CONTINUITY OF LEBANESE ENTERPRISE

For many Sierra Leoneans the typical Lebanese is the man who has run a particular shop for as long as they can remember. In terms of operations, location, and environment, his continuity is taken for granted.

The Lebanese trader has shown endurance and tenacity, and his optimism has carried him through bad years. These qualities of character were backed up by certain features of the family and the group. But the continuity was not all a result of virtue. A Lebanese – and his family – normally had no alternative to sticking it out. He could not return to Lebanon, either because he had no money or because he was loath to admit his failure.

However, his continuity was of benefit to Sierra Leone. In the process of development many new economic activities are necessary; and though most of them are only just viable, if they are not supplied, no development will take place. If they are supplied, their reward will depend on the pace of development; that is, rapid development will make them prosper, but slow development will cause them losses. The Lebanese were willing to move into such areas of marginal profit, sticking to them even when they were discouraged. It was fortunate that development in Sierra Leone has been rapid enough to bring prosperity to the majority of the Lebanese. The fate of those who struggled for many years in the wrong field or an unsuitable location must not, however, be overlooked.

The continuity of the Lebanese was in one sense superior to that of the Europeans. As organizations the European companies were of course more stable and reliable than the Lebanese, but they did not offer continuity of personnel in their Buying Stations. The African farmer could not count on seeing the same man in the European store, but he was sure of seeing the same face in the Lebanese shop.

Continuity meant creditworthiness. The Lebanese qualified for credit from the companies because of their continuity, for a company manager would reason that a man who had run a shop for several years and had bought supplies regularly could be relied upon to go on doing so. Virtually all European capital that was passed on in the form of trade and bank credit has been given on a personal basis, that is on the basis of past performance. Mortgages and other collateral have been rare. It seems to me that the Europeans were not always aware why they extended credit to the Lebanese. Sometimes they said that they trusted a Lebanese because they knew him, implying a knowledge extending over several years. Another argument was that the Lebanese were not able to abscond in the bush but they failed to say that only irregular traders may benefit from such an action; no settled trader would consider giving up his shop in order to escape a debt.

Their continuity was also the reason why they were expected to give credit. An itinerant trader has to be paid in cash because he moves on, but a shopkeeper will be there on the next day to receive his payment. The Lebanese were probably reluctant at first to sell on credit, but they were drawn into it, after they had settled somewhere. Thus a Lebanese trader's continuity made him a channel through which European capital reached the African consumer.

THE LEBANESE ENTREPRENEUR

The primary quality of Lebanese enterprise has been continuity, and the dominant form the family business. Family cohesion is a firm foundation for small-scale, unincorporated enterprise. Many Sierra Leoneans have noticed the way in which the Lebanese support each other, but it is often wrongly interpreted as national solidarity, when in fact it is family solidarity. This ingredient would be valuable for Sierra Leonean businessmen, too.

Success in business has always been a strongly desired goal in the Lebanese community, not only because of the money, but also because of the social status it brought. In a community of traders with few alternative opportunities commercial success becomes the unquestioned aspiration of everyone, and indeed, social and economic pressures combined to spur the Lebanese on to extra efforts and new initiatives. There was also social pressure from relatives and friends back in Lebanon. The desire to make good, or to be as successful as compatriot X, explains the strong motivation of the Lebanese. The small European and the large African community did not experience this pressure to the same degree.

Commerce in Sierra Leone is not impersonal and purely economic, for most customers expect a personal element in their relations with a trader. This element is declining, especially in the towns, but it has been very important. The Lebanese traders have lived up to the expectations of their African customers, in most cases quite naturally because they were used to it in Lebanon. Indeed, they have been very open-minded towards the African society in which they lived. One can belittle this and say that they could not afford to be indifferent, but I would rather credit them with a tolerant attitude and a willingness to see the qualities of other people. In this respect they have been more receptive than most Europeans. The Lebanese still possess this advantage: they are better informed about Sierra Leonean customs and behaviour than any other group of foreigners.

The political situation of colonial Sierra Leone was favourable for the

Lebanese. Some Sierra Leoneans believe that the British actively supported the Lebanese and so promoted their success, but the real situation is more subtle. The Lebanese community has always appreciated and admired the rule of law which the British guaranteed in Sierra Leone. For them it meant protection against avaricious and oppressive officials, and the possibility of invoking the verdict of an impartial judge in a long-drawn-out conflict.

All observers who have paused to consider the success of the Lebanese have tried to find an explanation for it. Virtually all these observers have looked for an answer in the Lebanese character rather than in their circumstances. Naturally, they have advanced suggestions such as superior intelligence, a long commercial tradition, a knack for learning foreign languages, and great psychological talents. I think that most of these suggestions exaggerate the facts.

It is interesting that those who were hostile to the Lebanese, the Creoles for instance, accepted the hypothesis of superior intelligence, but were convinced that it was used in the wrong way. According to them their shrewd intelligence verged on dishonesty, but I believe that this trait has been exaggerated as a cause of Lebanese success. Since this point has not yet been discussed systematically in this book, it should receive special attention here. It is not difficult to draw up a list of the objectionable practices of which the Lebanese have been accused:

1. Cheating individual consumers or producers by using false weights, by misleading them about the quality or grade, or by making faulty calculations which the other party cannot check.
2. Contravening Government orders. We note in passing that this complaint became more frequent as Government intervened further in economic affairs.
3. Instigating Government officials to act knowingly against the public interest; the evil of corruption.
4. Exploiting consumers collectively by raising the price of a scarce commodity, and even by engineering a shortage.
5. Instigating employees of European companies to act against their institutions.

The first three practices are legally wrong, and some Lebanese have been convicted of them in court. The last two were often considered wrong, too, but were not illegal. Moral standards, it seems, used to be stricter than legal standards. It should not be assumed, however, that moral standards were equally high in all parts of Sierra Leone. For many years British and Creole standards were very strict, stricter than those of the other Africans. In that period it was possible for the Lebanese to be accused of dishonesty by the Creoles and some Europeans, while at the same time being appreciated as giving a fair deal in the interior.

The unification of moral standards has proceeded fast in the post-war period, but agreed standards have become laxer. Many of the practices listed above occur today and fail to arouse the indignation of the press and public. Virtually all Europeans and Africans with extensive first-hand knowledge of the Lebanese told me that standards of honesty varied greatly within the Lebanese community: there were the men who could be trusted and the dubious characters. Normally the former were the more successful. I am therefore inclined to think that dishonesty has accounted for only a small proportion of the prosperity of the Lebanese community.

Two other qualities of the Lebanese have been mentioned as causes of their success: their industry and their frugality. As long as the Lebanese were itinerant traders, their industry was conspicuous. After they had settled as shopkeepers, however, the basis for this reputation disappeared. My own observations indicate that slack and busy spells alternate for Lebanese shopkeepers and that they adjust their work accordingly. Most of them could not be called busy or hard-working men. There is, however, a minority of very active men, mostly middle-aged, who spend a great deal of time in managerial work, inspection, travelling, negotiating, and personal relations. The word 'busy' is more appropriate for them than 'industrious'. Frugality is a virtue of the past. The early Lebanese lived very frugally and saved as much as they could; the capital thus accumulated consolidated their position in the 1910s and 1920s. Today there is little evidence of frugality, except among a few older people. We gather that the change-over occurred in the 1950s, when the Lebanese began to build modern houses which led to a new style of life.

THE CONTRIBUTION OF THE LEBANESE

The economic contribution of the Lebanese has been that they took on marginal roles in the process of development which did not attract other entrepreneurs. The development of Sierra Leone, especially of the interior, would have proceeded at a much slower pace in the period 1900 to 1950, if the Lebanese had not been there. The nature of their economic contribution has shifted with the emancipation of the Africans of the interior and the withdrawal of the Europeans. At present their main contribution lies in the importing and retailing of textiles and fancy goods; in the motor trade; in diamond buying; and in miscellaneous activities.

8

FARMERS AND MIDDLEMEN: ASPECTS OF AGRICULTURAL MARKETING IN THAILAND

A. Siamwalla

The role of agricultural marketing in Thailand is a subject worthy of serious examination for a number of reasons. Since Thailand has never been overtly colonized and since a large and socially powerful landlord class does not exist and since the backwardness of Thailand's agriculture has to be explained somehow, middlemen are made into scapegoats used to explain away all the ills that ail Thai agriculture. This scapegoat role of the middlemen is useful for the political elite for the additional reason that they are predominantly Chinese and thus form a convenient target for the critics, who came largely from the bureaucracy, and who thus hope that fixing the blame on the middlemen may lessen the scrutiny on their own mediocre performance in the general area of agriculture.

A second, more serious, reason that makes a study of agricultural marketing in Thailand particularly interesting is the fact that, at least as far as internal marketing is concerned, the role of the government has been minimal, a relatively unfettered private enterprise mode of marketing has been allowed to develop. Since a great deal of the debates concerning agricultural marketing end up hopelessly tangled on the issue of whether the marketing problems in backward countries are due to thieving middlemen or to meddling bureaucrats, the case of Thailand where bureaucrats have meddled very little and where an almost completely private internal marketing system was allowed to establish itself would be useful even as a counter-example.

To give a causal explanation of the degree of competitiveness for any particular commodity, we shall introduce the concept of the 'shifting cost'. We take a farmer who has been producing a commodity for some time. He would have been dealing with a trader or a set of traders for some time. We then take the trader with whom he traded last as a reference point and ask

---

Source: Passages from the paper published in *Economic Bulletin for Asia and the Pacific*, ESCAP, Bangkok June 1978. Reproduced with permission.

ourselves the question: how much would it cost the farmer to shift his dealings from this trader to an alternative buyer? The cost may arise owing to the fact that he would receive a lower price from the alternative buyer or it may arise owing to a higher transport cost. In these cases the shifting cost may be easy to calculate, in other cases it may be more difficult. The farmer may, by shifting, lose the right to borrow from that particular middleman – this right, given the lack of access to other sources of credit, may be worth something to him, but it is difficult to measure.

COMMODITIES WITH LOW SHIFTING COST: RICE, MAIZE, CASSAVA AND KENAF

The farming methods employed are traditional, requiring relatively little purchased inputs. This is true even in cases where the commodities concerned have only recently been introduced into Thailand, for example, kenaf. The farmers' dependence on traders for inputs and new technology is therefore minimal.

The processing industries connected with all four of these commodities are all fairly small scale, whether they be the rice-milling industry, baling-plants for kenaf or the chipping and pelletizing plants for cassava. The likelihood is thus very small for the processors to become collection centers covering a very wide area and thus to acquire monopsony power. The farmers *and* the various intervening middlemen thus face relatively low shifting cost and do in fact sell to many different buyers over the years.

For these four commodities then, the barriers to entry into the 'middleman business' seem to be very low. The only impediment that the middlemen themselves set great store by is something they call 'experience' – experience in grading, but more important experience in being able to predict price fluctuations. These 'experiences' however do not carry a high training cost – many middlemen apprentice themselves by working as clerks or even labourers for some other middlemen. Entry is therefore easy. The margins that accrue as middlemen's profits are thus very low, as borne out by many studies. The only exception among the four commodities appear to be cassava, where a government study shows that the profits accruing to middlemen and processors are as high as 18.96% of export price and 39.04% of retail price.

There is however a problem with this study in that the capital cost for the processing plants has not been deducted as part of the cost. There is also another possibility. Cassava exports have, until about five years ago, been very hard hit by a complete absence of quality control by the exporters. There were many complaints of adulteration. Eventually, the main buyers (mostly Dutch and West German firms) decided to set up their own exporting companies in Bangkok. There is now only one Thai firm actively exporting cassava to Western Europe which is its main market, the others are all European. This growth of foreign owned oligopolies with their notoriously high overheads may be partly responsible for the very high export margins for cassava. At the local and wholesale levels, the margins appear to be not excessive.

In general, it must be concluded that the internal marketing for all these commodities with the possible exception of cassava, is characterized by a high degree of competitiveness, and the price received by farmers for their products, compared to the export or retail prices, are quite high. I contend that this high share is due to a very low shifting cost.

COMMODITIES REQUIRING LARGE AMOUNTS OF PURCHASED INPUTS: TOBACCO AND EGGS

Lack of credit may be a major bottleneck to the adoption of modern inputs, thus the reluctance of rice farmers to apply fertilizers may, in many cases, be traced to their failure to obtain loans cheaply. There are, however, a number of commodities for which there are certain peculiarities which allow middlemen to take over the role, not only of traders in outputs *and* inputs, but also of a provisor of credit and even of an extension agent. Tobacco and eggs, fall into this category.

The curing house is to tobacco-growers as sugar mills are to sugarcane growers. The reason for this is not, however, the large scale of the processing industries, for the investment required for an economical curing plant is relatively small (a $50 000 investment would suffice for a medium-scale plant). The hold that they have over the growers arise more from their role as the provider of inputs.

Poultry feed is required on a regular basis; any withdrawal on the part of the egg-producers from this source would be disastrous for their hens. Egg producers cannot even shift their source of supply from one to another, because each manufacturer uses a different formula. The hens are very sensitive to any change in the formula, which if it occurs, would lead to a very sharp decline in egg production at least for a week or so. The shifting cost in this case is obviously quite high. The feed producers' agents are thus in a position to exercise a good deal of control over the producers.

CREDIT AND MARKETING TIES

The main conclusion to be derived concerning the relationship between credit and marketing ties is I believe somewhat different from much of the writings on the subject. The reasons for farmers being tied to middlemen must be sought in factors other than the latter's credit operations. Usually, it is the technology of the crop production or of the processing that enables the middleman to assume control over the marketing operations of the farmers. Once this marketing control is assured, then and only then are middlemen willing to extend credit to farmers. In this view, the credit tie between farmers and middlemen is a *consequence* and not the *cause* of the marketing tie. Where the latter is difficult to enforce, the farmer would not be forthcoming.

Much of the writings on agricultural credit have ignored collection problems inherent in any credit system, but peculiarly acute in a rural setting (a notable exception is Bottomley, 1970). Private credit operations are successful if and only if the borrowers are tied to the lenders in some ways. Landlords are very well-placed for such activities; so are professional moneylenders if borrowers feel that any delinquency on their part will be penalized by a cessation of future lending. To the extent that middlemen are engaged in competitive purchase of commodities, they are rather poorly equipped to engage in moneylending on a broad scale.

EFFICIENCY OF THE MARKETING SYSTEM AND THE ROLE OF THE GOVERNMENT

To provide a continuous link between producer and consumer is a target required of any marketing organization. The further question that needs to be asked next is: is this link provided efficiently? This question, in turn, presupposes the answer to the following question: how is efficiency to be measured?

If the target is as given in the previous paragraphs, the efficiency of a marketing system must be measured in terms of the resources consumed by the distribution sector in achieving that target. This sector may be privately or publicly owned; it has to feed itself by diverting resources from other sectors of the economy. The private trader may do it by charging a profit on the commodities that pass through his hands, the public distribution organization may subsist on government revenues derived from taxes. A price has, in short, to be paid somehow for the marketing services, and this price measures the efficiency of the marketing system.

With a private marketing system such as exists in Thailand, a crude measure of efficiency in this sense is a look at the margin obtained by the various participants of the marketing system, not only by the middlemen themselves, but by other firms that supply inputs into the distribution sector, in particular the transport sector.

The margins for three out of four of the major commodities and for rubber are quite low. In the case of rice the seasonal profit from storage is also low on the average. A study done on private trucking – a major mode of transportation in Thailand – tends to prove that the margin achieved by truckers is also low. All these point to the conclusion that the marketing system for the major crops in Thailand exhibits almost no super-normal profits, is competitive and thus permits very little room for improvements in efficiency.

That the marketing system is competitive at least for the crops discussed in the previous paragraph is, I believe, a defensible proposition, and fully justified by the facts. To conclude from it that therefore it is efficient is not justified, however. There is one major lesson that Chamberlin (1962) has left for us and that there may be 'too much' competition leading to excess capacity, and therefore higher costs than optimal. In agricultural marketing such a situation may arise from too many firms competing in a given area, pushing up costs. Looking at middlemen's profits may not be sufficient in such cases, as they would be making no super-normal profits. A more efficient system could theoretically be achieved by eliminating some middlemen, allowing the others to expand to their optimal sizes and thus cutting down marketing costs.

No empirical study has been completed which will throw light on this Chamberlinian inefficiency. Two studies in progress at Thammasat University on rice-milling tend to lend some credence to the view that the small rice-mills do have substantially higher costs than the large mills. However, these mills have the advantage that they are very dispersed. Transport costs are thus lessened. Of course, if this saving in transport costs is larger than the higher milling-cost of the small mills, then the existing distribution may still be efficient. No definite answer can yet be given pending further work, but it appears unlikely that the saving in transport costs is large enough.

For other crops, such as sugarcane, tobacco and eggs, it appears likely that the marketing system is statically inefficient, because of the quasi-monopolistic nature of the system. Even then, in tobacco and eggs marketing the static inefficiency may have to be weighed against the role played by the middlemen in transfering knowledge of modern methods to the producers, an activity in which the performance of the government has been at best mediocre.

It will surely be argued that efficiency is not the only criterion by which one should evaluate marketing systems. We now turn to another social objective which is important in considering an agricultural marketing system, namely price stability in the face of random factors which normally would lead to price gyrations. The most important agency, in this respect, is the government which has consistently attempted to dampen price fluctuations in the domestic market by means of export taxation or export control, with varying success.

This entry by the government into this area would be justified only if it can be proven that the domestic marketing system magnifies the world price fluctuations. For the major crops this judgment cannot be supported.

By and large, the marketing system is a neutral transmitter of the Bangkok wholesale market and the latter in turn is a neutral transmitter of f.o.b. price minus export tax and quota profits (when export control is in effect). This is true for most of the major export crops except sugar and is a corollary of the competitive nature of the markets for these commodities. For sugarcane and tobacco the situation is somewhat different and the fluctuation is probably less on account of the greater non-competitive elements. Egg production is entirely a domestic consumption item. The price fluctuations in this market follow the price fluctuations in the feed market which is in turn linked to the world market, as Thailand is an exporter of feed also (cassava, maize, soyabeans and fishmeal). Indeed, one may go even further. The only subsector of domestic market into which the government has entered, i.e. the distribution of cheap rice to consumers, largely in the Greater Bangkok area, has definitely magnified the price fluctuations. Because of the very loose administrative controls over the volume to be released through this channel, a rise in free market prices usually leads to a significant diversion of demand into this channel. To meet this demand, the government is then forced to draw on the supplies from the free market driving the latter's prices further up. This in turn leads to even greater pressures on the concessional sales, and so on in a destabilizing vicious circle. The government succeeds in keeping the price sold through the concessional market steady, but the quantity available is unable to keep pace with demand when free market prices are high. The inevitable result is long queues at the stores and chaos in the rice market as occurred in mid 1973 and in March 1974.

Thus, to conclude on the issue of price stability, the primary instrument in the hands of the government to achieve its target is export policy. This in itself is adequate for the major crops (except sugar). There is no need on this ground to replace the private marketing system. Indeed in the one area where the government has attempted to do so, the results have been negative.

The area in which the private marketing system can be said to be inimical to the interests of the farmers is the entire issue of quality variations and grading. Indeed the issue extends even further to the question of short-weighing. In the case of cassava and kenaf for exports, adulteration has also been a problem. Complaints on this issue are rife. Without a strong governmental action on this front, it is difficult to see any means of controlling the middlemen, yet governmental action in this area is strongly hampered by the general scepticism regarding the honesty of its personnel. This scepticism remains valid whether the government is to take over the entire marketing function or takes a less radical measure by sending its officials to inspect the scales used in weighing. The government is in fact empowered to do the latter, but the results have been negligible.

The last and most important issue that needs to be discussed is the question of income distribution. The actors that I shall be mainly concerned with are the poor farmers, rich farmers, and middlemen. I exclude urban consumers and the issue of the country–city terms of trade from consideration altogether as being too gigantic to be considered in a section of a short paper.

Whether middlemen make 'excessive' profits has always been a politically explosive issue. Their margins vary from crop to crop, and for the technically more 'sophisticated' produce, e.g. tobacco and eggs, it may be quite high on account of the monopsony power of the middlemen. Whether this profit is 'justified' or not is a difficult question which cannot be answered by a mere economist. Were it to be answered negatively, a much more careful examination of the alternatives has to be considered, given the fact that at this time, the middlemen for these products are important conduits for new technology, and they take over this function precisely because they can capture the benefits from the new technology. A proper design for an alternative scheme must be able not only to transmit the technology but also to pass on more of the benefits to the farmers themselves. Cooperatives appear to be superbly suited to do this, but how cooperatives can transmit the new technology whilst remaining basically a farmers' own organization with the farmers' interests uppermost in its mind is a question seldom raised, let alone answered.

When we come to the major crops, we reiterate again the earlier position that some middlemen do acquire considerable profits, but those profits are largely speculative profits arising out of price fluctuations. The atemporal margins for those commodities are typically small, and generally 'unsqueezable'. The issue of income distribution between farmers and middlemen thus reduces to the question of price fluctuations.

It has also been argued that the same individual may act as money lender and middleman. This doubling up of functions may enable the middlemen to obtain a stranglehold on the farmers' choice of production techniques and may end up successfully preventing new technology from being adopted at all (Bhaduri, 1973). Furthermore, such a semi-feudal relationship as it is called, tends to be more common with smaller farmers than with the better-off large farmers. Introduction of new technology may thus lead not only to small growth, but also to worsening of income distribution. This argument is relevant to agricultural marketing if it can be shown that large numbers of middlemen are also moneylenders. The empirical evidence for this is mixed. For the major crops, again excepting sugar, it can safely be said that there is very little overlap between the marketing and the credit operations. Credit operations require the ability to enforce contracts, either through violence or other economic threats, which middlemen simply do not have. In cases where middlemen do provide the loans, it is not true that smaller farmers depend on them more than large farmers. In cases where for technical reasons the marketing system is more structured and shifting costs therefore high, as in sugarcane, tobacco and eggs, this has enabled middlemen to extend credit to their farmers. But as pointed out, the credit tie followed the marketing ties. From the point of view of income distribution among farmers, for most major crops, the marketing system may be said therefore to be neutral.

REFERENCES

Bhaduri, A. (1973) A study in agricultural backwardness under semi-feudalism. *Economic Journal* 83(329), 120–137.
Bottomley, A. (1970) *Factor Pricing and Economic Growth in Under-Developed Rural Areas.* Crosby Lockwood, London.
Chamberlin, E.H. (1962) *The Theory of Monopolistic Competition*, 7th Edition. Harvard University Press, Cambridge, Mass.

9

WOMEN'S ACTIVITIES IN FOOD AND AGRICULTURAL MARKETING

FAO

Nearly everywhere women are busy turning small quantities of processed food and agricultural crops (including animal products) into cash or barter on an irregular and casual basis from their own homesteads. These sales often provide the savings and capital by which women acquire a greater stake in retailing their own produce. In addition, there are very many women marketeers and traders who engage in retailing as specialists, as committed even if part-time or seasonal participants. In some countries, such as Ghana where about 80% of all traders and nearly all fish traders are women, they dominate all stages of food marketing and agricultural trading. In others, while their participation is less dominant, it none the less forms an essential link in the distributive sectors of rural, urban and even regional economies. In a few areas, women play a major role in food staple markets, as in Manipur. In this small state in north-eastern India, women take the major part in the production and processing of rice and dominate rice retail trading. They also control important sections of the wholesale trade in commodities entering the state, such as cloth, and the catering industry. The market women of the state capital manage the central bazaar and handle its administrative relations with state authorities through elected managing committees.

There are very substantial differences in food market characteristics, commodity trading relations, and the nature of women's participation from country to country and area to area. However, it may be noted that women's participation tends to be highest where production, marketing and trading have been least affected by commercialization and industrialization. Not untypically, they are to be found concentrated in crowded assembly or retail markets which share common features across all countries. Less typically are they engaged in wholesaling or long distance transport or handling.

Source: Based on a paper by J. Jiggins prepared for FAO, Rome, 1984. Reproduced with permission.

ENTRY INTO MARKETING

The women are most often the sole or major income earner in their households and participate in marketing full time and with a long-term commitment. They are somewhat older at entry into market trading than women entering other income earning or wage jobs and are schooled to a lower level than male marketeers. They usually have had previous, though often discontinuous experience as wage employees or have been 'apprenticed' as unpaid family workers in another business (such as hawking or roadside vending) before setting up in the market.

Women often take up market vending under pressure of an immediate and pressing need. Start-up finance is borrowed from friends and relatives. The initial final profit from vending is low, and is mostly spent to meet immediate consumption needs, because a proportion of the turnover must be quickly reinvested in new stock and because it takes a considerable time to pay back the start-up capital. Their profits are also affected by price control, the numbers of other women selling similar items, and the poor distribution of these items to the vendors. Although a few have their own wet season fields outside the capital, most acquire their stock of produce by one of four means:

1. From farmers selling wholesale direct to the vendors outside the market site.
2. From the official marketing wholesale organisations at known weights, some quality control but at official retail prices.
3. Direct from farmers at the farm site.
4. Direct from wayside wholesalers who bring vans or lorry loads to the edge of the city and sell produce by the sack or box, at uncontrolled weights, quality or wastage.

They sometimes extend crop season credit to producers and often have to sell on credit (even though they are usually short of liquidity themselves), owing to the competitive nature of the market. In Zambia, as in many other countries, 'women' are not a homogeneous category. Market women see themselves to be in direct competition not only with each other but quite often with women who are street vendors of the same items, with women selling from their houses and from the unlicensed traders selling just outside the market boundary or adjacent open site.

Why are women concentrated in markets like these and on these terms? There are as yet few studies which have explored the question. Descriptive research suggests that the following are among the more important reasons:

1. The nature of women's home and maternal duties and responsibilities.
2. The desire for a degree of autonomy, at whatever cost in self-exploitation.

3. The effective barriers to access on equal terms with men to higher profit marketing (resulting from the procedural characteristics of formal service institutions, lower initial resource base, historical educational or legal inadequacies, etc.).

4. The pressure to survive and protect their children on whatever terms, at moments of family instability and crisis.

SPECIAL DIFFICULTIES FACED BY WOMEN

Women face especial difficulties, as women and as the result of the historical and current disabilities arising therefrom. For example, although perhaps the majority of peddlers and hawkers at some time, whether licensed or not, face harassment from public authorities, in women's case there may be the additional hazard of sexual harassment in return for the 'favour' of being allowed to continue trading. or there may well be barriers both structural and cultural to the acquisition of skills which might earn women higher profits. Or it might be that, because of historical restrictions on women's participation in public domain activities, they lack the confidence and knowledge and contacts to find and negotiate successfully with public officials and, often, the somewhat illegal or outrightly criminal circles which control trading and marketing activities.

Women's special responsibilities for the household and dependents in terms of constant daily care may pose especial problems for their participation in retailing and trading, particularly where the nature of the work, the location of the work place, and the hours of business cannot easily be combined with these responsibilities. The multiple nature of women's roles partially accounts for their concentration in local trading, in activities with flexible hours of work or market business in which child care and household security can be covered by others, usually by other women, in the home locality or workplace.

Husbands or male relatives may be a greater restraint, in their demands on women's time, energy, attention and income. The desire of women to protect their profits from their husbands or other relatives (usually to the greater good of the family) often leads them to seek group action, whether or not organized formally as cooperatives, or, where public law permits, as limited liability companies. A similar impulse may also lead women to prefer informal savings and credit clubs to banks or other institutional sources of finance which most usually demand husband's or male relative's sanction of an account or guarantee of a loan. But in highly fragmented and crowded markets the evidence suggests that peer-group organization among women may do little to meet their needs. Each one is afraid of jeopardizing carefully built up relations with her suppliers and access to their credit.

REVIEW BY COMMODITIES

In many countries (though less frequently in Africa) women are the main operators in the traditional dairy sector. In some cases, as in Kenya, they have become major participants in the modern dairy sector, in programmes to promote stall-fed smallholder grade dairy cattle. The best documented experience is that of the dairy cooperatives formed under Operation Flood in India. A recent review, however, shows that there is scope for increasing women's benefits through greater attention to how the income and other benefits are distributed by the formation of women-only cooperatives and by employing more female staff at every level, from the spearhead team right up through the technical and managerial grades.

Women are also often engaged in trading and in the marketing of goats, sheep and goat and sheep products. One of the better organised major markets in Togo (Gbo-Simé goat market in a suburb of Lomé) is dominated by women traders. In the southern highlands of Ethiopia both men and women of the Gujji own, manage and dispose of sheep and goats but it is women who sell their milk and butter in the market.

Women in Africa in areas where fresh fish forms an acceptable item of diet figure mainly as retailers although – especially in inland fisheries – they may regularly catch freshwater species for home consumption and local trade. They work often as street traders as well as stall owners and open market sellers, usually as retailers but, especially in West African coastal areas, also as intermediate assemblers. Typically they buy from fishermen a 'standard' basket or bundle of mixed varieties of unknown quality, weight or number, which they sell individually or in twos or threes, for a fixed sum, most often according to size. The smaller retailers often purchase according to their idea of what they can spend on stocking, rather than according to the state of the market, thus sometimes acquiring more/better quality fish and sometimes less for the fixed amount they are willing to lay out.

By far the largest part of the fresh fish catch in Africa is immediately processed. The main methods are: salting; smoking or drying or by various combinations of smoking and drying; fermentation; or by combinations of these (and other) processes. Modern freezing and canning factories are still rather few. Although the preference for fresh fish is strong, since most fisheries are dispersed, small scale and often located far from centres of potential custom, the bulk of the market is for processed fish.

The main challenge to marketing is thus to develop a system of collecting big enough volumes to make fish a commerically viable trade commodity and to move it to consumers along organized channels within the relatively short shelf life of the product. One example where the challenge has been met is offered by the Irepader Ilaje Fish Dealers' Association at Igobkoda in Nigeria, whose 231 members in 1976 were all women. It operated a marketing hall where fish were consigned to members, who sold it on to local buyers and traders who in turn delivered it on to the big consumer towns. Sales were made on a commission basis, ranging from 2 to 5%.

Women in many parts of Africa are predominant in small-scale horticultural production and retailing. Their participation in wholesaling is less strong but in some countries – especially for the less perishable items such as bananas, potatoes, yams – such as Kenya, Uganda, Cameroon, Ghana and Nigeria they play a strong role in wholesaling too.

In various parts of Africa women participate directly in input marketing in competition with men. They sell fertilizer in small quantities from retail stores in Kenya. In Lesotho they have built stores of 10–15 tonnes capacity on their farms to take in a full truck load of fertilizer. This is then re-sold on a 10% commission basis. Some cooperatives handling seed and fertilizers in Swaziland are managed by women. They buy inputs collectively for their vegetable plots in villages in Gambia. In Nigeria and Ghana they supply inputs on credit to farmers who contract in return to sell their output through them.

10

NEW TYPES OF MULTINATIONAL FIRMS IN THE AGRIBUSINESS SECTOR: IMPLICATIONS OF THEIR EMERGENCE IN THE LEAST INDUSTRIALIZED WORLD

J.C. Dufour, G.Ghersi and R. Saint-Louis

> First and foremost the food and feed industries, more concerned with low cost and high quality than country of origin for their raw materials, have shown an impressive facility for searching out foreign sources of supply. They have forced agricultural interests to play 'catch-up' taking advantage of new trade opportunities until they become an embarassment to domestic farm policy.
>
> Tim Josling

This tentative paper brings forward the following hypothesis: it may be that the best interests of Multinational Food and Feed Corporations (MNFFCs) are coming to rest more and more upon their ability to act as oligopolies seeking consenting partners rather than on their stubbornness to keep on behaving as fiercely competing ones, at least for the remainder of this decade. Three sets of features are presented on this issue. The first one throws in to relief a few recent trends in international trade of food and feed products. It raises questions concerning their suitability from the perspective of national food and/or nutrition strategies in some countries. The second deals with selected structural changes of MNFFCs during the 1980s. Finally, the third one deals with their 'new' ways of doing business deals.

INTERNATIONAL TRADE OF FOOD AND FEED PRODUCTS: PROTECTED MARKETS UNDER THE INFLUENCE OF FLOATING EXCHANGE RATES

The outstanding importance of the highly industrialized world in inter-

---

Source: Paper presented at the XIX Conference of the International Association of Agricultural Economists, Malaga, 1985.

national trade of food and fibre products is already well known. Less evident are the roles and the importance of specific firms supplying agricultural inputs, as well as that of those transforming raw food and feed products, despite the fact that more and more research is being conducted on them.

Rates of growth of volumes of food and feed products traded on international markets during the 1980s, which were apparently fostered by newly emerging solvent national markets since 1960, have been somehow surprising, especially in view of a parallel surge of sometimes vicious non-tariff trade barriers between the most relevant countries. These seem to become more and more efficient trade impediments.

Authors, such as Schuh (1979) and Ruttan (1980), have stressed the relationships between various degrees of changes in exchange rates between national currencies of some countries, the flows of international funds and the changes in the respective national food and fibre sectors both in the industrialized world and in the less industrialized countries (LICs) ever since the system of floating exchange rates has been restored. In other respects, Ruttan also documented the extreme weakness of economic situations in LICs, and in particular those which import a high proportion of cereals to meet their consumption needs.

The alleged ability of most major corporations, including those that do not belong to the MNFFCs' group, to take advantage of highly fluctuating exchange rates, to get to grips with the changing cost of living throughout the world, and finally to make the best out of selective trade barriers between countries, is put in the forefront by various authors. Others argue that MNFFCs stand to gain or lose the most by making wise decisions regarding the geography of their new investment capital and its congruence with various international trade patterns. There seems to be a more and more noticeable intention on their part to match their self-interest with national objectives even in countries more deeply committed to their own agricultural and/or food policies. Beyond these trends stands the real possibility that new types of MNFFCs might be in the making. More specifically, this paper suggests that significant structural changes in MNFFCs, some of which have a lot to do with the ways they spread out their investment capital throughout the world and in the LICs in particular, have been taking place. It also suggests that, in the meantime, MNFFCs have enhanced their ability to adapt their marketing strategies to worldwide changes as well as to local characteristics of the LICs, where perennial and rationally consistent food and feed policy objectives are sometimes wanting.

STRUCTURAL CHANGES OF MNFFCS: SOME PAST TRENDS RECONFIRMED

Growth and concentration of the leaders

Exceptionally rapid growth of MNFFCs is deemed surprising by some and troublesome by others. In 1981, the top hundred leaders that operated in market economies (according to AGRODATA), reported a cumulative sales value of 333 billion $US. This is close to one-third of the total production value of hundreds of thousands of firms operating in the food, feed and beverage sector. Their dominant importance can also be depicted by comparing them with the size of some national economies where they are present. For instance, in 1981, 17 out of 24 South American countries and 36 out of 40 African countries displayed a GNP (gross national product) a little different or even much smaller than the yearly sales of Unilever.

Since the turn of the century, leaders among MNFFCs have gained increased importance. In the last 20 years, cumulated sales volume of the top hundred has increased the most rapidly, going up at a rate of about 7 to 8% a year. This is, on average, from 2 to 3% more than the comparable growth rate of the industry throughout the world. Moreover, our own calculations show that, despite the recent crisis, the proportion of worldwide production which is accounted for by those leaders has gone from 28% in 1978 to a little more than 33% in 1981. This trend is likely to continue in the near future, in parallel with greater concentration, as well as with increased volume of production in some socialist countries and in LICs which are emerging as the new markets of the 1990s.

Stable and praiseworthy way through the crisis

According to recent trends, greater concentration may lead to the top hundred MNFFCs gaining worldwide control of about 40% of the whole sector by the turn of the 1980–1990s. The main reason for this is the outstanding performance of the mammoth firms. Indeed, there is some kind of a cumulative impact which is going on and that takes either one of the following forms. On the one hand, fiercely competing firms with exceptionally good self-financing capacity swallow up smaller and weaker ones and/or merge with kins. On the other hand, the tops among the top seem to attract other giants outside MNFFCs' groups, mostly because of the stable performance of the former.

Apparently, however, only the leaders were capable of such performances, which may explain the growing gap between the top and the bottom groups. Although significant changes are not expected in the membership club of the chosen few at the top, it is felt that market penetration by a second generation of firms of mixed breeds (Europe, Japan, Canada and eventually from LICs experiencing rapid growth), as well as the dynamic behaviour of specific firms, combined with the impact of research and development (R&D) efforts, may cause shifts in the order of firms appearing on the list as well as in their relative dominance in the MNFFCs' group (Hymer & Rowthorn, 1970).

Growth focused on specific targets

Research conducted in the last decade indicates some loss of enthusiasm by MNFFCs for the fluid and the industrial milk, as well as for the feed subsectors. By 1981, however, they were showing increased interest in cereals and cereal products, as well as in high value-added items such as brewery products. But the most striking change in this trend towards diversification is undoubtedly the highly significant entry of giants in bakery and fishery products. Despite this, it is not in the food and feed processing sectors that trends towards diversification in downstream direction seem to have been the most significant in the MNFFC sector. Wholesale trade and restaurants, caterers and tavern operations seem to attract the greatest share (13%) of new investments made by the leaders. In contrast, entry of MNFFCs in food retailing is perceivable but much quieter. It goes on mostly by way of food retailers integrating food processing.

In the restaurants–caterers–tavern subsector, fast food was undoubtedly the fastest growing area in the Western world in the past decade. Some 20 MNFFCs appearing in the AGRODATA catalogue are deeply involved in fast food outlets. Their activities also stretch out to management of chains of hotels and/or restaurants. This type of diversification sometimes leads to the development of very large conglomerates, within which the food and feed divisions are no longer dominant.

REGIONAL STRATEGIES: LOOKING OUT FOR DEVELOPING SOLVENT MARKETS

Along with diversification, the tendency for food and feed sectors to become multinationalized is growing. According to our own statistics, numbers of firms under the control of MNFFCs but not located in the same country as the head office, have gone from 2070 (in 112 countries) in 1978 to 2330 (in 127 countries) in 1981.

From the early 1900s until the mid-1940s, multinational firms were mostly transforming raw materials in colonies, exporting raw food and feed materials to European and North American markets. MNFFCs such as Swift-Esmark (1885), Brooke-Bond (1892), Castle and Cooke (1894), United Fruits (1899) and Unilever (1929) were born during that period. One may rightfully call this period the golden age of *input supply strategies* by MNFFCs.

Dominance by access to materials, technology and/or financial power is replaced by new forms of business intelligentsia. Intertwining of information and grouping of interests become closer and closer. Except for such things, however, firms stubbornly cling to their independence. The linking rings are chief executives, technical innovation channels and human know-how transfers. Those aspects seem to outweigh control of capital in cementing the newest groups.

In the manner of Japanese firms, infant international groups encompassing large, average as well as small organizations may have taken a substantial lead over non-linked firms in keeping in the closest touch with the various moods of a rapidly changing worldwide environment. Information systems of such groups are exceptionally rich. Highest-level executive meetings on linking matters strengthen their efficiency in fact-finding roles. Long-term objectives (within time horizons of about 15 years) then tend to fit in nicely both with overall market trends as with the group's self-evaluated capacity to successfully meet challenging opportunities.

Chief executives attend, more and more often, 'linking' meetings where long-term objectives are scrutinized whereas staff management personnel is led to think of practical strategies with other members of the group. Product lines marketed by these firms can therefore be viewed as the result of forward looking and strategically planned (not to say pruned) goals set up by the group, and having to do with financing, technical services, group image, merchandizing and foreign investments.

A suggested taxonomy for those newest groups already exists. It breaks them down into three sub-groups: (i) merger types of traditional international consortia (Unilever, Nestlé); (ii) types of groups highly supported by major banking institutions (DKB in England); (iii) types of groups specialized in highly industrialized product lines (Nissan, Komatus, Nippon Steel). All of them have at least two things in common: *diversification* into new product lines and/or in technological changes is the first; and selection of an *optimal basket* of lines of products conducive to the group's development, from input supply all the way to retailing (Burns *et al.*, 1983).

Links between firms within a group are fathered by highest-ranking executives. Integration may be two-pronged: it takes place either on a horizontal scale between chief executives of top MNFFCs and/or on a vertical scale. Networks of sequential meetings are institutionalized. Various levels of staff convene in such meetings episodically. There are of course some structural financial links which are woven within the group. In rare instances though, some firms are totally owned by others. The average degree of control by the leading firms is in the vicinity of 20–30%. Degrees of control tend to be the highest in those groups supported by major banking institutions, extending them from 40 to 80%. Lastly, other strong intermingling links exist between members of these groups, such as for availability of short-term and intermediate-term financing under special conditions, for shared access to input suppliers and/or to chains of buyers, for market prospection circles and/or lobbying services.

The building pressure for LICs to pay much greater attention to the structural changes that are going on in MNFFCs must be reckoned with. The choice of determining whether or not LICs ought to join in by ways of lobbying groups and/or by using other linking rings will be a crucial one. LICs declaring themselves out of such groups might be deprived of some degree of power in making MNFFCs espouse their long-term objectives.

THE NEWEST TECHNOLOGICAL PROCESSES: NEW SOURCES OF POWER?

R&D budgets allocated to research in the food and feed sectors indicate that discovery of new processes is of the utmost importance for MNFFCs. It is felt that structural changes will undoubtedly result from further major technological steps forward if capacities of MNFFCs to compete with one another are then intruded upon or impaired. For instance, bio-industry and the production of food and feed from non-agricultural materials might create a still greater dependence of agriculture and/or of food and feed industries production processes from progress in bio-engineering, chemistry, enzymetics and genetics (Table 10.1). Recent analysis of MNFFC strategies with regard to bio-engineering and to the newest food and feed production processes indicates that the dividing lines between relevant sectors are becoming more and more shallow. Sectoral stratification is coming to rest upon technological processes which are closer and closer akin. The food and feed sector, pharmaceutics, chemistry and energy tend to become almost every man's land from the point of view of the previously more specialized MNFFCs.

Recent studies have revealed the degree of interest of top multinationals for biotechnology. Some 225 uses of such processes, either under testing and/or under normal conditions, were recorded out of a relevant sample of 95 leading corporations operating in the food and feed, in the chemistry, in the pharmaceutic and in the energy sectors in 1983. The food and feed sector is on top of the list according to declared number of applications. Production of food and feed from unicellular organisms, of food additives and of sweeteners is also reported. More than a third of all

Table 10.1. Number of uses made of biotechnology by multinationals

Sector	Number of applications				Total number of uses recorded	Number of firms reporting
	Food and feed technology	Chemistry	Pharmaceutics	Energy		
Food and feed	47	14	13	10	84	28
Chemistry	22	13	13	1	49	28
Pharmaceutics	31	7	27	3	68	25
Energy	2	5	1	4	12	7
Other sectors	5	1	3	3	12	7
Total	107	40	57	21	225	95

Source: Our calculations from data collected and reported by CFCE (Centre Française du Commerce Extérieur).

items made known by sampled firms enter in these specific categories of uses. Chemistry and energy also seem to attract much attention. Peculiarly, some of the utilizations discovered thereon are no full strangers with food and beverage production. Indeed some of these firms extract various components out of food and feed by-products and/or from biomass.

AGRODATA statistics tend to corroborate such findings. However, top MNFFCs display a noticeable restraint by limiting themselves to the most traditional applications of biotechnology. Biotechnology-related production processes accounted for 36 production activities of the top hundred MNFFCs. Makers of cereal products and of fats are frequent users of enzymes in processing sweeteners and yeasts. Brewers of alcoholic beverages take the keenest interest in fermentation processes. The newly developed techniques with the latest uses are for production of dehydrated proteins and amino acids. According to these figures, a few new trends are in the making. First, significant attraction seems to emerge in chemistry, energy and pharmaceutic sectors for MNFFCs, especially for those under Japanese control. Perhaps of still greater interest is the fact that multinationals which were aliens to the food and feed sector are finding their way into it.

Tomorrow's food and feed production processes will undoubtedly come to rest more heavily upon innovative technology. That will bring along deeper changes in MNFFCs. Considerable matters are at stake for MNFFCs and for countries, some of which are not even aware of their consequences. Both the apologists and the fiercest denunciators of these trends tend to agree about at least one thing: what might otherwise become a promising stream of hope for a significant part of the world, it is feared, might make a large contribution to placing a few already dominant and highly diversified MNFFCs, endowed with the most advanced technology, in fuller control of world production of food and feed.

SUMMARY AND CONCLUSION

International capitalism has proved its capacity to adapt to local conditions. MNFFCs have played an active role in this matter. This paper suggests that the birth of the so-called 'third type' of firm has already been set going in the food and feed sector. Increased demand, which stirred up new flows of international trade particularly between countries of the western hemisphere, has led MNFFCs to seek new ways of making profits. Established strategies are being adjusted to or substituted for by new ways for firms to compete with one another, which are overall less aggressive and which come to rest more and more on technological advantage. LICs that are confronted with firms of this new generation might court serious disappointments, such as those of becoming still more deprived and/or more dependent, if they fail to coordinate their efforts right now. The case is pressing for scientific questioning of the likely impacts of new strategies which are being put forward by dominant MNFFCs on national policies, at least to the extent that food and nutrition programming remains a high-priority issue in most countries of the world.

From 1945 to 1960, MNFFCs extended their zones of influence to LICs with the greatest agricultural development potential. MNFFCs kept increasing the volume of their most basic activities but they gradually appeared also interested in supplying LICs' urban markets with high valued-added products with strong demand. The golden age of *input supply strategies* became overlapped by the age of *marketing strategies*. Competition did not seem to be as fierce in these infant markets as it was in already established ones. Moreover, the fact of being there first, together with the cumulative process of learning how to operate in these countries, somehow hid the first comers from the sight of the competitors.

During the 1970s, MNFFCs' presence in national economies then became more dominant and more integrated, by way of diversified local investments. Use was made of local financing and of subsidies of various types. They even complied with nationalization of some of their activities. It seems as if they were following a two-pronged strategy: on the one hand, they set up *marketing strategies* based on production in both *autonomous plants* and *assembly-line plants*, but on the other hand, they tried to improve their *production strategies* by better linking some of their strategic factors as multinational producers.

Nowadays, most giant firms display high proportions of multinational activities. AGRODATA statistics indicate that one-third of the leaders do more than 30% of their activities in foreign markets, and a little more than 40 of them operate plants in more than ten countries. Notwithstanding these trends, production rationalisation built into marketing strategies remains a regional management technique for MNFFCs. Indeed, in 1981 three plants out of four, among the 4200 branches operated under the control of the top hundred MNFFCs, were located in North America or Europe. The trend towards across-the-border ventures in countries of the northern hemisphere has thus significantly increased. But MNFFCs seem to favour countries or regions with high population growth rates, with the least abated economic growth rates and with stable social environments.

MNFFC: Newest Trends

Facing worldwide crisis and unpredictable political environments, some MNFFCs may already be in the process of changing somewhat drastically their management processes. The terms '*Third Type*' have been used in a rather loose sense to qualify firms which are openly trying to eliminate plant breakdown, delays in filling orders, construction defects, stocks of products and red-tape. Centralization is a key word but open-mindedness, flexibility, sharing in new ideas and looking out for relevant and critical facts and figures also come into prominence. Top executives get along with that by connecting themselves with quality circles, think-tank sessions, project partnerships as well as through their various links with other firms.

REFERENCES

AGRODATA, a bank of data compiled by the Groupe de Recherche Agro-Alimentaire of the Institut Agronomique Méditerranéen (I.A.M.) and by the Groupe de Recherche en Economie Rurale de l'Université Laval in Quebec (G.R.A.A.L.).

Burns, J., McInenvey, J. and Swinbank, A. (1983) *The Food Industry, Economies and Policies*. CAB International, Wallingford.

Hymer, S. and Rowthorn, R. (1970) 'Multinational corporations and international oligopoly: the non-American challenge'. In: Kindleberger, C.P. (ed.) *The International Corporation*. MIT Press, Cambridge, Mass.

Ruttan, V.W. (1980) Food strategies for grain deficit poor countries. In: Tyrchniewicz, E.W. (ed.) *The New Era in World Agricultural Trade: Perspectives for the Prairies and the Great Plains*. Occasional papers no. 12. Department of Agricultural Economics, The University of Manitoba.

Schuh, G.E. (1979) Fluctuations in foreign exchange rates: implications for agricultural trade. In: Tyrchiewicz, E.W. (ed.) *The New Era in World Agricultural Trade: Perspectives for the Prairies and the Great Plains*. Occasional paper no. 12. Department of Agricultural Economics, The University of Manitoba.

11

COOPERATIVES – EFFECTS OF THE SOCIAL MATRIX

G. Hunter

It is a platitude that the way in which cooperatives perform will largely depend upon who their members are, what attitude and tensions exist between the members, between the whole unit and neighbouring society, and between the unit and Government. The social matrix from which a cooperative is formed both permeates its structure and envelops it as environment. But unfortunately far too little work has, until very recently, been done to fill out this generalization by specific examples of exactly what effects on performance are produced by what structures and attitudes in particular societies.

This paper has two purposes: first, to indicate that some major hypotheses can be put forward relating the success/failure of cooperatives to the social/political stage of development of local society; second, to indicate that clear and important policy alternatives would flow from the establishment of these hypotheses at a level where they have at least *prima facie* predictive value.

GENERAL APPROBATION OF COOPERATIVES

The chief reason for lack of critical analysis has been the almost universal approval of cooperatives as a tool of choice for many kinds of rural development. The 1966–1970 Kenya Development Plan states: 'There is only one course of action open to the nation and that is to strengthen cooperatives to play their role adequately'. A typical statement of the International Cooperative Alliance is:

'Co-operation provides its own motivation under conditions in which

---

Source: Commentary on papers presented at the Seminar on Agricultural Cooperatives, Scandinavian Institute of African Studies, Uppsala, 1970.

it is sometimes extraordinarily difficult to enlist interest and active participation. It appeals to the self-interest of the rural producer in a way that he can understand and demonstrates, through tangible results, how he can pool his efforts and resources with others in a similar situation, in order gradually to lift himself out of poverty and stagnation. By enrolling him as an active participant in decision-making, the co-operative form of organisation stimulates initiative and gives him the will and the means to shape his own future.'

This emphasizes two benefits – self-interest and the stimulation of initiative and participation. There are other arguments in favour. One is the 'replacement of the middleman', with allegedly higher returns to growers, another is that 'They serve as a useful means of extending credit to farmers and thereby enforcing programmes of technical assistance and advice'.

Finally, the cooperative appears to draw upon traditions of cooperative behaviour in pre-colonial traditional society, and to chime with concepts of African socialism, Ujamaa in Tanzania, etc. It symbolizes an egalitarian, socialist society in contrast with a class-stratified, capitalist one. This is a formidable list of advantages: self-help, participation, democratic control economic advantage, defeat of capitalist middlemen, convenience to Government, egalitarian ideals, continuity with indigenous tradition.

Against these theoretical benefits has to be weighed the actual experience of cooperatives in developing countries. 60–70% of cooperatives started in East Africa (and a similar proportion in India) have proved unviable or semi-viable; a large proportion of those which have succeeded have emphasised stratification and resulted in rewards for the 'notables' and 'big men' of rural society as distinct from the humble farmers. 'Cooperatives are for big men'; 'cooperatives are for the rich', are remarks to be found all too commonly, not only in East Africa but in India, and in Latin America.

Finally, on the theme of approbation, it is generally the Rochdale model which has been approved, with variants from Denmark and Israel. Up to 200 different forms of 'pre-cooperatives', 'para-cooperatives' and small, simple forms of cooperative action that have succeeded in various areas of the developing world which have not followed the Rochdale model or come within the administrative straight-jacket of the 'Registrar of Cooperatives' are neglected. No one doubts that some form of cooperation is a good thing: many doubt the full-blown standard cooperative.

TRADITION

The notion of a full carry-over from traditional cooperative action (common house-building, weeding, path-clearance, etc.), usually based on extended family or clan relationships, to participation in 'modern' marketing cooperatives, is largely an illusion. Thus, many scholars trying to specify what makes the African peasant 'tick' tend to treat the existing peasantries as some exotic social type characterised by a basic benevolence and by being relatively undifferentiated. But these assumptions do not obtain in actuality. Socioeconomic differentiation and kinship–neighbourhood loyalties emerge into prominence among factors affecting differential participation by individuals in the cooperative organization. At the same time other variables such as ecology, transport and availability of market for agricultural goods affect the development of cooperatives. Many of these factors clearly did not pertain under the traditional system. In addition, they represented an expansion of the scale of the activities in which the African peasants are involved today. These factors alone point to the fact that there is no direct continuity between the autochthonous cooperative forms and modern marketing cooperatives.

THE EUROPEAN MODEL – VERTICAL AND HORIZONTAL RELATIONSHIPS

Nor is the European cooperative model easily applicable in the early stage of modernization. This new spontaneous form of cooperation was made possible by the absence of strong social ties based on kinship. The nuclear family system had already been sufficiently established to make the peasants realise that their strongest allies were not their relatives but the other peasants who shared the same economic fate.

This difference between a society already moving out of clan and extended family relationships and societies in which these are dominant can be seen as a difference between 'horizontal' and 'vertical' relationships. To this category belong teachers, priests, traders, administrators and politicians. They often get into leading positions, because the ordinary peasants believe them to be more able to defend the interests of the local community than they are themselves. The peasants give them full support in return for the favours or rewards that they can secure from the 'outside world'. Cooperatives in East Africa are thus very often ruled on the basis of already existing informal 'patron–client' relationships.

The importance of this 'vertical' relationship is reinforced by the fact that social ties based on kinship and other local institutions are still more important than the mutual loyalty between peasants in different village communities. The 'horizontal' ties of economic interest have not yet replaced the 'vertical' ties of social obligation based on such units as the clan, village, etc.

Obviously, the simple opposition of 'vertical' and 'horizontal' is absurdly over-simplified. Traditional societies had horizontal as well as vertical organization (age-sets in Africa; caste in India). Moreover, the *form* of vertical organization differs greatly in particular traditional societies; for these particulars the advice of the (currently rather neglected) anthropologists is needed. But in the critical areas of most LDCs we are not dealing with 'pure' traditional societies but with farming communities who have already had considerable contact both with the money economy and with the political systems which have grown up with independence. It is probably very widely true to say that the amalgam of persistent elements of tradition (extended family, clan, etc.) with new forms of influence (local politicians, commerce, the possibility of capturing benefits from Government programmes) has led to vertical, patronage relationships in its present phase far more than to 'horizontal' economic relationships which cut across family, clan, caste, or traditional hierarchies.

REINFORCEMENT OF STRATIFICATION

A 'vertical' stratified relationship is both used and reinforced then when cooperatives are introduced to a society at this stage. The businessman cum cooperative leader is usually able to divert some of the money of the cooperative institution for purposes connected with his own private activities. It may be for further improving agricultural production on his land or it may be for building up political support in the area. The latter is not uncommon. It must be remembered that very few, if any, businessmen become cooperative leaders because it is commercially profitable. They belong to cooperatives because these institutions in the rural areas offer a convenient platform for political campaigning and the maintenance of social control over the population in the area. While the businessman acts in order to build up his economic position while in his business, he acts in order to boost his social and political position while in the cooperatives. This is one of the important aspects of current sociopolitical change. Cooperatives assist in institutionalising the power of the already economically privileged.

Moreover, these small vertical segments or factions compete violently with each other. What has been said above about the socioeconomic conditions in East Africa has definite implications for the management of cooperative institutions. The committee members in a cooperative society or union are usually elected on the basis of local constituencies – a village, location or division, etc. Unlike the situation in Europe, where elected committee members shared an interest in fighting a common enemy, the contradictions in East Africa are still regarded as those between different local communities: village against village, division against division, etc. A committee member is there to fight for his own interest or that of his constituents.

This renders the process of managing cooperatives extremely difficult.

Each committee member is pulling in his own direction or seeking coalitions based on non-economic principles. There is no real consciousness of the common objectives, certainly no strong will to realise these objectives. Favouritism and corruption and often the result of planting democratically managed cooperatives in this kind of environment. Committee members who deliberately use their position for other purposes than that of promoting the economic welfare of all members can easily be re-elected as long as they are socially acceptable to a majority of the electorate. The idea has been put forward that socialism is necessary to cooperatives rather than cooperatives to socialism.

MIDDLEMEN

The idea that 'middlemen' exploit peasants, which has, of course, some justification in some circumstances, in Africa has had mainly a racialist background. We should be aware, however, that the forces in operation in East Africa have not been the same as in Europe. Those who took the initiative in challenging the Asian merchants were not the ordinary peasants, but the African 'entrepreneurs' in the countryside, who found the Asians standing in their way. These entrepreneurs, by appealing to the peasants and telling them that they were 'their men', as opposed to the Asians, who were 'outsiders', managed to use the cooperatives to strengthen not only their own economic position but also their political control over the countryside.

It is worth mentioning that detailed economic studies both in India and in Africa have largely failed to prove that middlemen give unduly low prices to farm producers; on the contrary, their real margins are usually small. There are also many occasions where merchants give either a higher price than cooperatives or a quicker, more acceptable service in payment, or collection – why else do cooperative members so often sell to merchants rather than to cooperatives, and why else does Government so often protect cooperatives by imposing a monopoly on their behalf? There are too many cases where cooperative crop purchase is proving disadvantageous to the membership, compared to the open market system.

PARTICIPATION AND SIZE

Another somewhat confused issue lies in arguments concerning efficiency, size and participation. The argument for size in primary societies and for building unions and federations on top of primary societies, is usually put forward on efficiency arguments – larger funds, better services to members, more power in the market, and so on. While some marketing and processing operations need considerable size for economic operation, effective organization and effective participation are governed by the rule of optimal size, as well as the processing of tea or coffee. Effective participation decreases sharply with increasing membership.

There are really three points here. First, the relation of size and participation; second, the fact that large size requires professional managerial ability; and third, that, whereas the initial reason for starting cooperatives is usually to provide a primary grouping of many individual farmers (for convenience of extension and credit services, and for sharing some kind of facility – tractor, pump, dairy, etc.), once this primary grouping is achieved, the higher, large-scale functions can just as well be done either commercially or by a semi-public corporation. Is the primary object of cooperatives to encourage participation and provide a grouping of a few score of individual farmers, or is it to create an organization capable of competing, in managerial efficiency and in rewards to the growers, with a commercial enterprise? Too often, cooperative marketing organizations end up with a six storey building and low prices to producers.

The interest of cooperative employees (and also to chairmen, etc. for political as well as economic reasons) lies in increasing the size and importance of the whole enterprise. The interest of employees is partly in preserving the size of their hierarchy and, if need be, expanding it, and, of course, in better conditions and pay for themselves. This is quite contrary to the interests of the farmer.

GOVERNMENT INFLUENCE

Is the cooperative a spontaneous, autonomous group of farmers who realize the need to cooperate and both see and accept the disciplines which cooperation involves? Or is it a tool of the administration, which needs a convenient channel for delivering services? Let us now turn to the basic problems of democracy, the involvement of the government and the effective involvement of members. In some cases the cooperatives are in fact manipulated by the wealthier peasants, and with 'traditional' cleavages. There is also the case in which the government has pressed or forced a cooperative upon a peasant community which has not asked for one. In both cases, the alienated membership may react with indifference and apathy and with a feeling that they cannot influence the decisions about their own cooperative and about their own future.

ECONOMIC BENEFIT

Finally, great care is needed in defining the tangible economic benefits which the formation of a cooperative may achieve. It is not so easy to find 'indivisibles' which can only be secured by cooperative organization. A tractor? But it may not be suitable even if used cooperatively. There certainly are indivisibles – a dairy, a powered boat, a large pump – and cooperatives built around such tangible needs have a higher chance of success. But by no means do all cooperatives have any such convincing purpose; some are simply established for administrative convenience or for ideological reasons.

SOME PROPOSITIONS

1. 'Communitarian' or 'traditional' forms of common action differ in motivation and structure from modern cooperative organization. The 'spill-over' from one to the other may be relatively weak.

2. Traditional societies with a fairly strong vertical type of social organization, which have already developed elements of 'modern' leadership, will tend to absorb cooperatives into this style (patronage, segmentary competition, etc.). In consequence, cooperatives will tend to re-emphasize and strengthen social stratification. The egalitarian and socialist aims of cooperatives will thus be, at least partially and temporarily, frustrated.

3. Scandinavian and British cooperatives were formed when extended family and clan organization had already broken down and horizontal relationships in face of a common economic enemy could be more readily formed. Thus, it appears to be in the intermediate stage between a purely traditional society and a more individualised society that cooperatives tend to be absorbed by, and to strengthen, the upper layers of a stratified system. Very large sections of LDC society are precisely in this intermediate stage.

4. The objectives of democratic participation may be impeded not only by the stratification but also by excessive size of the cooperative group and the tendency to transfer power upwards to unions and federations.

5. The objective of giving producers better prices may be impeded both by this tendency, supported by both staff and leadership, to create large organizations, but also by unproven objections to 'middlemen' sometimes racialist, sometimes ideological. If the main objective of instituting cooperatives is: (i) to have a convenient grouping at farmer level (ii) to encourage participation; and (iii) to obtain good prices, this might be better achieved by allowing organization (especially marketing and major processing) above the primary level to be carried by more professional organizations (whether public, semi-public or private).

6. The creation of cooperatives by government for administrative convenience is likely to negate active democratic participation, and may (particularly in the case of settlements) produce attitudes of dependence or apathy.

This is particularly likely to happen if there is not a clear, tangible benefit around which the cooperative is built.

This list of propositions tends to negate many of the advantages claimed for cooperatives, in certain circumstances:

- Self-help, participation, democratic control – not if stratification, size and heavy Government tutelage operate strongly.
- Economic advantage – certainly to the powerful few; often not in terms of good prices and benefits to rank and file membership.
- Defeat of capitalist middleman – yes, but often by substituting local indigenous entrepreneurs for racially alien ones; and often at the cost of low prices or less acceptable services to the grower.
- Convenience to Government – possibly, but often at the cost of dependence or apathy among ordinary members.
- Egalitarian ideals – seldom furthered, save in societies already far along the path of individualization and relatively emancipated from 'vertical' influences.
- Continuity with indigenous tradition – not true in most cases.

This is, of course, a highly critical and negative position; it would even justify an assertion that an obsession with cooperatives is one of the major impediments to agricultural development, because their frequent failure to achieve goals set for them results in a really catastrophic defeat of the attempted 'package' deal, discourages the extension service and gives opposite political effects from those intended by any egalitarian or socialist-minded government.

It remains, therefore, to suggest both what positive advantages and successes of cooperation may remain, and to suggest conceivable lines of policy.

First, the scope of the argument is, of course, limited; it suggests relative difficulty or failure in the pursuit of egalitarian/socialist ideals – and this is important to those Governments which are thus committed. From another standpoint it might suggest success in assisting the evolution of a more individualized commercial and political leadership at grass-roots level, and the growth of entrepreneurs has been a subject on which much ink has been split and much aid directed. It is worth mentioning that at least some of the entrepreneurs have come from the ranks of the illiterate and poor.

It could also be argued, on almost historical dialectical grounds, that the 'intermediate' stage of stratification will, in turn, be negated by horizontal cooperation of the under-privileged against the entrepreneurs. the policy implication here would be to control, rather than destroy, the new leadership, and to assist the lower levels to participate more fully. Schemes such as the Indian Small Farmers' Development Agency follow this line of thought.

Second, there are undoubted and major successes of cooperative organization in both East Africa and India at two levels. Some very big and flourishing organizations have been built up – the Lake Victoria Cotton Cooperatives (Mwanza, Tanzania), the Anand dairy (Gujarat, India), or some of the Maharashtra (India) sugar cooperatives. In such cases, the two major factors seem to have been: (i) a clear economic advantage, and an 'indivisible' (processing, by ginnery, dairy or sugar factory); and (ii) strong and competent business management, although sometimes marred by corruption.

At Mwanza there was a third factor – common interest against a (racial) enemy – the Asian ginners. At the lower level there have been many, less often recorded, successes at very low levels – extremely small cooperative groups built around a small facility (such as a tubewell shared by 20 or 30 farmers). Some of the Comilla cooperatives were extremely small, and there are a host of 'cooperative' ventures on a very small scale, with extremely simple organization, which have succeeded. Note that these are also built around a clear, tangible benefit.

In relation to small schemes, and particularly to the first attempt to ensure cooperation, policy could take far more care:

1. not to fly at once to the full Rochdale pattern, but to start with very simple groupings of like-minded farmers round a clear, tangible facility or benefit;
2. to take particular care to see that it is primarily those whom it is intended to benefit (small farmers) who are first approached and approached without preconceived ideas of what structure should be imposed upon them;
3. to retain a very open mind whether, above the essential 'Primary' group, a union, or a public corporation (e.g. Kenya Tea Development Authority) or merely a better controlled private commercial system should be entrusted with the higher level organization, whether of processing or marketing or both, with the recollection that if a small, genuinely cooperative, genuinely participatory, genuinely self-governing group is established at village level, most of the key objectives of cooperation have already been achieved; above this level efficiency may become the chief criterion. In the light of difficulties spelt out in this paper, of achieving horizontal, non-stratified organization, it may be necessary to avoid including whole communities (complete with their structure) in the Primary but to encourage deliberate, selective, horizontal groupings of just those farmers who see a common economic benefit. This group, may not include either the most backward or, indeed, the most powerful and active (who tend to join for prestige, patronage or political reasons even at some sacrifice of their purely economic freedom of action).

12

COMMODITY MARKETING THROUGH COOPERATIVES: SOME EXPERIENCES FROM AFRICA AND ASIA AND SOME LESSONS FOR THE FUTURE

COPAC

In the developing countries, where cooperatives are still in an early stage of their development, agricultural cooperatives have achieved some marked successes. Examples are:

1. the coffee cooperatives in Kenya which made the participation of small producers possible;
2. the societies and country-wide cooperative in the Kenya dairy industry which provide the basis of a successful dairy industry and milk production by small farmers;
3. the remarkable performance of the Anand type of cooperatives in the Indian dairy sector;
4. the rapid and successful expansion of the oilseed marketing and processing cooperatives and of cooperative sugar mills, also in India;
5. the substantial role of the Korean cooperatives in agricultural development and marketing;
6. the survival and continued operation of the coffee and cotton processing cooperatives in Uganda despite appalling adverse political and economic conditions.

KENYA

Cooperatives play a major role in Kenya's widely diversified agricultural economy. Until the 1950s, agricultural production and structures were mainly geared to large-scale European type farming. Since then, the small-scale farming sector has steadily developed and now contributes 53% of all

---

Source: COPAC Occasional paper, COPAC Secretariat, FAO, Rome, 1984.

commodities supplied to marketing boards. In line with this, cooperative institutions – primary societies and secondary unions – were established and the country-wide cooperative bodies were adapted to the new situation.

Cooperatives at all levels are essential to the production of coffee, Kenya's main export crop. Coffee arabica – unlike robusta – has to be processed immediately after picking. The larger farmer has his own machinery and facilities required to carry out this operation. The small producer has had to set up primary cooperative societies to carry out these functions because of the substantial investments required. The small grower delivers the coffee cherry to the cooperative factory and is paid according to the weight of the cherry. After processing, the society delivers the parchment – via the district union – to the Kenya Planters Cooperative Union (KPCU). The larger farmer delivers his directly to the KPCU.

The KPCU hulls, grades, classifies and stores the coffee in its modern and efficient factory. The coffee is then put at the disposal of the Kenya Coffee Marketing Board (KCMB) which operates a weekly auction to sell the whole Kenya crop. The proceeds are pooled and the primary society or the grower, large and small, is paid according to the grade (quality) delivered.

The KPCU, which handles the whole coffee crop, has kept operational costs at a very reasonable level. It pays the producer (individuals or small growers through their union) the proceeds of KCMB's sales after deducting costs for transport, processing and marketing. The cooperative societies and unions deduct their costs of initial processing and handling as well as interest on loans or amortisation of the investment of their coffee factories before paying the individual farmer. The efficiency and costs obviously vary greatly from society to society but to ensure that growers get a fair deal, government has ruled that 80% of the KPCU proceeds must be passed on to the producer. This is generally adhered to. Kenya is one of the very few countries where farmers actually receive something near the world market equivalent price.

Inefficiency and corruption at the society and union level has sometimes occurred and government has exercised its right under the law to remove cooperative boards and/or management after due investigations. Apart from these general supervisory functions, however, government does not interfere with elections of cooperative office bearers or the appointment of management staff. The same applies to the KPCU, where both cooperative society and union interests, as well as individual farmers, are represented on the Board.

The assumption that competition and private traders ensure the best services at the lowest costs – and hence best price to producers and consumers – certainly does not apply to the coffee industry in Kenya. When, in the 1940s, coffee growers voted in favour of the Coffee Marketing Board as the sole selling agent, private traders predicted quite wrongly a decline in coffee quality, prices, markets and popularity of Kenya coffee. Instead, the combination of cooperative processing and KCMB auctions has proven to be one of the most successful systems anywhere in the world.

Kenya Cooperative Creameries (country-wide) and the Marikani Milk Scheme (coast only) have a near monopoly of the sale of dairy products in all urban centres. The KCC owns and operates dairy processing factories, milk tankers and vehicles for milk distribution to retailers and handles over 240 million litres per annum. Farmers who produce sufficient milk can supply direct to KCC factories. The smaller producer has to bulk, cool and transport the milk to the KCC through a primary society. The success and level of payments to farmers depends on the quality of the milk supplied by members and general management of the society. The performance of cooperatives varies greatly and while some societies do a good job, others do not.

The whole price chain, from producer to consumer, is fixed by government. At times, producer prices were raised but the KCC was not allowed to pass on the increase to the consumer for some months, and hence incurred heavy losses. This, together with sometimes inadequate processing and marketing margins, has led to a large deficit being carried forward. This is now gradually being reduced. While there is room for improving efficiency in processing and marketing, the present system works well. Competition from private traders might result in better price for a few producers but it would certainly not ensure country-wide production and collection of milk.

ANAND COOPERATIVE, INDIA

The Anand pattern of cooperative development is an amazing and, unfortunately, almost unique success story and provides an innovative model of cooperatives for India and elsewhere. It started in 1945 with government entering into an agreement with a private firm to supply milk to Bombay. The arrangement was highly satisfactory to all concerned – except the farmers. The government found it profitable; the firm kept a good margin; milk contractors took the biggest cut. After a 15-day protest milk strike, the Kaira District Cooperative Milk Producers Ltd, Anand, was established and formally registered in 1946. It began with a handful of farmers producing 250 litres per day. By 1948, 432 producers had joined village societies and the Union handled 5000 litres per day. By 1955, the cooperative's first modern dairy plant was in operation. The plant was expanded in 1958 and again in 1960. In 1965, an additional factory was erected. By 1974, the Anand complex had a milk throughput of 750 000 litres per day and produced butter, milk powder, and other dairy products.

In 1964, a cattle feed plant was commissioned and cooperatives took responsibility to improve members' milk production by providing artificial insemination services, veterinary and extension services and cattle food supplies. More recently, Anand has contributed funds to general rural development.

Neighbouring states joined in the enterprise and the Gujarat Cooperative Milk Marketing Federation was set up. It now covers 850 000 members, has a throughput of 1.5 million litres per day and an annual turnover of Rp 1250 million per year. Through various development programmes, mainly Operation Flood, cooperative milk procurement increased to 3.9 million litres per day (1981) and enabled 20 million poor milk producing families to earn Rp 8 million per day.

The basis of the Anand pattern is the village cooperative society of primary producers. A milk producer becomes a member by paying an entrance fee and buying a share. He also undertakes to sell his milk only to the Society. Members elect a committee and a chairman, all honorary workers. The paid staff includes a secretary, milk collector, butter fat tester, clerk, accountant, inseminator, and so on. Farmers, or their wives or children, take the milk to the society milk collection centre twice a day. The milk is measured and tested and when the farmer delivers his evening milk, he receives payment in cash for the morning delivery. In addition to the daily cash, paid on the basis of quantity and quality of the production, the farmers also receive a yearly bonus based on the society's surplus.

The district union represents all the village societies and has a 19-member board of directors. The union purchases all the milk from its societies, processes it and provides the societies with back-up and production enhancement services. This includes artificial insemination centres and services, mobile veterinary units, supply of feed concentrates, and training facilities for society and union staff. Meetings with farmers to explain better production methods are also organized.

In addition to its economic contribution and the development of the dairy industry, the Anand pattern has had an important impact on Indian rural society:

1. The democratic process of election is breaking down social and economic divisions.
2. The equalizing nature of the queues at milk collection point is helping break down caste barriers.
3. The daily payments, often collected by women, has enhanced their status and income; it is also providing an income to the landless who are able to keep cows.
4. Non-formal education is being provided through training in better production techniques, veterinary and health care, and artificial insemination.

5. General rural development efforts are taking place which are improving income and rural life.

The success of the system is also due to inspired leadership and a loyal and devoted staff at all levels. Staff employed by the society are frequently paid incentive bonuses. An inseminator, for example, is paid an additional fee for each cow got in calf and others are paid a bonus according to the quality of milk delivered to the union. This extra payment is not only an additional income for the staff but also creates an interest in the success of the cooperative, an attitude often lacking in personnel employed by cooperatives elsewhere.

The Anand example is now spreading its influence to other commodities, the latest venture being groundnut marketing and processing. The oilseed cooperatives in Gujarat have grown from 34 societies and 300 members in 1978 to 946 societies with 70 500 members three years later, and handling nearly 70 000 tonnes of groundnuts. Members were paid about $460 per ton in 1981 or more than double the market price of the previous season.

The cooperatives now own two oil mills and plan to expand both membership and marketing volume. They, however, face bitter opposition from the 'oil kings' (13 powerful families controlling the oilseed trade and processing). The profits of this group has been reduced from $225 million to $60 million in 1980/81. The Gujarat Cooperative Oilseed Federation (GCOGF) is determined to break the traders' absolute control of the processing and marketing system and plans to erect five more mills, two solvent extraction plants and storage facilities at a cost of $850 million. The oilseed success shows that the Anand experience can be duplicated, given the right leadership, conditions and determination, and that farmers' interests can be promoted by cooperatives despite powerful opposition.

REPUBLIC OF KOREA

Korea has made phenomenal economic advances in the last twenty years and the cooperatives have been an important component. In 1961, a three-tier structure including the National Agricultural Cooperative Federation (NACF) was established to take over all functions of the cooperative movement. This consisted of primary cooperatives at the township level together with county or city level cooperatives federated to the NACF. Cooperatives were multi-purpose as well as specialized (livestock and horticultural). In 1981 the livestock cooperatives affiliated at the national level into the National Livestock Cooperative Federation (NLCF).

The overall objectives of the agricultural cooperatives are threefold: (i) to increase agricultural productivity; (ii) to improve the social and economic status of its members; and (iii) to promote the balanced development of the national economy. To achieve these objectives, cooperatives are engaged in banking, marketing, farm supplies, insurance, processing, farm guidance, research, education and public relations. Government has given its support in a practical way to these activities but also exerts considerable influence on cooperative policies.

A sophisticated system has been established and developed to carry out the cooperative marketing functions. Primary cooperatives assemble, sort, grade, inspect, transport, process and store the various commodities. They own and operate facilities such as collection and selling points, vehicles, rice mills, and warehouses. In addition, 24 townships and 25 horticultural cooperatives located near or in cities also operate marketing centres which function as wholesale markets.

Farmer members, unless they have entered into a special contract, are under no obligation to sell through the cooperative channels and are free to use private traders. Farmers are either paid outright, as is normally the case with rice purchases, or, when high risk commodities such as fruits and vegetables are involved, deliver on a consignment basis. An 80% advance payment of the expected price is made to farmers if they so wish.

1389 collection centres for farm products are operated by the agricultural cooperatives in the producing areas and 145 marketing centres have been established at the township level for the sale of various commodities. This extensive collection and distribution network makes the effective handling of the vast amount of produce passing through the cooperative channel possible. The sale of commodities by agricultural cooperatives reached 576.6 billion won in 1982, of which 21.4% were fruits, 26.0% food grains, 22.9% vegetables and 19.7% cash crops and others. The cooperatives purchase rice and barley from farmers at fixed prices on a commission basis. Cooperatives also store rice, transport it to the mills and distribute rice and other grain to the consumer on behalf of government.

Although agricultural cooperatives in Korea enjoy a privileged position as government agents for specialised products and as supplier to the military, they also face strong competition from private firms. Moreover, some of the tax exemptions hitherto accorded to the cooperatives were to be phased out in 1982. In these circumstances, it is increasingly recognized that the cooperative share in the market can only be expanded by improving efficiency as well as investment in new facilities.

Amongst the measures which have been, or will be, taken to achieve a better performance are:

1. Improving standardization of produce grading and packing.
2. Supplying packing material with the cooperative brand name.
3. Providing interest-free credit to cooperatives as a form of advance payment.

4. Improving market information – current and future – through telephones and the present 115 telexes.

SOME LESSONS

For cooperatives to be viable as well as independent, a large proportion of the funds should come from within the cooperative and its members rather than from external borrowing or government grants. It is a general rule that on joining a cooperative members must purchase one or more capital shares. This is usually a minimal amount, but even then small farmers find it difficult to raise the cash or have it deducted in installments from their deliveries. Further capital contributions are even more difficult to find but are necessary as the cooperative expands and requires additional funds. The small equity participation of members reduces their financial stake in the success or failure of their own organization and their interest to ensure that capital is used to the best advantage of all members is, therefore, not great.

The cooperative itself may find it difficult to accumulate funds through the retention of surpluses. Unless farmers receive at least as good a price as from other buying sources, the cooperative would lose members, so it is difficult to make substantial profits. Every opportunity must be taken to overcome this constraint. For example, when producer prices are high it will be easier to retain a small portion to build up a capital fund, and such opportunities should not be missed. Where external funding is required this is usually best obtained through cooperative banks or through development grants from donor agencies.

In many developing countries, agricultural cooperatives act as collecting or procurement agents on behalf of government marketing boards. Where there is a fixed producer price the cooperative also receives a fixed margin for its services. Where the price payable is only a minimum one there is of course some flexibility with the possibility for cooperatives to exercise market judgement, and hence seize opportunities for increasing their profits.

It might appear desirable for cooperatives to operate in high risk marketing sectors such as fruit and vegetables. However, it is doubtful whether cooperatives, with their inherent constraint of not being able to make up losses in one period with profits in another, are really competent to operate in such volatile markets. There are also difficulties coping with sudden market changes. They certainly must have very experienced staff and be able to build up an effective network functioning from the producer to the market outlet. Such ventures have been successful in Korea, but have failed in Kenya and the Philippines.

Farmers frequently sell their produce to traders, or on parallel markets, for cash even though they receive lower prices than from cooperatives. This is because selling to a cooperative usually means waiting until the end of the month, or even longer, for payment. Even more important is the consideration that women, who do most of the work, and usually deliver the produce, prefer being paid cash. Otherwise, they may never receive any money if payment is made through an account, which is in the name of their husbands. Cooperatives, therefore, should pay cash on delivery wherever possible. The success and the importance farmers attach to this procedure and its impact on the success of the cooperative enterprise is clearly demonstrated in the case of the Anand Dairy Cooperatives. Unfortunately in many cases cooperatives lack adequate working capital to implement cash payments and even more frequently marketing boards do not provide sufficient cash to enable cooperatives to effect prompt payment. In some instances, cooperatives have had to borrow from the banks, at high interest rates, to improve their cash flow.

Cooperatives which undertake primary processing as well as marketing seem to be better supported by members and are more successful (dairy, coffee, cotton, sugar, oil seeds). It is now recognized that this type of cooperative must be equipped with adequate equipment and facilities.

In many countries, cooperatives are either not willing or not able to pay salaries comparable with the private sector and hence can neither attract nor retain suitably qualified personnel. Sometimes this is due to lack of finance but often it is because of the idea that salaries should be in line with those in the government sector, where security of employment and pension rights compensate for lower salaries (Korea, Tanzania, Zambia, Indonesia, Philippines). A review of cooperative salaries should be carried out where those do not match the level in other sectors of the economy. Incentives and bonuses should be introduced to encourage initiative and efficiency as has been done in the Anand cooperatives.

13

ECONOMIC TASKS FOR FOOD MARKETING BOARDS IN TROPICAL AFRICA

W.O. Jones

The concern of this paper is the economic efficiency with which tropical African markets supply basic foodstuffs to consumers and direct farmers in their production. The burden of the paper is that governmental boards might perform specific tasks that would significantly improve the economic efficiency of private food-marketing systems, but that in order to perform these tasks the objectives and structure of existing marketing boards will have to change profoundly. It will not be easy to wrench management and staff away from their present preoccupation with handling and policing, but it can be done.

Uses of marketing boards for purposes other than to promote more efficient allocation of goods and resources are ignored. Prominent among these uses are taxation; income stabilization; subsidy of consumers, farmers, or other favored constituencies; ever-normal granaries or buffer-stocks; and attempts to manipulate world markets. Price stabilization is considered only in terms of persisting departures from equilibrium as determined by the interplay of domestic and international supply and demand.

The procedure will be to consider which economic imperfections identified in recent studies of African markets might be corrected or modified by some sort of agricultural marketing board. This attempt to match identified imperfections in private African marketing systems with services that might be rendered by marketing boards seems pertinent at a time when the virtues of existing free marketing systems are being demonstrated and when there is widespread disillusionment with marketing boards.

Economic efficiency considerations are much more telling for domestic food marketing than for export marketing. The latter is most often simply a

---

Source: Taken from *Food Research Institute Studies*, Stanford **19**(2), 1984 (footnotes omitted).

matter of moving crops from farm to port at a price that will induce farmers to grow the crop for market; it is essentially a mechanical or physical process. Furthermore, when export marketing boards were established for cocoa, coffee, cotton, peanuts, and palm oil they simply took over thriving private trades that had developed long before.

The food marketing system has a mechanical side, too, but in addition to assembling products from many growers it must allocate them among many dispersed consumers. The food marketing system does this by generating prices that reflect relative shortages and surpluses of supplies at various marketplaces throughout the system.

National integration of markets for export crops is likely to be better than for crops that are only consumed at home. Farm prices of export crops are simply port prices less costs of assembly, including trading profits. Export crop production tends also to be more concentrated than production of food crops, and this concentration in itself reduces marketing costs. (One benefit of increased commercialization of food crop production is similar concentration.) Parallel influences on market integration result when marketed production is shipped to a distant domestic market center (maize supplies for the Zambian Copperbelt, for example, and gari and cowpeas in Nigeria).

MARKETING BOARDS

Political sanction and support are essential for regular economic exchange. The necessity for some sort of state intervention in the marketing process itself is also generally recognized, but the nature of intervention is more controversial. It may vary from enforcement of a market peace at infrequent periodic markets, through investment in roads, telecommunications, and financial intermediation, to compulsory delivery of crops at harvest and their rationing to consumers.

Marketing boards comprise a varied set of governmental or government-regulated marketing institutions that lie closer to the authoritarian than the anarchic, voluntary-cooperative end of the spectrum. The earliest marketing boards had their origins in pressure from growers for higher and more stable farm prices. Hoos (1979) suggests that they may be viewed as compulsory cooperatives with their roots in Rochdale principles. The first national marketing boards were established to raise farm incomes during the Great Depression, and African marketing boards were established for similar reasons when exports were curtailed by enemy action during World War II. African food marketing boards in British East and Central Africa were also partly inspired by concern about food supplies for India and the Middle East. More recently, food boards have been advocated variously to assure the domestic food supply, to benefit urban consumers or other constituencies, and to implement government programmes for farming and rural settlement.

Among the more ubiquitous arguments for the establishment of national marketing boards have been those stemming from the belief that private marketing was disorganized, exploitative, and inefficient, that marketing margins were excessive, that farm prices were unnecessarily low, and that retail prices were unnecessarily high. The list of alleged imperfections is familiar. Jones (1982), however, suggests that allegations about market imperfections and the incompetence of private entrepreneurs may have been no more than an excuse for government control.

Abbott and Creupelandt (1966) define marketing boards as public bodies established by government to improve the marketing of agricultural commodities and empowered to exercise varying degrees of compulsion over producers, traders, and processors. They vary greatly in form and function. At one extreme are advisory and promotion boards that provide market information, engage in market research, promote sales, and do technical work on quality of produce. Their activities are financed by levies and taxes on growers, merchants, and processors. At the other extreme are boards with statutory monopolies over the foreign and domestic marketing of specified crops, and the fixing of domestic buying and selling prices. They administer equalization and stabilization funds, promote and operate marketing facilities, and control imports and exports. Abbott and Creupelandt identify the following types:

1. Advisory and promotional boards.
2. Regulatory boards.
3. Boards stabilizing prices without trading.
4. Boards stabilizing prices on domestic markets by trading alongside other enterprises.
5. Monopoly trading and price stabilizing boards for:
   (a) export crops; and
   (b) domestic crops.

PERFORMANCE OF PRIVATE MARKETING SYSTEMS

The notable finding of recent research into private marketing of agricultural products in tropical Africa is that most of what was said about these markets a generation ago was wrong, but the word seems to have taken a long time to get around. African markets are not disorganized. African farmers are not unfamiliar with commercial activity, and African farmers do not respond perversely to prices. Nor are African farmers improvident. Most marketed food crops are grown by farmers who sell only part of their crop at harvest, holding the rest of their own consumption, for operating expenses, and for the seasonal price rise. African farmers know what current market prices are and rarely are limited to only one or two prospective buyers. Private marketing is generally competitive and affords farmers a rather high share of the consumers' dollar while assuring supplies to urban areas at reasonable prices.

This is all the more remarkable when account is taken of the obstacles that private merchants must overcome. The physical task of assembling and distributing staple food crops is enormous. Most of tropical Africa is still thinly populated and road systems are poor. In many countries financial crises and political disturbances have led to deterioration of what roads existed and to extreme shortages of spare parts for vehicles. As a consequence, fragmentation of market systems has increased, even in such areas of well-developed economic exchange as the western Sudan.

Tropical Africa has hundreds of languages. Nigeria alone contains about 150 separate ethnic groups, each with its own tongue. Considerable ingenuity has been required to overcome the barriers to trade presented by language, custom, religious belief, and hostility to strangers, and to establish understanding, trust, and security of contract. Devices employed include trading partnerships, brokerage and safe houses, traders' enclaves, and magico-religious enforcement of market peace and behaviour.

Specialization in food crop production is in its infancy. The trading community contains few large merchants of foodstuffs. Not many traders are literate, keep written records, or have bank accounts. Traders operate most often in crowded, poorly equipped marketplaces, and the absence of any sort of trading floor or general meeting place for wholesalers complicates price discovery. There are great differences, of course, in the capacity of national farm marketing systems, depending on the nature of internal transportation, stability of government, population density, government policy regarding commerce, the general wealth of the community, and the extent of commercialization of staple food production and consumption.

Students of private marketing systems in tropical Africa in the 1960s and 1970s have been impressed with how well they operate despite these difficulties of space, demography, and technical development (CILSS, 1977; Gilbert, 1969). Lele (1971) with wide knowledge of marketing systems both in tropical Africa and in India, says, 'considerable evidence has accumulated with regard to the working of traditional markets dispelling stereotypes about the degree of oligopolistic tendencies and spacial and seasonal price differences' and Kriesel (1974), who has studied agricultural marketing on both sides of the continent, goes so far as to say 'indigenous private enterprise food marketing in developing countries of Africa is more efficient than statutory marketing of any agricultural product in such countries.

Perhaps recent students of farm marketing in Africa (and elsewhere in the developing countries) have gone a bit overboard in their praise of existing private marketing systems. In their defence, it is fair to say that all had been taught the same stereotypes about the behaviour of people in developing countries, and most were surprised, and many were pleased, to find that they were not so: that private African markets can be expected to respond in orthodox fashion and reasonably well to changing demand and supply, and that they can be employed with confidence to implement government distribution policies that are adopted for political or social reasons.

This mild euphoria should not blind students of African markets to their many deficiencies and imperfections. Timmer *et al.* (1983) observe of markets generally that they 'do not always function in the best interests of a broad cross section of society, especially in poor countries where communications and transportation facilities are poor, markets highly segmented, and access ... is greatly restricted. The efficiency and economic gains potentially available from successful market coordination of a society's food system are an empirical issue, not a matter of faith and logic.'

DEFICIENCIES AND IMPERFECTIONS

It is useful to distinguish between mechanical, technical, or engineering deficiencies in physical execution of marketing services on the one hand, and economic imperfections that impair market allocations of products and resources on the other. Both reduce potential national economic product. Under the first category may be listed deficiencies in transportation, communication, market-place facilities, standards and grades, units of measurement, banking services, security in the marketplace, and security of travel. Imperfections may be most easily classified in terms of the requirements for the hypothetical perfect market, conditions that would lead to optimum allocation of goods and services and resources under existing technical conditions. Allocation over time, space, and form is improved, of course, as engineering efficiency improves and physical costs are reduced. Ideally, differences in prices between markets, between seasons, and between forms of the product should approximate costs of transporting, storing, and processing.

Deficiencies

In most tropical African countries road systems are incomplete, road surfaces tend to be poor, and many road are impassable in the rainy season. Road transport is frequently handicapped by extreme shortages of imported spare parts, including tyres, and infrequency of repair shops and sometimes of fuel stations. Railroads carry a considerable volume of agricultural products from hinterland to port cities and between major regions, but have almost no feeder lines. Telecommunications within most countries are extremely unreliable. Banks are infrequent in the hinterland, and almost all transactions are in cash, although trade is also facilitated by rather personalized forms of informal credit (see, for example, Southworth *et al.*, 1980 and Franke, 1982).

The most overt consequence of physical deficiencies in the marketing system is increased cost of moving crops from farm to consumer. High physical marketing costs might be expected to result in large marketing margins, but in fact, gross margins in tropical Africa sometimes appear to be small compared even with those in the highly competitive United States food industry. This may, of course, reflect no more than the complex allocative functions performed by sophisticated marketing systems in the advanced economies. After all, margins are smallest when consumer buys from producer at the farm gate. Reliable information about net trading returns – net margins – is to get; such evidence as there is indicates that returns are modest.

High transportation costs may also reduce the effectiveness of allocation over space. If it costs 5 cents to move a pound of grain from one market centre to another, and the average price of the grain is 25 cents, then the difference in prices between the two centres can change by as much as 10 cents, or 40% of the average price, before the grain will move from one market to the other. The situation with sorghum in Sokoto and Kano, Nigeria, from 1953 to 1966 provides an example. Poor correlation between changes in the price of sorghum in Kano and in Sokoto was the consequence.

As regards storage, evidence is that most cereals, pulses, and staple root crops are stored on the farm at reasonable cost. The widespread practice of farm storage has more serious consequences, however, for temporal and spatial allocation; these are discussed below.

Information about commodity supplies and requirements is peculiarly deficient. Crop reporting services have been minimally developed and standing estimates prepared in offices in the capital may differ widely from values reported by infrequent sample censuses. Crop estimation from marketings is impaired by widespread speculative holdings of crops on farms. Rate of release of stocks is determined by expectations for the new crop year and variation in other farm income as well as price expectations, so that marketings may sometimes be larger when crops are smaller.

Lack of information about crop condition and unreliable estimates of the harvest are major causes of food crises in countries with monopoly food marketing boards like Kenya, Tanzania, and Zambia. The problem is made worse by the fact that a much larger proportion of the marketed crop moves through private channels in years when crops are poor, i.e. when board prices are below the market clearing price, and a much smaller proportion in years of bumper crops when board prices are above market clearing prices.

Imperfections

Markets and marketing systems are economically efficient to the extent that the following conditions are met:

1. Commodities traded are fungible and divisible.
2. Buyers and sellers act in an economically rational way.
3. Each buyer and seller accounts for a small part of the volume traded and behaves as if his actions had no influence on price.
4. Buyers and sellers have entry to all activities in the marketing systems on the same terms.
5. Buyers and sellers have full knowledge of all forces likely to affect prices and requirements.

Not often mentioned in lists of imperfections but also important is the ability of market participants to respond to trading opportunities when they know about them. This requires not only market access, but also the financial means to carry out transactions that look attractive, and implies both the possession of funds and the ability to employ them.

Against this set of criteria, free African markets score moderately well.

Fungibility and divisibility. A frequent problem is the lack of standardized measures of quantity and quality. Sale of cereals and beans by the bag, pan, or cup and of root crops by the piece makes price comparison difficult, especially between markets. Quality standards are generally lacking, but they seem less important to customers.

Rationality. No problem.

Smallness – may be oversatisfied in retail markets, but in many remote and inaccessible places the buyer often can influence price. Most farmers, however, can choose among numerous independent buyers.

Access. Ethnic dominance of a trade seems not to reduce competition among members of the dominant group, but outsiders are often at a disadvantage. Local trader associations, made up of members of different cultural, language, or religious groups, have similar characteristics. Problems of access may be overcome by trading through brokers or trading partners or in separated marketplaces or market sections.

Knowledge – a major problem that is compounded by the lack of reliable information about the amount and location of supplies. Retail prices of staples are easy to learn – every household buys them – but knowledge about wholesale prices is more difficult. The only effective way to overcome this difficulty is through trading in the market directly, or through establishing close commercial relationships with a resident merchant who does. Wholesale trading floors where open bids and offers are made are rare, and the second stage of price discovery may be difficult.

Response. Response may be limited not only by impaired market access, but by poor communications and underdeveloped financial intermediation. Banks are few, and to carry large amounts of currency is hazardous. In most African societies, an elaborate network of credit among farmers and market intermediaries up and down the marketing chain eases the response for ordinary transactions, but it might not be able to support large increases in volume of trade. Outsiders are again at a disadvantage because they must bear the risk of price decline alone and find it difficult to raise the money needed to take advantage of a bargain.

Various arrangements have been devised by west African traders to overcome problems of information and response. Trading partners in supply and terminal markets provide access and information but most often only between pairs of markets. (The ancient trans-Saharan trade featured a few family firms with members in numerous markets.) Eastern Nigerian merchants in Umuahia ship palm oil and gari to trading partners in the north who may once have been their apprentices, and couriers travel with the loads, four shipments each way a week, conveying orders, payment, and market news that is posted for all to see. Wholesalers in the Onitsha market order from their rural suppliers and receive market information through letters carried by truck drivers and messengers. Saharan and Sudanic trade has long been characterized by dependence on brokers, permanently located in the large markets, who act on behalf of merchants from a distance and provide accommodations for trader, goods, and livestock.

These measures, some of them actually countervailing imperfections, help to reduce imperfections of knowledge, access, and finance, but they fall far short of the kind of market integration characteristic of the industrialized countries.

MARKET PERFORMANCE

The consequences for a free marketing system of high physical costs, imperfect knowledge, difficulties of market access, and primitive financial intermediation can be seen by examining the effectiveness of allocation of products – of arbitrage – over time and space as indicated by price behaviour. Prices set by the market or by government fiat are the primary allocative devices in all societies. (Compare Lukinov, 1971, on marketing in the Soviet Union.) Only when prices depart too far from equilibrium do governments resort to rationing and compulsory delivery. The effectiveness of seasonal arbitrage can be judged by the closeness with which average annual increases in price approximate the cost of storage. The efficiency of interyear arbitrage is crudely indicated by the occurrence of crises of supply; if the system is working well, shortages will be anticipated and stocks provided to meet them either by carrying over larger domestic stocks or by obtaining supplies from foreign sources. The effectiveness or arbitrage over space can be judged roughly by the extent to which prices of a commodity in various markets move together, most often measured by some sort of bivariate correlation coefficient.

Intraseasonal arbitrage

Contrary to common belief, intraseasonal arbitrage appears to be rather good when nonseasonal influences are removed. This evaluation is not as firm as it might be, however, because of deficiencies in information about costs of storage. Maize prices, in particular, show a rather large increase for a cereal, but whether because of large storage losses or because the supply of maize and competing staples is persistently underestimated is not clear. In general, however, seasonal arbitrage is good, particularly in urban centres, with modest increases for cereals and beans and increases for yams that approximate or are less than the cost of storage.

Interseasonal arbitrage

The evidence about interseasonal arbitrage is not clear, but there is some indication that it is impaired by tardiness of information about the size of the old crop. Because crop estimation depends greatly on observed flow to market and because most storage occurs on farms, accurate appraisal of production may be delayed until a month or two before the new crop comes in, but it has not been much studied. Interseasonal arbitrage may be a serious problem in countries with monopoly farm marketing boards.

Spatial arbitrage

Arbitrage over space is another matter. Markets for many commodities appear badly fragmented even in countries where internal trade in domestic staples across district boundaries is legal. (It is not over most of English-speaking east and southern Africa.)

Consider Gilbert's calculations of the correlation of monthly prices between 29 millet markets and 43 sorghum markets in the northern states of Nigeria from 1952 to 1965 (Gilbert, 1969). These two cereals are commercially well behaved: they store well, are relatively inexpensive to transport, are major components of local diets, and are widely traded. That

Economic Tasks for Food Marketing Boards in Tropical Africa 153

Fig. 13.1. Millet and sorghum price links in northern Nigeria.

major markets are interrelated is demonstrated by maps of price links between them at a correlation value of 0.70 or greater (Fig. 13.1). But of the 903 pairs of sorghum prices, only 40 (4%) are 0.70 or more and out of 406 pairs of millet prices, only 30 (7%). Furthermore, no pair of sorghum or millet markets was correlated at 0.90 or higher. A correlation of 0.70 is not very high. Lele (1971) reported weekly price correlations of 0.90 to 0.97 between each pair of six major grain markets in India, 1955 to 1965. Mohammad (1983) reported similar values for the Pakistan Punjab. A value of 0.70 means that less than half of the month-to-month change in prices in city A is associated with similar price changes in city B.

Perhaps this is as much as can be expected, when account is taken of costs of transport and difficulties of communication and finance. (Access is less a problem in northern Nigeria.) The integration of millet and sorghum markets was much less, however, than that reported for 25 gari markets in the southern states and 29 cowpea markets in the north and the southwest. Of 300 pairs of gari markets, 71% were correlated at 0.70 and 15% at 0.85; of 406 pairs of cowpea markets, 30% were correlated at 0.70 and 5.9% at 0.85. Both gari and cowpeas move long distances to provision the large Yoruba cities of southwestern Nigeria.

Comparisons of this sort should not be pushed very far. The price series are of varying length and reliability, and half to three-fourths of the correlation coefficients are below 0.50. These correlations of prices between free African farm markets were not calculated originally to determine whether the bear danced well, but whether it danced at all. Quite clearly, it does, but equally clearly, not very well.

Weak intermarket connections are also demonstrated by detailed analyses of commodity trade flows, which tend to be to major cities, not through them, although increasing specialization in production is beginning to establish multiple market connections to the specialized supply areas. In Nigeria, Umuahia for gari and the country around Shaki for maize are examples, but terminal markets in general are not redistributive centres.

The picture of free African markets that emerges is of a set of two-level systems, each focused on a different consuming centre, with adjustments at the margins of the supply areas but with long-distance arbitrage only when supplies within the separate areas fall badly out of balance.

The causes of this weak integration of staple food markets lie partly in deficiencies of communication, transport, handling, and finance, but they are also related to the wide dispersion of population in many areas and to heavy reliance on own production for basic foodstuffs. Most rural people probably still obtain at least half of their food energy requirements from their own production, and this is both cause and consequence of small and irregular supplies of food crops in remote markets. If thinly dispersed populations could rely on the market for their staple food supply, they would be more likely to enlarge their production of crops for sale, a first step toward the development of specialized production areas and consequent decline in costs of moving produce to and from farms, thus raising the prices of what farmers want to sell and reducing the prices of what they want to buy. Continued reliance on own food production is partly caused by the thin market. Another major cause of reliance on own consumption is internal insecurity or fear of it, a matter not within the province of marketing boards. Concern about security of holding reserves in the form of commodities rather than cash, however, is a matter that a marketing board might help with.

Better integration among town markets could contribute to integration of bulking markets in addition to stabilizing supplies and prices in the towns and facilitating specialization in production. This is a sure way to reduce the risk of food shortages and hunger. Of the first importance, then, are measures to reduce obstacles to spatial integration. Almost equally important is the need for better knowledge of changes in national supplies.

Although African markets for domestically consumed food crops work well without the marketing board poultice, many in spite of legal monopolies of state agencies, they could work better. It is certainly worthwhile to examine ways in which some sort of marketing board might assist them.

GOVERNMENT'S ROLE IN OVERCOMING DEFICIENCIES AND IMPERFECTIONS

Many of the defects in Africa's food marketing can only be overcome by governmental action, and others can be overcome more quickly with such help. An important question is when, if ever, the statutory marketing board can be considered an appropriate instrument for reducing marketing costs and improving the allocation of farm products and resources.

Physical plant

Roads, bridges, telecommunication, mails, and marketplaces are usually recognized as governmental responsibilities in tropical African countries. But there is no reason why their construction or maintenance should require the establishment of a marketing board, although private merchants have built such facilities when governments neglected to do so. Government most often enters only negatively into vehicle maintenance, to the extent that it restricts supplies of fuel, lubricants, parts, and accessories. Public road and rail transport are, of course, important for marketing in some parts of Africa, but they are not appropriate marketing-board responsibilities.

Regulation of trading

There is considerable precedent in tropical Africa for governmental intervention in market behaviour. Some sort of government sanction appears to be necessary for regular market meetings to occur, and general supervision of marketplaces and market meetings is usually by government officials or their delegates. Standards and grades, weights and measures, hygiene, market entry, honesty in trading, prevention of collusion, and enforcement of contract are certainly matters of public concern and merit government action. They might properly be responsibilities of a department of marketing charged with rendering the same services to agricultural marketing that departments of agriculture render to agricultural production.

Deficiencies in roads, vehicles, and telecommunications are major causes of poor market integration, of surpluses and low prices in one location and of shortages and high prices not far away. Improvement of communication should have a high priority in national budgets, but it is costly and takes time. Nor is it the only cause of poor spatial integration. Even if roads are much improved, problems of information and ability to respond to market opportunities – to correct market disequilibria – will persist. Gradual improvement of communications will reduce obstacles to spatial arbitrage but will not eliminate them.

Information

Crop acreage, crop condition, size of harvest, and size of stocks determine interseasonal movements of prices within boundaries set by import and export parity, but knowledge of their magnitude is neither necessary nor sufficient for effective spatial arbitrage – not sufficient because they tell nothing about demand, and not necessary because competitively determined prices can substitute for them. Furthermore, crop statistics tend to be costly and unreliable, whereas the general level of prices is easy to learn. Collecting and disseminating such information is an appropriate task for a department of marketing, and some African governments now record monthly wholesale prices in selected markets, though often only for official use. To collect daily price information is likely to be costly and ineffective unless it is already being compiled by the trade, as it is in some places (the Umuahia Food Traders Union in Nigeria is an example). Under African conditions its dissemination is also likely to be long delayed. Wholesale prices in African markets typically are not arrived at openly, so that the second stage of price discovery, the price at which goods change hands, can only be determined precisely by trading. Yet the difference between general level of prices and the prices of actual transactions may make the difference between profit and loss in interspatial arbitrage.

The best way to obtain reliable information about prices in other markets is the way most African wholesalers do, by buying and selling through local trading partners, brokers, or agents (compare Hays and McCoy, 1978). The greater the trading capacity of such a merchant, the better his market information and the more effective his spatial arbitrage. This in fact has been a principal advantage enjoyed by the largest European and American firms trading on world markets.

There seems to be more involved in imperfect spatial arbitrage, too, than costs of transportation that can be lowered by better roads and vehicle maintenance and knowledge about prices. Access and finance may be even more important. The Ibadan wholesaler who knows that prices are lower in town B than in town A where he usually buys, but doesn't buy in town B because he doesn't have a 'customer' (trading partner) there, is thinking about availability of funds and protection from being cheated by local merchants, matters of less concern when trading with a merchant from whom he buys regularly.

Although competitive market prices summarize all information in the system at any moment, they may mislead about future prices if stocks information is inaccurate, for accurate knowledge about size of carryover is essential for interseasonal arbitrage. The position of stocks may also affect spatial arbitrage by its influence on the speed with which supplies can be brought to deficit areas.

TASKS FOR A MARKETING BOARD

These two aspects of agricultural marketing in tropical Africa – arbitrage and location of stocks – might benefit from a marketing board of Abbott and Creupelandt's type 4. Arbitrage over long distances and off-farm storage of stocks could be undertaken by a well-funded national trading company that was required to compete alongside private firms but enjoyed a competitive advantage in interspatial arbitrage because size and number of transactions provided reliable market information and because its consequent trading position assured access to many markets and ability to respond to market opportunities.

Spatial arbitrage

Profits from effective intermarket arbitrage increase with the information assembled and with the number of transactions, themselves an important source of market information. Movements of prices in an efficient free market are the surest guide to shifts in local supply and demand relationships.

An adequately financed national trading company, charged with earning profits from arbitrage among major market centres, could contribute a great deal to efficient market integration, but only if it traded alongside independent private firms. If it were set up as an effective monopoly, prices would no longer convey information about local supplies and requirements or serve as indicators of market conditions.

Such a marketing board differs from Abbott and Creupelandt's type 4 in its definition of stabilization as reducing the magnitude and duration of deviation of market price from equilibrium price. Perhaps Abbott and Creupelandt are thinking of the same thing when they speak of 'wide' and 'sharp' price changes, but there must be at least a suspicion that by stabilizing prices they mean making prices constant, in which case arbitrage will not take place (see Bressler and King, 1970).

Price stabilization defined in terms of persisting departure from market equilibrium should be distinguished from other kinds of market operations that might be carried out by such a board, like price support in underdeveloped areas with good potentials (the infant-industry approach to economic development) or buffer-stock programmes to protect against errors in interseasonal arbitrage, or consumer subsidies. When such distortions of equilibrium prices are considered necessary to satisfy development, welfare, or political requirements, the board should be compensated specifically. If it is not, the primary task of spatial arbitrage will be compromised, as it is now by monopoly food marketing boards that engage in territorial pricing.

A national trading company charged with trading between principal bulking and terminal markets could solve problems of information, access, and finance that now handicap the private staple-food trade. Buying and selling through trading agents (brokers) in principal markets would provide it with necessary information about prices and local supplies, information that could be conveyed by couriers when telecommunications are inadequate; market access could be assured by contractual arrangements with resident trading agents; and financial intermediation could also be assured by resources of the state. (This is not as big a problem in urban markets as it can be in rural markets that have no banking facilities at all.)

Storage

Though spatial arbitrage is the most troublesome defect in market performance, storage is probably the most troublesome aspect of market behaviour, not because of the costs of storage but because the location of stocks makes access costly and knowledge about their magnitude imperfect. Farm storage is not costly and farmers need pay only small out-of-pocket costs in order to hold produce from harvest to sale. They are also likely to regard produce as a safer store of value than cash or bank accounts.

For a marketing board to undertake to store the entire marketed share of a crop from harvest to harvest can be a costly proposition. If the Kenya National Cereals and Produce Board in 1980/81 had been required to store the marketed half of the crop, about 1 000 000 tons, in order to meet seasonal requirements, it would have had to make an average investment of at least $65 million for eight months. Calculations are similar for Zambia and Tanzania. Such a burden may be more than simply interest foregone; there may be an absolute problem of raising the money at all.

Another function that might be undertaken by a department of marketing or a national trading corporation would be the operation or licensing of secure warehouses for merchant or farmer storage of grains, pulses, and other storable staples, and issuance of negotiable warehouse receipts that could be used as security for bank loans. (It is probably too much to hope that stocks could be hedged on the Chicago Board of Trade, but it is not impossible.) Government operation of crop storage facilities has not been very impressive in Africa, but perhaps some marketing board funds could be used to finance construction of warehouses for private management.

If establishment of such warehouses and availability of loans against warehouse receipts were accompanied by convincing official approval of merchant storage, it might be possible to reduce merchants' fear that they will be prosecuted and their stocks confiscated if they engage in this socially useful speculative service. If merchants felt that their stocks were secure, they would welcome the trading flexibility it would provide. It will not be easy to persuade merchants in most African countries that it is safe to admit to owning stocks greater than are needed for current transactions. The effects of 20 years of enlightened performance by government can be cancelled by one misguided seizure of stocks in a 'crisis'. Confidence might be restored more readily if the marketing board or marketing department undertook to insure stocks in storage against such an occurrence.

Persuading farmers to convert their stocks of food crops to cash at times of harvest so that the amount and position of stocks can be estimated better will also take time, but evidence from central Ghana in the 1970s indicated that a few more prosperous farmers with easy access to markets were already doing so (Southworth *et al.*, 1980). Farmers who hold stocks for the seasonal rise cannot be sure that it will occur – prices are influenced by more things than the cost of storage – and may prefer not to incur risk. It is conceivable, of course, that some farmers might be willing to trust their speculative stocks to a bonded warehouse when inflation or domestic unrest makes them reluctant to hold cash.

A proposal

A national trading company could assist in correcting certain major deficiencies in the marketing of storable staple foodstuffs like cereals and pulses in various tropical African countries – deficiencies in spatial arbitrage caused by problems of information, access, and finance and difficulties in interseasonal arbitrage caused by ignorance about the size of the stocks. Such a board would be charged with competitive trading in designated staples between major terminal markets and with establishing, perhaps operating, bonded warehouses where merchants and farmers might store these commodities. It probably should not initially be authorized to engage in international trade; it will have its hands full conducting long-distance internal trade economically at the outset. It will also be easier to maintain its competitive character in internal trade than in foreign trade, where marketing boards have proved to be such a convenient taxing device. It might be associated with a department of agricultural marketing charged with fostering marketing efficiency (not with raising farm prices).

Accomplishment of the proposal would not be easy. Marketing boards as they now exist in almost every African country have quite different objectives, and their staffs are ill-suited to aggressive merchant activities. In fact many of them, particularly the food boards, find it difficult if not impossible to accomplish the simple tasks of buying, assembling, and storing a few hundred thousand tons of grain. The poor performance of marketing board employees is a convincing argument in favour of Abbott's recommendation for 'maintaining and building up the services of private firms in the produce trade' and 'in continuing to use these sources rather than by extension of the board's own organization' (Abbott, 1974).

But the central direction of the proposed board would differ completely from that of existing boards, and this would call for a basic reorientation of staff. The task will not be 'to provide a good buying and collecting system for farmers' crops' (private traders do that), 'to market them economically' (private traders do that too), or 'to sell them on the best possible terms overseas' (the task of export boards). Instead the task is to follow market prices in major domestic markets and to move produce from one market to another when a profit can be made by doing so, thus stabilizing prices among markets and over time by preventing large or persistent departures from equilibrium.

There is a great difference between price stabilization that consists of buying and selling at fixed prices over a period of weeks, months, or years and price stabilization that attempts to reduce lags and restrain overreaction to changing market conditions. The first transfers the risk of price change from farmers and merchants to government at the risk of progressive distortion of supplies from requirements. The second increases the reliability of prices as indicators of the relationship between supply and demand, and in this way it enhances their adjustment. The first destabilizes, the second equilibrates.

The perfect market, as conceived by Working (1958) reflects instantly all that can be known about supply and demand; a change in prices reflects a change in supply or demand that could not be foreseen, hence prices in a perfect market pursue a 'random walk' through time. Some price uncertainty is thus inevitable in the perfect market if it is to achieve an economically optimum allocation of resources; uncertainty will be lessened by improved information and capacity of traders to respond. Any attempt to reduce price uncertainty by fiat is likely to reduce the efficiency of allocation, although it may sometimes be necessary to impose limits on the rate of change in prices in order to dampen hysteria or price manipulation.

As the tropical African economies mature, it is expected that private trading firms will develop with national capacity comparable to that of the national trading company. When this happens, it is appropriate that activities of the governmental agency, with its ever-present threat of political intervention, should be brought to a close. The risk of private monopolies seems slight. As Ehrlich (1982) said 'long before African countries became independent there were measures to eliminate competition' by the establishment of farm marketing boards in Kenya and Zambia, perhaps elsewhere; this agitation was strongest among European farmers who feared the competition from African farmers. It will be a long time before a principal problem in private African food marketing ceases to be that merchants are too small.

Could it happen?

Directors and staff of existing marketing boards are likely to find it difficult to achieve this change in objectives. The notorious inability of existing food marketing boards to perform the tasks assigned to them is only partly a consequence of politically inspired overstaffing and of managerial incompetence. It stems in large part from the fact that the price-fixing task assigned to them is an open invitation to bribery and corruption that becomes stronger the further prices depart from equilibrium. Even if it were not, and cadres were incorruptible and conscientious, enforcement would still require massive and equally incorruptible police action. (Wartime experiences with price control in the industrial nations give some notion of the difficulties.)

The task of the proposed national trading company is much simpler. Although the market can be a stern taskmaster, the staff would be working with it rather than against it. As Temu (1983) has said, if private enterprise is not only tolerated but actively encouraged, it can serve 'as a yardstick against which to measure the economic performance of public operators'.

There are competent, alert staff at all levels, some with commercial experience and more with commercial ambitions. Unfortunately there are not many, and of those not a few already have their eyes on more remunerative employment in private commerce and industry. Nevertheless, there are probably enough in most African countries to provide cadres that, with proper support, could conduct profitably the kind of commercial activities described here. It cannot be done all at once. Acquisition of trading skills will take time, even for those who have already learned how to requisition, transport, store, and invoice. One of the surest ways to sort the sheep from the goats is to make rewards depend on achievements – on trading profits. After all, to buy cheap and sell dear is not entirely foreign to tropical Africa.

REFERENCES

Abbott, J.C. (1974) The Efficiency of Marketing Board Operations. In: Onitiri, M.A. and Olatunbosun, D. (eds) *The Marketing Board System: Proceedings of an International Conference.* Nigerian Institute of Social and Economic Research, Ibadan, Nigeria.

Abbott, J.C. and Creupelandt, H.C. (1966) *Agricultural Marketing Boards: Their Establishment and Operation.* FAO Marketing Guide No. 5, Food and Agriculture Organization of the United Nations, Rome.

Bressler, R.G. Jr. and King, R.A. (1970) *Markets, Prices, and Interregional Trade.* John Wiley & Sons. New York.

CILSS, Club du Sahel (1977) *Marketing, Price Policy and Storage of Food Grain the Sahel: A Survey.* Working Group on Marketing, Price Policy and Storage. Prepared by Center for Research on Economic Development, University of Michigan, 2 vols.

Ehrlich, C. (1982) The marketing of cotton in Uganda 1900–1950. PhD dissertation. University of London, London, England.

Franke, C. (1982) The Kumasi cattle trade. PhD dissertation. New York University, New York.

Gilbert, E.H. (1969) The marketing of staple foods in Northern Nigeria. PhD dissertation. Stanford University.

Hays H.M. and McCoy, J.H. (1978) Food grain marketing in Northern Nigeria: spatial and temporal performance. *Journal of Development Studies* 14(2).

Hoos, S. (ed.) (1979) *Agricultural Marketing Boards: An International Perspective.* Ballinger Publishing Company, Cambridge, Massachusetts.

Jones, D.B. (1982) State structures in new nations: the case of primary agricultural marketing in Africa. *Journal of Modern African Studies* 20(4).

Kriesel, H.C. (1974) Some economic performance problems in the primary marketing component of statutory marketing systems. In: Onitiri, M.A. and Olatunbosun, D. (eds) *The Marketing Board System: Proceedings of an International Conference.* Nigerian Institute of Social and Economic Research, Ibadan, Nigeria.

Lele, U. (1971) *Food Grain Marketing in India: Private Performance and Public*

Policy. Cornell University Press, Ithaca, New York.

Lukinov, I. (1971) The methodology of forming prices of farm produce: history of price formation in the U.S.S.R. *Economic Policies, Planning and Management for Agricultural Development.* Papers and Reports, Fourteenth International Conference of Agricultural Economists. Minsk, U.S.S.R.

Mohammad, F. (1983) An analysis of the structure and performance of agricultural markets in Pakistan. PhD dissertation. Simon Fraser University, Burnaby, British Columbia, Canada.

Southworth, V.R., Jones, W.O. and Pearson, S.R. (1980) Food Crop Marketing in Atebubu District, Ghana. *Food Research Institute Studies* **17**(2).

Temu, P.E. (1983) Marketing Board Policy in Tanzania and Kenya: A Comparative Analysis. Paper presented at the Leiden Conference.

Timmer, C.P., Falcon, W.P. and Pearson, S.R. (1983) *Food Policy Analysis.* Johns Hopkins University Press, Baltimore, Maryland.

III
PHYSICAL INFRASTRUCTURE FOR MARKETING

By its nature agricultural production tends to be dispersed, so access to markets and inputs when needed is often a constraint. An adequate road network is essential for development. J.S. Spriggs (1977), 'Benefit cost analysis of surfaced roads in the eastern rice region of India', *Journal of Farm Economics* **57**(2), 375–379, concluded that the returns were substantial.

For many years the World Bank's main line in marketing improvement was financing the rehabilitation of trunk and feeder roads. Awareness of the difficulties governments face in maintaining them has pointed to greater user responsibility (see Mittendorf's paper below). The World Bank, however, is sceptical, certainly regarding voluntary maintenance (see J. Riverson, J. Gaviria and S. Thriscut, 1991, *Rural Roads in Sub-Saharan Africa: Lessons from Bank Experience*, World Bank, Washington).

The development planners of the 1960s sometimes thought it was enough to put down a processing structure: development would follow. Frequent failure in practice sparked Mittendorf's paper presented here. C.P. Timmer (1975), 'Choice of technique in rice milling in Java', *Bulletin of Indonesian Economic Studies* **9**(2), 52–76, also stressed the distinction between technical and economic efficiency. Von Oppen and Scott's 1976 paper is the lynch pin of processing plant location theory. M. von Oppen was then with ICRISAT in India; J.T. Scott was professor of agricultural economics at the University of Illinois. Abbott's *Agricultural Processing for Development* Avebury, Aldershot, 1988, provides case studies and economic experience. The EDI Working Paper (World Bank) 'Agroindustrial Investment and Operations' by J.G. Brown with Deloitte and Touche (1992), details market identification and financial analysis, also providing commodity profiles.

Observation of rural markets in India during von Oppen's ICRISAT assignment brought from him their rural development rationale – that they help most of all the smaller farmers who cannot transport produce to more distant outlets. H.J. Mittendorf expanded on this role, e.g. *Indian Journal of Agricultural Economics* **28**(1), 101–119, 1982.

Periodic markets and the routes followed by produce moving to consumption centres have also been studied by geographers. R. Bromley, then at the University of Swansea, was a prolific writer in this area (*Periodic Markets, Daily Markets and Fairs: a Bibliography*, Department of Geography, Monash University, Melbourne, 1974).

Promotional literature on storage reached its apotheosis with Kissinger's call for a food loss prevention programme at the World Food Conference of 1974. Bourne's *Post Harvest Food Losses in Developing Countries*, Cornell University, 1977, noted the exaggeration in many of the loss estimates. He was followed by E.R. Pariser (*Postharvest Food Losses in Developing Countries* National Academy of Sciences, Washington, 1978).

A.W. Shepherd's paper, summarized below, ties postharvest management firmly into marketing. B. Berman sets out priorities for investment in storage. Packaging and packing operations have been treated mainly by aid organizations, e.g. *Guide to Establishing Small Packing Stations for Fruit and Vegetables in Rural Areas*, FAO, Rome, 1984.

14

MARKETING ASPECTS IN PLANNING AGRICULTURAL PROCESSING ENTERPRISES IN DEVELOPING COUNTRIES

H.J. Mittendorf

It is logical that most of the developing countries should be seeking to industrialize on the basis of their agricultural production. The output of crops and livestock is both their main source of wealth and their most obvious raw material for industry. In many important cases, products which were formerly exported in their primary state are now being processed in varying degrees before shipment, so that the national economy gains the value added by processing. In other instances, locally processed products are being sold on the domestic market as substitutes for imported goods. In both cases, the implication for a country's balance of trade can be highly important. At the same time, the new processing industries help to make more effective use of labour, and thus reduce unemployment. Agricultural development itself may be notably helped if the processing enterprises provide a new and reliable market for production. Also, major processing projects have significant 'linkage effects' by promoting new business for service companies, transporters, traders and various others who are affected by its operations. Finally, there may be welfare benefits, such as the improvement of public nutrition that should follow the successful establishment of a milk plant.

Although the importance of stimulating agricultural processing industries on these grounds is well appreciated, there is not always a full understanding of the marketing and economic factors which vitally affect the success or failure of a project. It is a particular feature of most agricultural processing that the value added by the process is low in relation to the value of the primary commodity and the other materials used. In a complex manufacturing industry, raw materials generally form a very minor part of the ex-factory cost of the product; in the simple processes of agro-allied industry, they generally form the major part. In some cases of primary processing, notably oilseed crushing, the value added in processing

---

Source: *Monthly Bulletin of Agricultural Economics and Statistics* **17**(4), 1968. FAO, Rome.

Table 14.1. Marketing problems encountered in utilizing full capacity of selected agricultural processing plants established in recent years in developing countries

	Type of problems encountered		
Processing plant	Supply	Demand	Marketing methods
Slaughterhouses and meat processing plants			
Africa: Eastern			
1			X
2	X		
3			X
4			X
Africa: Western			
5	X		
6			X
7	X	X	
8	X		X
9			X
10			X
11			X
Africa: Central			
12			X
Southern Europe			
13			X
Latin America			
14		X	
15			X
16			X
17			X
18			X
Dairy plants			
Far East			
1		X	
2	X		
3	X		
4	X		
5	X		

Marketing Aspects in Agricultural Processing Enterprises

Processing plant	Type of problems encountered		
	Supply	Demand	Marketing methods
6	X		
7	X		X
Africa			
8			X
9		X	X
10	X		
Near East			
11	X	X	
12		X	X
Latin America			
13	X		
14			X
Fruit and vegetable processing plants			
Africa			
1 Fruit and vegetables	X	X	
2 Tomato paste	X		
3 Citrus	X	X	
4 Pineapples	X		
5 ”	X		
6 ”	X		
7 Fruit and vegetables	X	X	
8 ” ” ”	X	X	
9 Onion drying plant	X		
10 ” ” ”	X		
Asia			
11 Tomato paste	X	X	
12 ” ”	X	X	
13 Packing stations, dried products	X		
14 Figs and raisins	X		
15 ” ” ”	X		
16 ” ” ”	X		
17 Fruit and vegetables	X	X	X
18 ” ” ”	X	X	X
19 Pineapples	X		
20 Fruit and vegetables	X	X	
21 Packing station for raisins	X		
Latin America			
22 Fruit and vegetables	X	X	
23 Juices	X	X	

Table 14.1 continued

Processing plant	Type of problems encountered		
	Supply	Demand	Marketing methods
Southern Europe			
24 Tomato paste	X		
Other plants			
1 Groundnuts	X		X
2 "			X
3 Sugar	X		
4 Palm oil	X		X
5 Cocoa		X	
6 Cashew nuts	X		
7 Rice	X		
8 "	X		
9 "	X		
10 "	X		
11 "	X		
12 "	X		
13 Feed-mixing plant	X	X	

may be less than 10% of the value of the raw materials. In this kind of situation it is clear that the terms on which the enterprise procures its raw materials and the terms on which it distributes and sells its products are of dominant importance in determining success or failure. Efficiency in the factory processes, accounting for a minor part of total operational costs, cannot save the project from failure if it is not fully competitive in its buying, transport, marketing and selling operations.

The object of this article is to emphasize the marketing and economic aspects which determine the success or failure of a processing enterprise and to highlight some of the topics which should be exhaustively studied before an investment decision is made.

FAO marketing advisors were asked to identify processing plants that had failed or faced serious problems. Detailed studies or dossiers based on interviews of directors and managers were assembled for 70 plants. They featured:

1. Definition of the processing project (ownership, scale, capital outlay, arrangements for raw material supply, prospective markets, management arrangements, and whether a proper economic feasibility study had been carried out before a decision to establish the project was taken).

2. A brief account of the operations and results of the processing project (e.g. period of operation, whether the plant was idle or running uneconomically, utilization of capacity (per year, future prospects).
3. Reasons for failure and how far attributable to failures of basic planning.

Table 14.1 summarizes the data obtained. It is emphasized that they refer to plants known to be in difficulties, not to a representative sample of agricultural processing enterprises. All of the 70 processing plants appeared to be well designed from the engineering point of view. The problems could be grouped under three main headings, namely:

1. *Market demand*, including overestimation of prospective demand, misjudgment of tastes, preferences and habits of the consumers; underestimation of competition from alternative marketing channels and substitutes, and of obstacles to entering foreign markets.
2. *Raw material supply*, which include overestimation of supply, lack of suitable varieties for processing, insufficient incentives to farmers and lack of supporting services to farmers, such as extension and credit.
3. *Management*, in particular marketing managements: lack of adequate marketing facilities, i.e. for collection, storage and distribution; lack of consumer education and sales promotion; inefficient internal management, inadequate financial resources, especially for working capital; overstaffing; inappropriate government intervention.

These deficiencies resulted in excessive marketing and processing costs compared with competing enterprises. They prevented the plants from making their intended contribution to national development.

ASSESSMENT OF MARKET DEMAND

One of the main reasons for the economic losses of a number of plants listed was the overestimation of the capacity needed. In some cases, no proper market outlet studies had been conducted before a decision on the investment had been taken. In other cases, the study was superficial and led to inaccuracies in the assessment of actual and future demand.

Export markets

As far as the demand for well-defined primary processed commodities is concerned, as for instance fats and oils, cotton, cottonseed, etc., information on prices, quality preferences and conditions of sale may be obtained rather easily on wholesale markets where these products are sold for further processing. Trends of production, imports and consumption on foreign markets can be identified in broad terms by an analysis of official statistics. However, the assessment of the commercial demand for processed consumer items such as canned goods is more difficult and requires further accurate and detailed information on packaging, pricing, credit, transport, storage, sales promotion, trade rules and procedures, quality, access to markets – in particular information on tariff and nontariff barriers. It should also be known which are the likely groups of consumers to be supplied (higher, medium or lower income consumers, institutions), their location and their tastes. One onion drying plant, for instance, could not sell its products derived from red variety onions because consumers preferred the white variety. Another important aspect is the substitution between fresh and processed products, as in the case of fruit and vegetables. Planning sales of canned fruit and vegetables requires, therefore, information on seasonal availability of fresh products. All these data on consumer preferences and commercial sales conditions (particularly needed for marketing of commodities ready for final consumption) can, in most cases, only be obtained by detailed market studies, which often, in the past, have not been given adequate attention in feasibility studies.

Local markets

The assessment of demand for processed products on local markets requires particular attention to the consumer groups to be supplied and their tastes, preferences and purchasing power. One of the major reasons for the failure of some processing plants, particularly with regard to fruit and vegetables, has been the overestimation of demand for processed products on local markets. Often there is not full awareness of the low purchasing power of the majority of consumers, who are simply not able to pay for elaborately processed and packaged foods. A typical example is canned fruit and vegetables which appear excessive in price to low-income consumers in comparison with available fresh, dried or pickled fruit and vegetables. Often the can costs more than the raw material. The convenience factor, which is so important to consumers in high-income societies, is much less significant in developing countries and canned goods are not widely substituted for fresh products until the prices are competitive. This often means that the demand for processed fruit and vegetables is largely seasonal, since it is affected strongly by the availability of fresh products. In Ghana, for instance, tomato paste is demanded mainly in the rainy season when fresh tomatoes are not available. Such factors have important implications for production planning and the financing of stocks.

Adequate attention is often not paid to local tastes and preferences, and the existence of traditional foods which have been well established on the local market for decades. Some dairy plants failed to operate at full capacity because their operation was too much oriented to sales of liquid milk and the product mix did not include yogurt, cultured buttermilk and salted cheese, which would evidently have been more popular with local consumers. This is also one of the reasons why well-processed vegetable oil produced by modern oil mills has not been able to compete on some local West African markets with oil supplied by the traditional small-scale process, which has a higher content of free fatty acids and a flavor preferred by the indigenous consumers. 'Sometimes it is recommended to install parboiling equipment in rice mills, for nutritional reasons, without studying carefully whether the consumer is accustomed to parboiled rice. There are cases where such equipment has been installed and never used, since consumers were not willing to accept the product.

ASSESSMENT OF AGRICULTURAL RAW MATERIAL SUPPLY

The availability of an adequate supply of raw material in terms of quantity, quality, price and regularity is a major concern of all agriculture-based industries, and lack of supplies has been one of the major shortcomings of a great number of plants. In many instances, assumptions on the quantities of produce that would be purchased for processing from a local growing area appear to have been excessively optimistic. For example, the underutilization of slaughterhouses has often been due to the fact that the supply of slaughter cattle was overestimated. Livestock statistics in most developing countries are only broad estimates, or even guesses, and careful surveys on slaughter cattle supplied to markets and sample surveys on the composition of herds are, therefore, needed to obtain more reliable data. Available data on the supply and export of hides and skins may be helpful in making more realistic estimates of the number of cattle slaughtered in recent years.

There is often no thorough understanding of the productivity of a cattle herd, that is, the number of cattle which may come forward for slaughter annually from a given cattle population. The availability for slaughter depends on the reproduction rate, the number of calves reared per year, and the speed with which these calves will be raised and fattened. In many African countries, the number of calves surviving annually per 100 cows is less than 50, as against 80–90 in developed countries, and it often takes between five and seven years before cattle are ready for slaughter. Consequently, slaughter rates (number of cattle slaughtered per 100 cattle kept) 7–11% are common, instead of the 20–30% achieved in more developed countries. Even where cattle are available for slaughter, it has to be taken into account whether the cattle owners are prepared to sell. There are still many areas where nomads or semisubsistence farmers are reluctant to sell cattle for slaughter. In such cases, ways and means of encouraging sales have to be considered in order to ensure a regular supply to a slaughterhouse.

A large part of the problems of the dairy plants was due to difficulties in getting sufficient quantities of milk. The expected increase in the milk supply had certainly been overestimated during the planning stage.

Particular difficulties have been faced with regard to the supply of fruit and vegetables to processing plants. There are various reasons for this. In some instances, the suitability of existing varieties of fruit and vegetables for processing had not been carefully checked as, for instance, in the case of pineapple, tomato, orange and peach processing. In one case, the existing varieties of tomatoes had an excessively high acid content and their yields in processing were too low. Even if culled qualities of these varieties are available for processing, and may appear cheap in relation to prices obtained for good and average quality on the fresh fruit and vegetable market, they may still be rather expensive for processing in view of their low processing yields and the irregularity of supplies. The assumption, still widely held, that processing plants could operate on a raw material basis of culls and second-grade produce may only be true in limited cases as, for example, on small-size domestic markets protected against outside competition. It is certainly not true for larger size markets with full international competition.

A point of particular concern is to estimate realistically the period of the year during which fruit and vegetables will be available for processing, particularly in those cases where farmers have the alternative of selling their products either on the fresh market or for processing. In cases where there is a rather high degree of substitution between sales on the fresh market and sales for processing, an answer to this problem can only be found by a thorough analysis of the seasonality of supplies, demand and prices on the fresh fruit and vegetable market. The processing plant can, to some extent, secure the supply of raw material by contacting supplies in advance, but this will only be successful if the price stipulated is in line with the market situation and offers long-term incentives to producers.

In assessing future supplies of the raw material for processing, a careful study is needed of the price level necessary to provide adequate incentives to farmers. The returns expected by the farmer from a crop destined for processing has to be compared with alternative crops. In one case, farmers abandoned the production of tomatoes for processing when they found that tobacco production was more remunerative. The farmer's returns depend to a large extent upon his ability to produce efficiently and to fully utilize the production potential. Particularly at the beginning of a processing project, it is sometimes observed that crop yields are low due to deficient production practices and, especially, insufficient inputs. Tomatoes, for instance, can be offered at much lower prices when the yield amounts to 100 tons per hectare than at yields of 20 tons per hectare.

Since in many cases technical know-how is not fully developed, adequate arrangements have to be made for introducing the knowledge which is lacking.

The right varieties of fruit and vegetables have to be tested for processing purposes before the plant is established, and farmers have to be trained in planting these varieties. In order to ensure a continuous supply of the raw material, it may be useful in some circumstances for the processing enterprise to establish its own farm or plantation which would also serve for production experiments and as a place of demonstration for farmers.

It is easy to overestimate the primary producer's response to the establishment of a new processing plant and his willingness to supply it. In practice, the advantages of leaving traditional markets in order to sell to the new project are often not obvious to the local farmers, especially when the processor requires unaccumstomed standards of quality, punctuality and regularity for deliveries. Particularly where perishable products are concerned, a processing plant must be prepared to devote considerable resources to organizing its supplies and maintaining relations with producers. Provision may be to be made not only for transport units to collect produce from a network of collecting points, but also for field extension staff to make detailed arrangements with growers, give advice and possibly provide improved seeds and credit for the growing season. Without such initiative it has often proved impossible for processors to ensure their supplies.

The need to make the best use of by-products has been rightly stressed on various occasions. The feasibility of processing by-products should, however, be carefully studied from the economic point of view. There are, unfortunately, a number of cases where investments in by-product processing units have not yielded an adequate return, since the raw material supply and the demand for processed by-products had been overestimated. This refers particularly to processing units installed in some slaughterhouses in West Africa, where most of the raw material which would be available for processing in countries with a higher income is directly consumed by the local population, and the demand for meat, bone and blood meal is small because of the limited existing commercial poultry and pig industry in these areas.

Along with technological factors, the assessment of the present and future demand for processed products and the available raw material must determine the economic capacity of the plant. It is often observed that the capacity of processing plants is larger than the scale of operations can justify, so that overhead costs are excessive. The possibility of planning a much smaller plant and operating in two or three shifts during the peak season is often not fully explored. Capacity should be planned, wherever technically feasible, for the immediate future and, where appropriate, provision may be made for expansion at a later stage when increased demand justifies it. In this way, the burden of overhead costs can be kept at a minimum while the project is in its development stage.

NEED FOR OPERATIONAL EFFICIENCY

The importance of operational efficiency is often underestimated. Many enterprises were not able to compete with other suppliers because their operating costs were too high. This was due to various reasons: the plants were either expensively constructed, inadequately organized, wrongly located or inefficiently operated. Slaughterhouses financed by outside assistance programmes included rather expensive equipment, material and layout, and no serious attempt was made to keep expenditure at a minimum consistent with the requirements of a functional and economic plant. Often no careful study is undertaken on the needed capacity of refrigerated rooms attached to slaughterhouses. Excessive cold-storage facilities based on experience in developed countries have been installed in west African slaughterhouses but are hardly ever used, since consumers prefer fresh meat.

Identifying competition

A new processing plant will only be economic if it is able to compete on the market against other suppliers. The planning of a new processing enterprise should, therefore, not be based only upon information on present and future market demand and raw material supply but should also take into consideration the type and degree of competition on the market, to provide a guide to the most effective marketing strategy. The assessment of the present and potential supply of competitors requires an analysis of the quantities supplied, the market share, the financial strength, the market strategy covering prices, promotion, sales practices, marketing channels, agents, margins, etc.

A point often neglected with regard to the supply of local markets is the competitiveness of small-scale rural industries. They have often been disregarded because of their outdated technical equipment and the apparently low quality of their products. Experience has shown, however, that they can be keenly competitive on local markets, since their equipment is often amortized and labour is paid on a level nearer to opportunity costs, while large-scale mechanized factories have often to pay government-fixed minimum rates. Also, the products of rural industries, although seemingly of low quality according to the standards of developed countries, may be preferred to the indigenous consumers because of their traditional style and flavour.

It must also be borne in mind that it is relatively simple for small rural industries to obtain their raw material supplies and to market their products. They can rely on local suppliers and local customers and can easily adjust their output to suit current conditions. Large plants, on the other hand, must often assemble their supplies and distribute their products over a wide radius, involving special organization, substantial transport, and higher buying and selling costs. Sometimes, large processing enterprises have to buy their supplies at official prices while their small-scale competitors can avoid controls and buy more cheaply. This is a feature which has caused problems for soap manufacturers in Eastern Nigeria for many years. Small plants with simple equipment and minimal labour forces also do not experience the difficulty of maintaining employees all year round. Machine operators in highly technical processing plants require considerable training and constitute a valuable asset to the plant. This type of skilled labour has to be maintained throughout the year, whether the plant is operating or not.

Selection of marketing channels

The strategy of marketing of the processed products plays an important role in the success of an enterprise. Particular attention has to be given to the selection of marketing channels and agencies in order to maximize sales volume and returns. Margins offered to wholesalers and retailers should contain an inducement element for them to expand their sales of the processed products. Where the factory employs its own sales agents, proper training and supervision of the agent is necessary in order to ensure full exploitation of the existing markets. Attractiveness of packaging material, size, type and price of packages are important aspects in the marketing of food ready for consumption, and particular attention has to be given to the limited purchasing power of consumers in developing countries who are not able to pay for sophisticated packing material. One milk plant, for instance, had great difficulty in selling its product because the size of the package was too large in relation to what consumers could afford to pay.

Need for sales promotion

There is often insufficient awareness of the importance of sales promotion for canned products, particularly on highly competitive markets in industrialized countries. The sale of canned products can be successfully promoted under brand names which, however, require considerable financial resources to establish and promote. In the United States, for instance, large-scale fruit and vegetable canners often spend about 4–5% of their annual turnover on sales promotion. The establishment of national or even regional brands requires a considerable volume of sales, often far beyond the quantities sold even by a large-scale canning plant. This means that a canning plant to be established in a developing country which intends to export to industrialized countries should study carefully the possibilities of making use of existing brands and established sales services through arrangements with the firm owning the brand.

Need for sound location

Another reason why new processing plants were unable to compete successfully on the market was wrong location. The decision on location is sometimes influenced by political considerations rather than economic factors aiming at minimizing marketing and processing costs and maximizing marketing advantages – such as, for example, convenience for ports or wholesale distribution centres.

The location of slaughterhouses has proved to be a major issue. While in industrialized countries, where long-distance transport is involved, it has been more economic to locate them in producing areas, because meat transport under refrigerated conditions is cheaper than transport of livestock, this may not always apply to the conditions prevailing in developing countries. In cases where a careful study was conducted, as in Madagascar (Lacrouts and Tye, 1962) and Southern Africa (Mittendorf and Wilson, 1961) it was found that meat transport was not as economic as the transport of live animals. The operation of a refrigerated transport chain is often more expensive in developing countries and meat prices usually lower, so that the economic incentive to minimize tissue shrinkage of live animals during transport by means of refrigeration is less than in industrialized countries.

Another important aspect is the combination of different commodities for processing in one plant. Since the demand for individual processed products is usually very limited in developing countries, processing plants have sometimes to handle a variety of products in order to achieve a utilization of the capacity and keep overhead costs within limits. Therefore, an important problem is to decide on the right combination of products, such as fruit and vegetables along with meat and others. Experts from industrialized countries, where large-scale specialization in processing is an obvious means of rationalization, are in many cases no longer familiar with the optimum combination of relatively small quantities of different products on economic lines.

Arrangements for raw material supply

The efficiency of the arrangements for raw material supply is a determining factor in the success of a processing plant and the assembly and transport of perishables for processing are often a particular problem. Some milk plants failed to attract sufficient supplies because their milk-collecting system was expensive, inefficient and not flexible enough to meet competition from peddlers. In the case of the supply of livestock for slaughtering and processing, an important task is to ensure that the right type of animals are collected by traders at the agreed price. The determination of a clearly understandable grading system is an important requisite for efficient purchasing.

Working capital

Particular attention has to be paid to the financing of raw material supplies, as well as the marketing of processed products, especially in view of the lack of institutionalized credit in most developing countries. In many cases, farmers are tied to traders through credit received. Credit arrangements may have to be made with them to ensure the supply of the crop to the factory. Corresponding arrangements may also be needed for the distribution of processed products, since merchants and retailers may be unable or unwilling to carry stocks on the scale necessary to develop sales unless they are allowed generous credit terms. At the same time, finance is needed to store products processed during a few months until later in the year. The terms on which working capital is obtained have therefore an important bearing on the economic success of a processing enterprise.

Competitiveness requires also proper internal organization of the processing operation and, in particular, the labour force. There is often strong pressure to employ as many local people as possible in the processing plant in view of prevailing unemployment. This has jeopardized the economic success of some plants and should be resisted. The operational efficiency of a plant often suffers from the use of outdated accounting systems; they may provide a check on the flow of funds but do not make it easy to identify major shortcomings in efficiency. Lack of an incentive system to encourage staff to improve its work performance and productivity has been a major shortcoming.

The importance of marketing means that managers of processing plants should not only understand the technological process but be able also to analyse markets, forecast supply and demand, and organize marketing operations at all stages. Great flexibility is needed in operating the business in view of the strong seasonal and annual fluctuations in supply, prices and market conditions of agricultural raw material and processed goods. Therefore, more time and energy is often needed for marketing than for the supervision of the technological process.

The discussion of the problems encountered in organizing the marketing and processing operations of processing plants indicates the importance of all round management skills. Much more emphasis has to be placed on training in marketing and managerial skills for persons directing processing enterprises than is now often the case; only then can a higher rate of economic success of new processing plants be expected.

GOVERNMENT MARKETING POLICIES AND FACILITATING SERVICES

The successful development of sound agricultural processing industries depends also upon government marketing policies and services. The implications of government export and import policies for processing industries are well known and need not be discussed here in more detail. Certain enterprises studied faced difficulties in organizing exports of processed products, since some of the inputs imported, i.e. cans, salt, sugar, chemicals, etc., were not exempted from import duties even though they were destined for re-export. In another case, imports of cheap milk powder hindered sales of local fresh milk seen as the main product of the dairy plant.

The point to be stressed here is that governments can provide useful services in marketing to promote processing industries, and in this respect there has been little or no headway in most developing countries. The development of new processing industries and the progress of existing ones would benefit from applied market research, regular market information, assistance in organizing trial shipments to new markets, economic feasibility studies, sales promotion activities at exhibitions and fairs, establishment of national brands where appropriate, advertisements, a quality control service, a grading and inspection service for processed products and for the raw material to be processed.

Since agricultural processing projects involve a number of different groups of people, such as farmers, processors and traders, as well as government agencies, municipalities, transport and road construction agencies, it is essential that the various complementary activities are coordinated for the benefit of the enterprise as a whole and it is in this field that governments have to take the necessary initiative and action. Where governments have been unable to provide these services or to take sound and quick decisions in a reasonable time, this has led to a slowing down of a project's implementation, or even to its termination.

REFERENCES

Lacrouts, M. and Tyc, J. (1962) *Etude des problèmes posés par l'élevage et la commercialisation du bétail et de la viande à Madagascar.* Ministère de la Coopération, Paris.

Mittendorf, H.J. and Wilson, G. (1961) *Marketing of Livestock and Meat in Africa.* FAO, Rome.

15

A SPATIAL EQUILIBRIUM MODEL FOR PLANT LOCATION AND INTERREGIONAL TRADE

M. von Oppen and J.T. Scott

The general theory pertaining to the spatial location of economic activity can be divided, according to Hoover (1970), into the following classes: location, regional analysis, and interregional trade. Location theory describes the economic analysis involved in comparing alternative spatial locations for a specified kind of activity, while regional analysis is 'concerned with groupings of interrelated economic activities in proximity within certain specified areas'. Interregional trade refers to the buying and selling of inputs or products and their movement among two or more delineated areas.

Highly sophisticated methods and approaches have been separately developed in each of these fields of spatial economic theory. However, apparently few attempts have been made to integrate these theoretic fields, especially the theory of location and interregional trade analysis, or to integrate their associated models (Candler *et al.*, 1972; Leuthold & Bawden, 1972; Rao, 1973).

There are an increasing number of cases in which public decision makers, especially in developing countries, are concerned with the question of implementing a new, efficient agricultural industry (from raw materials production to processing and marketing of final products) so that social welfare can be maximized. To solve such a problem requires investigations in two fields: (i) an analysis of the interregional trade in order to predict the flows of the new goods, the quantities demanded regionally, and the resulting price level; and (ii) a determination of optimal locations and sizes of individual processing plants. Therefore, we believe that an approach integrating the theory of location and interregional trade analysis will be of general interest. The model presented here was developed to represent the emerging soyabean industry in India, but it could be adopted to almost any agricultural production and processing industry.

Source: *American Journal of Agricultural Economics* 55, 437–445, August 1976.

THE MODEL

The spatial equilibrium model for location of processing plants integrates all of the following important economic functions: (i) transportation of inputs and products; (ii) average processing costs related to plant size; (iii) size of market area per plant; (iv) regional supply functions of inputs to be processed; and (v) regional consumer demand functions for processed products. The model is constructed in two parts: the plant location is determined with the help of a single equation optimization model and the interregional trade of inputs and products is analysed by means of a quadratic programming model. From the optimum solution of the plant location, model regional average processing costs are derived and fed into the interregional trade model. From the optimum solutions of the interregional trade model, the quantities to be processed and distributed by processing plants are derived and fed into the plant location model. If applied in this fashion, both models are linked by an iterative solution procedure and jointly form a spatial equilibrium model for plant location and interregional trade (Fig. 15.1).

Plant location – assembly, processing, and distribution

The location equation is based on the following assumptions: (i) as the number of plants within a region increases, average processing costs will increase at a linear rate; (ii) both the average assembly and distribution costs of the major inputs and products in the market area of a plant decrease at a decreasing rate as the number of plants in a region increases; (iii) the assembly and distribution road network fits approximately a 60° triangular grid such that hexagonally-shaped plant market areas within regions are appropriate; and (iv) within a region, producers of inputs for the plant and consumers of products from the plant are evenly distributed.

[This assumption is perfectly fulfilled only in the case of perfectly homogeneous (i.e. for all practical purposes, very small) regions. However, even in the case of larger regions with urban population concentrations, the assumption may still be acceptable and realistic as long as these urban areas as such are evenly distributed.]

Furthermore, as the number of plants is initially increased (from the number at which plant size is large enough that average processing cost is a minimum), the sum of average assembly and distribution costs decreased faster than the increases in average processing cost. As the number of plants continues to increase, however, processing costs begin to increase faster than the decrease in assembly and distribution costs. The optimum number of evenly distributed plants is determined by the minimum of the sum of these average cost functions.

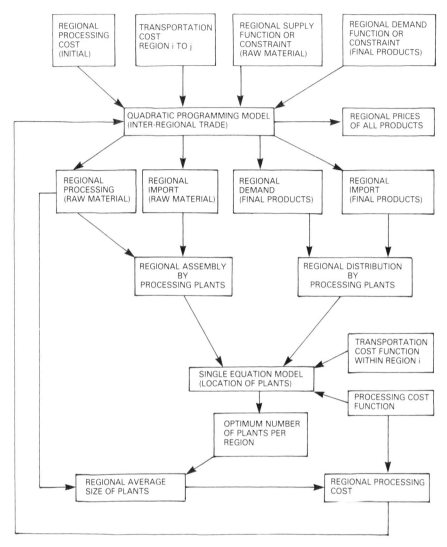

Fig. 15.1. Flow chart of a spatial equilibrium model for plant location and interregional trade.

Given these assumptions, the plant location equation can be derived as follows:

$$B = 6\frac{r^2}{4} 3^{1/2} \quad (1)$$

where B = the hexagonal market area for a plant and r = the circumscribing radius of the hexagon. If A is the total area of the region for which the optimum number of plants is to be found and N is the number of plants, the area per plant can be explained as

$$B = \frac{A}{N} \tag{2}$$

Solving equation (1) for r and substituting equation (2) into (1) gives

$$r = 0.6204 \left(\frac{A}{N}\right)^{1/2} \tag{3}$$

for the radius of each of the N hexagons representing the market areas of the plants in the region. The average distance from all points evenly distributed within the hexagon to the central point for assembly and distribution can be shown to be $0.7r$.

We assume that assembly of raw material and distribution of final products have the same per unit costs and these are a linear function of the average distance from the plant with an initial fixed cost for loading or unloading and vehicle use, so that

$$T = f + d(0.7r) \tag{4}$$

where T = the transportation cost per unit of input or product, f = the fixed cost, and d = the per unit cost of transportation per unit distance. [Where assembly and distribution are carried out by truck and as long as the commodities concerned are not extremely bulky, perishable, or otherwise difficult to be transported by truck, their freight rates normally are based on the same per quintal charge.]

The quantities of raw materials (R) to be assembled for processing and the quantities of final products (M) to be distributed for consumption within the market area of the plant can be expressed in relative terms using as a common denominator the quantity regionally available for processing (Q). [Note that Q includes regional production adjusted by imports and exports of the region as determined from the interregional trade portion of the model (see Fig. 15.1).]

Relative weights of the quantities transported within the market area of a plant are required to form the sum of the transportation cost functions for both of these quantities before adding the processing cost function. Thus, after substituting equation (3) into equation (4) and multiplying by the relative weights, the regional assembly cost function for locally assembled input R becomes

$$T_1 = T\frac{R}{Q} = f_1 \frac{R}{Q} + (0.43) d \left(\frac{R}{Q}\right) A^{1/2} N^{-1/2} \tag{5}$$

and the regional distribution cost function for locally distributed final product M becomes

$$T_2 = T\frac{M}{Q} = f_2\frac{M}{Q} + (0.43)d\left(\frac{M}{Q}\right)A^{1/2}N^{-1/2} \tag{6}$$

The average regional processing cost is assumed to be

$$T_3 = f_3 + c\frac{N}{Q} \tag{7}$$

where N/Q = the reciprocal of average plant capacity, and the processing cost is assumed to be a function of the reciprocal of the quantity processed. [This functional form is suitable for estimating average cost functions of production processes where economies of size imply that with increasing capacity processing costs are asymptotically approaching a minimum value.]

Then summing the three functions (5), (6), and (7) and setting the first derivative of this sum with respect to N equal to 0 gives the number of plants (N^*) that according to the second order condition minimizes assembly, processing, and distribution costs:

$$N^* = \left(\frac{0.43}{2}\frac{d}{c}(R + M)\right)^{2/3}A^{1/3} \tag{8}$$

This shows that the optimum number of plants is a linear homogeneous function of degree one in the two variables – the sum of inputs assembled and products distributed and the area of the region. This optimum number is then theoretically located by evenly distributing the plants within the region.

Interregional trade – regional supply and demand functions

The basic concept of this model is a special case of the general problem of nonlinear programming, which is solved for the saddle point (Takayama & Judge, 1971). In this case the objective function is a quadratic function constrained by linear inequalities, a problem that has a unique solution (Takayama & Judge, 1964).

The major assumptions and the economic environment are as follows. A given country is divided into n regions. Each region is represented by one base point at which the supply of primary product, the demand for the final products, and the processing capacity are assumed to be concentrated. All possible pairs of regions are separated by known transportation costs per physical unit for each product. The processing capacity in each region is assumed to be unlimited. Processing is performed as per unit costs that are assumed to be given for each region (from above). Demand for each of the final products is assumed to be known for each region and to be a logarithmic function of price. [For programming purposes this nonlinear function is assumed to be represented by its linear tangents. The incorporation into the model of the procedure for selecting the representative tangents is described below.] For each of the n regions a nonnegative quantity of the primary product is given.

In this environment it is possible to formulate a model that accounts for the interaction of the spatially separated economic units. The model determines the level and location of processing of the primary product into final products, gives volume and direction of all product flows that will minimize the aggregate transportation and processing cost, and finally determines the pricing system of all products that accompany the optimum allocation system.

Regional demands for final commodities Y are assumed to be represented by functions dependent on price P. These are of the general form $P = aY^{-b}$. Their tangents represent linear demand functions for the final products in each region. By applying the familiar technique of formulating the 'net' social benefit function as the sum of the line integrals of individual demand relations minus the total costs of processing and transportation, a quadratic spatial equilibrium problem can be set up (Takayama & Judge, 1971). The objective of this problem is to maximize the 'net' social benefit subject to a set of constraints on primary commodity allocation and flows; final commodity production, flows, and consumption; and processing and joint production. All prices and quantities are restricted to be nonnegative.

If the assumptions of the net benefit function and the constraints are satisfied, then a necessary and sufficient condition for an optimum solution of the problem is that the corresponding Lagrangian forms a saddle point. The appropriate Kuhn-Tucker conditions ensure that the solution is a maximum (Takayama & Judge, 1964).

Combination of the model parts

When the two models are applied jointly, it is assumed that interregional flows take place between regional centres (reference points), while in reality, of course, shipments would flow from all plant centres or local assembly points (in case unprocessed raw products are shipped) to consumption centres. This assumption is justified because it implies realistically that the sum of the costs of all shipments between all pairs of plant centres and assembly points in region i and consumption centres in region j does not, or only to a negligible extent, differ from the costs of shipment between the regional centres of regions i and j.

In combining the two parts of the model into one spatial equilibrium model, the following concept is applied. Assume that there are two systems that are mutually dependent in that the output of one constitutes part of the input of the other. Then, as these systems are optimization models, they form a combined equilibrium model. Their interdependence generates a sequential set of objectives and restrictions, the objective of the plant location model forming a restriction to the interregional trade model and the objective of the interregional trade model forming a restriction to the plant location model. In other words, for each of the two models the optimum solution is to be found subject to the constraint that the other model be optimized. Since part of the input of the model is generated by the other model, a sequence of alternating applications of both models – one in the objective, the other in the constraint, and vice versa – asymptotically approaches an equilibrium solution.

[The spatial equilibrium model consists of the following programs: (i) quadratic programming model (QP), (ii) single-equation plant location model (SE), (iii) derivation of regional demand functions from a tangent to the exponential demand function (DT), and (iv) testing of the tangent for representativeness and if rejected finding new tangent (TT). In applying these programs in the empirical case given below. The most efficient sequence was found to be

$$SE_1 \searrow$$
$$ \nearrow QP_1 \to SE_2 \to QP_2 \to TT_1 \to QP_3 \to SE_3]$$
$$DT_1 \nearrow$$

We assume that the solution is unique because each of the two parts of this model has a unique solution. Application of the model with empirical data showed consistent convergence of these solutions. [The uniqueness of the solution depends upon the character of the set of the various parameters included. In the unlikely but conceivable case in which the parameters included generate non-convergence of the results (e.g. when the sum of the costs of transportation of the raw materials equals the sum of the costs of the transportation of the final commodities), there would be no unique optimal solution to the problem. By submitting the model to appropriate sensitivity tests, the likelihood of such cases can be evaluated.]

AN EMPIRICAL APPLICATION

While there is a sophisticated oilseed processing industry established in India, soyabean production, processing, and consumption is a new developing sector in this country. After their introduction in the late 1960s, cream-coloured varieties of soyabeans covered about 10 000 hectares in 1970–71 and about 43 000 hectares in 1973–74. In 1974, five solvent-extraction plants with a total capacity of 265 tons per day and five screw-press-expeller plants with a total capacity of 11 tons per day were reported to be processing soyabeans in India (see Table 15.1).

In view of this development it seems worthwhile to review the results

Table 15.1. Soyabean processing facilities reported in different regions in India in 1974

Plant type	\multicolumn{8}{c	}{Region}
	1	2	3	4	5	6	7	All-india
Screw-press expeller								
Number	—	1	—	2[a]	—	1	3	5
Average capacity (tons/day)	—	4	—		—	4	1	2.2
Annual capacity (tons)	—	800	—		—	800	600	2200
Solvent extraction								
Number	—	1	—	1	—	2	1	5
Average capacity (tons/day)	—	60	—	60	—	10	125	53
Annual capacity (tons)	—	18000	—	18000	—	6000	37500	79500
Total annual capacity (tons)	—	18800	—	18000	—	6800	38100	81700

Source: Rathod & Motiramani, 1974.
[a] One extrusion cooker and one soyamilk plant, both still in the experimental stage.

that were generated in 1971 by the above spatial equilibrium model of a future Indian soyabean economy based on data collected in 1970–1971. The following is a summary of the major assumptions made on supply demand, transportation, and processing of soyabeans.

Based upon certain agronomic assumptions about the general production potential for soyabeans in India, about the extent to which a selected number of crops and fallow land might be substituted by soyabeans, and about yield, the potential production of soyabeans by districts was computed (see Fig. 15.2); coterminous districts of homogeneous potential production densities were grouped into seven regions and the share of each region in the total potential production was calculated.

[Formulation of the interregional trade of a soyabean economy (one raw product, two final products) into a quadratic programming model as specified above requires a matrix for n regions of $n \times 7$ rows and $n \times 23$ columns. Seven regions were chosen because on the one hand a larger number of regions would have required excessive computational costs and on the other hand less than seven regions would not have allowed a satisfactory division of the country into reasonably homogeneous areas.]

The two joint products, soyameal and soyaoil, were assumed to be substitutes for gram flour and groundnut oil, respectively, for each of which demand functions were fitted using time-series data on prices, quantities consumed per capita, and income per capita. The per capita demand for gram flour was estimated, and after conversion into the required form it became

$$P_g = 1.5716 \ Y_g^{-0.84048}$$

where P_g = price of gram in rupees per kilogram and Y_g = the quantity of gram consumed in kilograms per capita. [The function was estimated by using seventeen years of All-India per capita net availability and annual average prices of gram (Government of India, 1968, 1970, and 1973). The standard error of the exponent is 0.1812, the coefficient of multiple correlation is 0.76, and the F-ratio is 21.5.]

The per capita demand for groundnut oil used by the vanaspati [cooking margarine] industry was estimated, and after conversion into the required form it became

$$P_o = 0.001955 \ Y_o^{-1.8298}$$

where P_o = the price of groundnut oil in rupees per kilogram and Y_o = the quantity consumed in kilograms per capita. [The estimation was based on monthly observations for five years (1964–1969) on per capita groundnut oil consumption (Q_o), groundnut oil prices (P_o), and per capita income expressed in vanaspati consumption per capita (V_r) (Government of India,

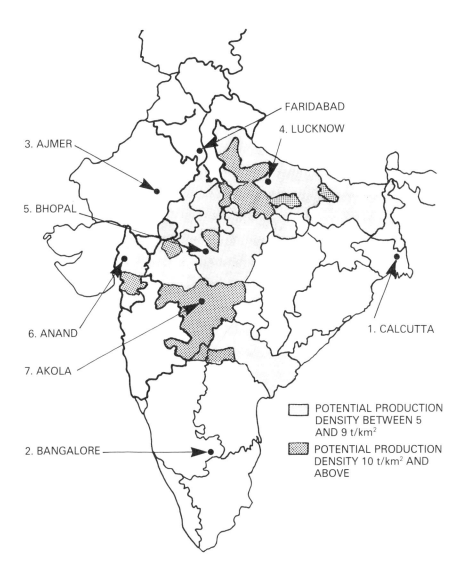

Fig. 15.2. Estimated soyabean production potential and demarcation of regions and reference points in India.

1964–1969). After adjusting for an autocorrelation of 0.6, the logarithmic function fitted gives

$$\ln Q_o = -0.8590 - 0.5465 \ln P_o + 0.653 \ln V_o$$
$$(0.4004)(0.2213)(0.254)$$

Standard errors are given in parentheses. The coefficient of multiple correlation is 0.37 and the F-ratio is 5.3.]

The demand functions entered into the model were defined as tangents uniquely representing these exponential functions. Regional demands were then derived by dividing the slope of the soyameal demand by regional population, by dividing the slope of the soyaoil demand from the vanaspati industry by regional population proportional to vanaspati industry prevailing in each region, and by adjusting the intercept of the soyaoil demand function according to the postulated per capital income. [Division of the slope of the soyameal demand by population implies multiplication of per capital quantity and thus aggregation of the demand function.]

The average costs of processing in solvent extraction plants were estimated from engineering data and found to be $T_3 = 59.95 + 1383.5\ S^{-1}$. [Where S represents annual capacity in tons, note that $S = QN^{-1}$.]

Assembly of soyabeans from the farmer to the plant and distribution of soyameal from the plant to the consumers was assumed to be carried out by truck; truck rates were found to be a linear function of distance. The intraregional transport was assumed to be carried out by rail according to freight rates for the different commodities in 1970–1971.

Table 15.2 summarizes the major assumptions made with respect to time. The years 1970–1971, 1973–1974, 1978–1979, and 1988–1989 are labelled model years A, B, C, and D, respectively. Estimates of the Indian population are taken from official sources. Estimates of income per capita are based on an average annual income between 1964 and 1969 and a compound ground rate of 2% annually during the entire period is assumed. Table 15.2 also gives the most recent data available on the actual development in 1971 and 1973.

Comparison of the actual development with the model years indicates that in 1973 population, income, and soyabean production have developed about three years past the model year A and are about five years 'away' from model year B. Therefore, the results (Table 15.3) for model years A and B should be comparable with the actual location and size of soyabean processing plants; if so, the results for the later model years C and D would allow one to draw useful conclusions.

Average processing capacity of soyabean solvent extraction plants was calculated by the model to grow about 50 tons per day in year A to about 140 tons per day in year B, eventually reaching up to 350–400 tons per day in year D. Present capacities of the five solvent extraction plants for soyabeans of 53 tons per day seem to be quite in line with this result, even

Table 15.2. Assumed and actual development in population, income, and soyabean production

Item	Assumed development in model years				Actual development	
	A	B	C	D	1971	1973
Population in millions (1968–1969 = 100)	550 (105.0)	592 (113.0)	662 (126.3)	787 (150.2)	551 (105.2)	574 (109.5)
Income at 1960–1961 prices in Rs per capita (1968–1969 = 100)	337.0 (104.0)	357.6 (110.4)	394.8 (121.9)	480.9 (148.6)	345.8 (106.8)	337.5 (104.2)
Soyabean area in ha. (or tons)[a]						
Region 4	7280	145600	436800	1820000	4909	16820
Region 5	4190	83800	251400	1047000	5122	17077
Region 6	2610	52300	156900	653000	2935	8827
Region 7	5920	118300	354900	1480000	11254	532
All-india	20000	400000	1200000	5000000	24220	43256

Source: Rathod & Motiramani, 1974; Government of India.
[a] Assuming yields are 1 ton/ha.

Table 15.3. Theoretical optima of number, average capacity, and average radius of market area of soyabean processing plants in four model years

Model year	Region				All-India
	4	5	6	7	
Number of processing plants[a]					
A	0.5	0.3	0.2	0.4	1.4
B	3.8	1.9	1.2	2.4	9.3
C	7.7	3.7	2.3	4.7	18.4
D	19.3	9.5	5.9	12.0	46.7
Average capacity (tons/day)					
A	47	45	47	51	48
B	127	147	145	163	143
C	189	224	224	250	217
D	314	368	370	413	357
Average radius (km)					
A	507	531	504	471	—
B	187	215	197	189	—
C	132	153	141	135	—
D	83	96	89	85	—

[a] The fractional numbers present here as theoretical optima should for practical purposes be rounded upward, such as to give 'recommended' numbers.

though they range from 10 to 125 tons per day (Table 15.1).

The existing processing plants are located in regions 2, 4, 6, and 7, which is partly in line with our predictions. In contradiction to the model result, there is no processing capacity for soyabeans in region 5 as yet, while in region 2 soyabeans are processed. This can probably be explained by the fact that a reclassification of rail rates in 1973 resulted in a reduction of the rates for soyabeans and a substantial increase in the rates for soyameal and in making soyabean processing feasible also in non-producing regions such as region 2. [The ratio of the sum of the costs of transporting soyameal and soyaoil over the sum of the costs of transporting soyabeans increased from 0.52 in 1970–1971 to 0.75 after 1973.]

Without entering into further discussion of the many reasons why or why not reality differs from the model results in some cases, it can be concluded from the evidence gathered so far over the short period of three years that the soyabean economy in India appears to be generally allocating plant capacities and sizes in line with the optimality criteria developed in the model. It should be mentioned here that for several years plans existed to establish one single government-operated processing plant with an initial capacity of about 100 tons per day and a provision to later step up the

Table 15.4. Comparison between costs of processing and transportation of the optimal solution and of an arbitrary solution

Costs[a]	Optimum solution (million rupees)		Arbitrary solution (million rupees)		Optimum solution divided by arbitrary solution	
	Year A	Year B	Year A	Year B	Year A	Year B
Processing cost	2.64[b]	27.85[c]	1.60[d]	23.80[e]	0.61	0.85
Transportation cost of interregional trade	0.14	7.46	2.04	42.13	14.10	5.60
Total cost	2.78	35.31	3.64	65.93	1.30	1.86

[a] Based on 1970–1971 processing costs and rail rates.
[b] Assuming one plant in each production region with capacities as indicated: region 4: 25 tons/day, region 5: 15 tons/day; region 6: 10 tons/day; and region 7: 20 tons/day, i.e. in total, 70 tons/day.
[c] Assuming average processing costs.
[d] Assuming one plant with a capacity of 70 tons/day.
[e] Assuming one plant with a capacity of 1 333 tons/day.

capacity. This plant was planned to be located in Faridabad, south of New Dehli (Rathod & Motiramani, 1974). If only one plant had been available to process the quantities of model years A and B so that interregional trade would have been forced to move soyabeans to Faridabad and if soyaoil and soyameal were distributed among regions according to the quantities demanded in model years A and B, then this arbitrary solution would have cost the soyabean economy in year A about 130% and in year B about 186% of the costs for the optimum solutions (see Table 15.4).

It should be emphasized that starting a soyabean economy in India imposes the 'chicken-and-egg' problem of simultaneously developing raw material production and processing facilities. If left to itself, this problem will probably be solved by an industry that especially in the initial stages is expanding at very low rates and at the costs of likely mistakes resulting from trial and error approaches. However, with the help of a tool such as the above, despite its many limitations some of the major principles involved in the spatial allocation of the soyabean processing industry are recognized and can be applied for future planning by both public and private decision makers in order to foster the growth of the soyabean industry.

CONCLUSION

With the aim to contribute to the integration of location theory and of interregional trade theory, a model for plant location and interregional trade was developed. Application of the model on data relevant to a projected soyabean industry in India produced feasible results. There is evidence that the actual development in the soyabean processing sector up to 1974 has followed the earlier projected path fairly closely. It can be shown that an arbitrary plan of processing soyabeans in one single plant located in a consumption region would raise costs to the industry 30% above the optimum level for 1970–1971 and 86% above the optimum level in 1978–1979. Application of the spatial equilibrium model for plant location and interregional trade may help decision makers to recognize the principles involved in the spatial allocation of agriculture-based processing industry and thereby reduce costs and time required for its development.

REFERENCES

Candler, W., Snyder, J.G. and Faught, W. (1972) Concave programming applied to rice mill location. *American Journal of Agricultural Economics* **54**, 126–130.

Government of India (1964–1969) *Oilseeds Situation* (monthly). Mimeographed. Government of India Press, New Delhi.

Government of India (1968, 1970, and 1973) Ministry of Food, Agriculture, Community Development and Cooperation, Directorate of Economics and Statistics. *Bulletin on Food Statistics* (18th, 20th and 23rd issues). Goverment of India Press, New Delhi.

Hoover, E.M. (1970) The Partial Equilibrium Approach. In: Dean, R.D., Leahy, W.H. and McKee, D.L. (eds) *Spatial Economic Theory*. The Free Press, New York, pp. 3–14.

Leuthold, R.M. and Bawden, D.L. (1972) *An Annotated Bibliography of Spatial Studies*. Exp. Sta. Res. Rep. 25, College of Agriculture, University of Wisconsin.

Rao, A.S. (1973) Buffer storage location under economies of scale. *Indian Journal of Agricultural Economics* **28**, 14–29.

Rathod, K.L. and Motiramani, D.P. (1974) *Status of Soybean Industry in India*. All-India Coordinated Soybean Research Project, Jawaharlal Nehru Krishi Vishwa Vidyalaya, Jabalpur, India.

Takayama, T. and Judge, G.G. (1964) An inter-regional activity analysis model for the agricultural sector. *Journal of Farm Economics* **36**, 349–365.

Takayama, T. and Judge, G.G. (1971) *Spatial and Temporal Price and Allocation Models*. North-Holland Publishing Co., Amsterdam.

16

INFRASTRUCTURE FOR FOOD MARKETING: SOME INVESTMENT ISSUES

B.W. Berman

GRAIN

Normally, the following justifications for a grain distribution project are presented to an international or bilateral development agency when seeking funding.

1. Reduction in grain losses.
2. Lack of capacity in storage, processing, distributive infrastructure.
3. Growing deficits in the operating results of the grain parastatal.

Usually, some broad macroeconomic indicators are available to back up such justifications.

From the World Bank's perspective in the early 1980s a major incentive to tackle grain distribution projects is to provide an operational entry point to addressing the issues of grain price levels, and the subsidies associated with parastatal grain distribution. Often the parastatals operated at a financial loss, due both to the inadequacy of the operating margins allowed by the government set farmgate and urban price levels, and the parastatals' own inefficiencies. Projects aimed at improving the parastatals efficiency are seen as the first step in a strategy of addressing grain pricing and subsidy issues.

This is in fact the reverse of an effective project oriented strategy. Pricing policies should, ideally be adjusted prior to any project lending. Realistically this is not possible. Instead the worst case assumption should be adopted: projects normally have to be appraised and implemented with no change in the pricing levels. To do this requires a cost and risk minimizing strategy for project design.

Source: World Bank working paper, 1980.

Demand for and supply of marketing services is dependent on price levels. If marketing margins are inadequate no services will be provided to those areas that cannot be accommodated within the margin. As under- and over-serviced areas emerge in response to the controlled price levels, so too will the demand for infrastructure investment. Attempting to relieve apparent infrastructure bottlenecks, through marketing investments, may well simply cement the distortions into permanent structures. If, therefore, the ultimate intention of these marketing projects is to correct price levels, the very act of building infrastructure may work against the programme objectives. Faced with only very macro level data and uncertainties about price, just how can a project be put together? This paper will attempt to outline as risk-free an approach as is possible. This approach generally is disappointing in lending programme terms because it demands relatively limited levels of initial investment with an intensive institutional focus to build a database for a future systematic investment programme.

The transition from macro-indicators to strategy and project formulation is hampered by a lack of data. Within a normal project cycle time frame it will not be possible to obtain such data, and especially so because grain distribution systems are dynamic and have complex matrix type interactions. The operationally relevant questions that must be faced are:

1. Is there a way to prepare and justify a relatively riskfree initiating project on the basis of only rudimentary data?
2. How can a data base be established for future subsector and project work?
3. What form of analysis framework would be necessary to identify, design and justify future projects?

In other words, is there some defensible way of initially investing in this subsector, thereby alleviating some immediate short-term problems, while setting the stage (and not preempting it) for a future systematic investment programme based on a reasoned subsector strategy?

Two approaches would typify investments:

1. Infrastructure for the parastatal grain authority.
2. A credit line for private sector investment in storage and processing.

In a situation where the database does not allow for confident assessment of the infrastructural and parastatal investment needs, the private sector approach is easier to justify using the macroindicators. This is especially so because over-investment is prevented by the regulating mechanisms inherent in a credit programme: private investors normally are prudent in assessing the market, and do not risk their capital in over-investment.

The private sector credit line approach has some limitations too. The first is that it is directly responsive to current (distorted) price levels, and will allocate its resources on this distorted basis. If large credit lines are proposed, there is a risk of seriously distorting overall investment patterns in the same way as parastatal investments would in the absence of a systematic investment strategy.

Another limitation to the private sector approach is that the investor tends to follow the grain production trends rather than anticipate them, and for sound reasons. Thus, most credit line utilization would be for expansion of existing capacity in already serviced areas. New or increasing production areas would experience a processing capacity lag that may result in depressed prices. It would be interesting to calculate the cost difference to the economy between the alternative of temporarily depressed prices and temporarily underutilized capacity.

Under a number of the current price control regimes it is not profitable for the private sector to invest in custom storage (storage-for-hire) and storage built under credit programmes is usually captive to a processing facility. If, therefore, storage is identified as a major constraint, and pricing policies cannot be changed to make custom storage profitable, only a public sector investment approach can be adopted. Private sector investment in transport, and working capital for stocks have not usually been addressed in past credit line projects, and although this would be desirable it is not very easy to include in formal credit programmes because of collateral consideration (assets – trucks and stock – and owner area very movable and not easy to attach!).

Infrastructure investment for the parastatal is more difficult to assess. Because the parastatal is not a profit maximizer, and often not a cost-minimizer either, investment proposals are based on demanded levels of service. The first question to confront a project mission is 'why is the capacity needed?'. This soon translates into questioning the assumed market share of the parastatal. Because of the fact that this is an initiating project, this question is difficult to assess without more data collection. An operationally useful rule of thumb is:

- Under initiating project conditions, incremental capacity should be limited to the level necessary to maintain the parastatal's current average market share, but not to increase it.

However, even a limited parastatal investment programme in an initiating project runs the risk of misallocating resources. Reducing this risk requires knowledge of the outcome of the systems analysis of the parastatals investment requirements. In the absence of this analysis a risk averse initiating project could be guided by the following:

- Avoid new construction, concentrate on storage loss reductions and pipeline efficiency through rehabilitation of existing plant and needed information systems.
- Initiate manpower development for design skills for new storage and

distribution investments as well as requisite planning and supervision skills.
- Design and institute improved inventory control, cost control and distribution scheduling systems, and the concomitant manpower training.

If the project has a short enough implementation time horizon (and no more than three years should be considered for initiating projects), then any investments can be limited to filling immediate obvious needs, while additional data gathering and planning is undertaken to establish an investment master plan. However, interpreting obvious needs presents some difficulties. Some examples are given below.

A number of investments are proposed for countries that have experienced recent large increases in production. All usable storage space is filled, and excess grain is being stored in schools, other public buildings, and in the fields. Large losses result, and the proposal is an emergency building programme for more storage. Operative rules of thumb are:

- If the storage shortfall is related to an acknowledged bumper crop, only temporary storage (e.g. plinth and tarpaulin type) should be considered.
- If the storage shortfall is related to 'normal' crop levels, assess if there are any structural reasons (price, bad access roads, lack of trucks, etc.) at those locations to account for the large unmarketed surplus, and invest in alleviating those constraints.
- If a true storage shortfall is identified, priority should be given to rehabilitation and efficiency improvement in existing storage, with new construction concentrated at transport terminal points.

Often the parastatal will wish to increase the capacity of the transport system between the production and consumption areas. This would relieve storage constraints in the production areas, by transferring surplus to consumption area storage during harvest season. (This also implies increased consumption area storage.) The rule of thumb here is:

- It is normally more cost effective to increase production area storage capacity than to increase railroad capacity (and often also truck fleet capacity); or, storage to flatten transportation demand peaks is more cost effective than increasing transport capacity to level out those peaks.

Requests for an increase in consumption area storage are the most likely to need modification after a systemwide examination, because of the interaction of storage and transport capacity. Therefore, the following rule of thumb is recommended:

- The most cost effective consumption storage level is that which allows for a steady inflow delivery stream matching the steady consumption demand, modified by considerations of batch size and delivery schedules, and the need for a buffer to cushion delivery delays. Large consumption area storage is generally less cost effective than production area storage because of higher urban land cost, and urban wage levels than in the production areas.

This rule of thumb must be modified if physical conditions (e.g. impassable roads) for the transfer of stocks makes it impossible to provide a steady delivery stream.

For grain imports for urban consumption, two investment decisions are normally required: size of port facilities, and types of grain handling equipment.

1. If port storage is more costly than other urban storage, it may be cost effective to (i) periodically evacuate dockside facilities and place the grain in other urban stores instead of increasing port storage, (ii) invest in increased evacuation capacity (e.g. trucks and rail).

2. Bulk handling for evacuating ships is probably justified, but bagging at the terminal is initially more cost effective than early introduction of bulk transfer systems. Ability to modify later to bulk transfer would be an advantage.

For requests to increase captive (i.e. owned by the parastatal) processing capacity, in a system where the private sector is also maintaining the commercial activity level, the following is recommended:

- It is more cost effective to promote private sector capacity, with custom processing on behalf of the parastatal, than to build additional public sector capacity.

However, for those systems with little or no private sector processing capacity, location of parastatal processing capacity is dependent on transport cost comparisons between processed and unprocessed grains, and storage characteristics of processed grains.

- The determinant of processing capacity at the chosen location would be the steady state demand level of that location. Steady capacity utilization across the entire crop cycle is preferable to a low capacity installation with post harvest peak loading.

This rule of thumb implies ancillary investments in storage capacity and delivery systems.

For a request to provide storage capacity specifically for security reserve stocks (i.e. stocks of grain as insurance against shortfalls in production) the following rule of thumb normally implies injection of the request unless the rule can be disproven:

- It costs less to import and handle grains for an unforeseen shortfall in production than it does to operate a security reserve stock.

The cause of 'unforeseen shortfalls' are normally found to be a lack of information for timely determination of import requirements (and therefore timing of import orders), logistics capabilities and stock management. As the information needed to design an effective grain security programme is normally not available at this initiating project stage it is more cost effective to build up the information system, and continue the imports programme than to build stockholding capacity.

The rules of thumb given above are designed to minimize the impact of incorrect decisions under conditions of poor data and a distorted price regime. Alternative investments would require additional unavailable information to prove they are more cost effective. It is expected that additional subsector analysis would provide investment guidelines based on cost effectiveness criteria, that take into account the following:

- the impact of changing price levels on the total system;
- future market share of private and public sector;
- interaction between the location and capacity of storage, processing and distributive infrastructure;
- types of materials handling systems.

Grain distribution system investments must take into account the annual variability of grain supplies, changing urban demand pattern, variable rural demand patterns, annual changes in urban/rural terms of trade, absolute income levels, crop yields (especially other non-grain cash crops) and so on. To measure the impact of these factors on investment strategy and provide investment justification the following types of information are necessary:

- Supply and demand projections by region taking into account changing urban populations.
- Projected market surplus and market demand (i.e. demand by non-producers).
- Seasonal and crop specific private sector and parastatal market shares.
- Capacity inventory of storage and processing, both private and public sector.
- Assessment of rail, road, barge and port seasonal capacity available for grain.
- Estimates of efficiency of utilization of all infrastructure related to grain.
- Time series analysis of grain storage losses by type of storage.
- Analysis of variance of grain production trend data, and associated market surplus levels.

- Quantification of grain supply changes due to exceptional weather and other types of 'disasters' and 'windfalls'.
- Sampling of costs of grain procurement, storage, processing and distribution by location, over time and distance, for private and public sector.
- The country's performance in the grain import and export markets.
- Analysis of the grain price regime and grain price controls.

This data would provide the basis for identifying investment needs by region, and analysing the alternative strategies possible especially where location and transportation options are interrelated.

For those systems where the private sector is to play a role, any investment alternatives in the public sector can be expected to have an impact on the private sector's continuing operation. This in turn would influence the private sector's response to any project initiatives (e.g. credit) to encourage additional investment. The price regime too would influence this. For any interventions in the private sector the following information would be necessary as input to designing a private sector strategy.

- Role of private sector in each level of market and distribution chain.
- Relationships between various participants (family and social obligations, credit linkages, contractual obligations, etc.).
- Enterprise profitability (including byproducts sale, etc.).

In addition the 'investment climate' needs to be assessed – through a compilation of the government official pronouncements on the role of the private and public sectors in grain, as well as an analysis of the financial impact on the private sector of various government policy decisions, and regulatory actions of the parastatal grain authority. This compilation should be compared to the private sector's actual investment as well as operating decisions. If possible their private assessment would also be valuable, but can be expected to paint an unfavourable picture if price controls are in place.

The next most often encountered questions are capacity and location of new storage. The suggested systems data approach would use the data especially gathered to identify cost minimizing alternative systems. For example, in the production area, alternatives would be a series of small village level stores, a system of sub-regional stores or a large store at the regional centre. Economics of scale of the storage, expected distances of delivery to and from the stores, and economies of transport lot size would be used to assess which of these storage alternatives provided the least cost solution. A region with extensive redistribution of locally stored grain for local consumption would have a solution responsive to smaller local delivery lot sizes. A region serving as a supplier to other regions would have storage size and location decisions almost determined by the economies of transport. Similarly, a region served by rail would have economies of transport lot size requiring larger lots than a trucking system. These economies would in turn favour large storage installations, and probably faster materials handling operations. Locations would be determined by line of rail. The truck dependent system on the other hand would be more dispersed, with storage at locations that minimize the cost of delivery to storage and consumption points.

Another example of the dependency of storage decisions on transport infrastructure is that of the length of time that grain should be kept in the production areas before being transferred to consumption areas. This decision also impacts on the size of consumption area storage. In countries where rural storage is less costly than urban storage, minimum stockholdings in the urban areas would be expected. Determining the size of urban stocks would depend on the economies and timelags in deliveries from rural areas.

Often governments plan for large urban food storage, and relatively fast removal of grains from rural to urban storage. From what has been outlined above this would appear not to be a cost minimizing approach. However, use of scarce rail and trucking resources, impassability of rural roads at certain times of the year, and similar factors need to be taken into account in assessing the balance between rural and urban storage, and distribution infrastructure. The systems analysis framework would allow the mission to evaluate the government proposal for stockholding patterns, against the cost minimizing solution, or any modification thereto for protection from delivery disruption, etc. The difference in cost between the government proposal and the alternative mission views as optimal is the cost of the government's urban stockholding decision. Often this decision on urban grain stock levels is a political one, related to prevention of food shortages (especially so called 'artificial' ones created by the private sector) and any resulting urban unrest.

Bag or bulk storage

Two major reasons are given for proposing bulk storage.

1. Lower storage losses.
2. Lower transport costs in bulk.

Two types of bulk storage are available: flatbed and silo. Choices between these two are determined by site conditions, land prices and the level of turnover. The level of turnover is determined by the role of the storage facility at that point in the distribution chain. High turnover locations normally favour silos, while low turnover areas could utilize bulk flatbed or bag storage.

Objections are often voiced that bulk storage facilities displace labour because of mechanized materials handling. The storage technologists argue that mechanical handling is necessary to achieve fast and efficient turnaround time for ships, or unit trains, and even trucks. If the benefits of this fast turnaround are truly captured by the economy it may compensate for the labour displacement effects.

For port bulk handling it seems that these benefits are captured in less port congestion, faster turnaround and lower demurage. Often, however, the rail system, as an example, is so cumbersome that the impact of improved train turnaround is dissipated by inefficiencies elsewhere in the system. It is worthwhile to consider seeking for bulk storage designs that effectively combine the use of labour and mechanical handling in a mix that captures most of the benefits from both bulk handling and low cost labour usage.

Processing

Aside from increases in capacity to handle increased volume of grain in the market system, investments are often recommended to replace obsolete processing facilities. This obsolescence is measured primarily in terms of grain losses, but nowadays some attention should also be paid to fuel efficiency. Two issues will therefore face a mission:

1. Capacity and location of processing facilities.
2. Choice of processing technology.

The capacity and location decision for processing will be influenced by all the systems-type factors that are operative for storage decisions. In addition, the transportation cost differential between unprocessed and processed grains, the cost of storage of processed grain and its shelf life, economies of scale in processing, and delivery times, must be included in determining capacity and location decisions through the cost minimizing approach.

Choice of technology predicated on improved extraction rates (i.e. reduced grain losses) should also be assessed in terms of a number of system related aspects. Processing efficiency depends on grain quality, which in turn is dependent on all the steps in the distribution system from farm to the processing facility. If any one of these steps adversely affects grain quality, the introduction of improved technology at the processing end will have only small (if any) benefits. Often to actually capture the expected benefits the introduction of a new processing technology requires investment at some early point in the distribution chain (for example, grain dryers). An assessment of the feasibility of such an investment usually requires an examination of the systems implications.

As an example, the case of rice dryers adopting the system approach highlighted a number of factors. The cost of drying is borne by the farmer or primary trader. The price structure is often sufficiently noncompetitive that any benefits from both quantity and quality improvement are not passed down the chain to the farmer. Thus there is no incentive to use a dryer, and therefore no advantage in introducing the new technology, until the noncompetitiveness of the price structure is altered. Such system reforms are sufficiently removed from the immediate technically motivated decision about new rice mill and designs, as not to be in any way influenced by them.

Regretably in recent years policy makers have been persuaded of the need to 'modernize' the rice mills, but there has been no realization of the necessity to make changes in other parts of the system to ensure that the economy does actually benefit from this modernization.

PERISHABLES

For perishables, fast delivery to consumers is the primary requirement of a marketing system. This is the major difference from a grain system which is designed to store, and deliver to the consumer on demand. The perishables system is driven by supply.

Only certain relatively affluent countries can afford the costs of cold storage of essentially seasonal products. In some countries where fish and meat are now considered as the primary protein source, governments have been willing to subsidize cold storage facilities. The wisdom of this has not yet been questioned, but should be prior to extensive lending in this area: even more so in those countries advocating cold storage of locally produced fruits and vegetables for local consumption. Concepts of food substitution by season seem to have given way to satisfying possibly very expensive consumer preferences, through subsidy.

For most perishables, however, the traditional markcting channels are still operating. Very little parastatal activity is currently evident, although some countries are being tempted to consider it because of large differences between farmgate and consumer prices. New investments should concentrate on cost reductions in the private sector distributive system mostly through new and improved rural roads, new, better located market places, and improved market information systems.

Decisions on location of market places, their size and regulation of the system under which they operate can also be analysed through systems approach. Location should minimize transport costs and therefore the relationship of the location to both the supply transport grid, and to the retailers is important. The urban setting (e.g. traffic congestion, land values, complementary enterprise locations, the market place as a social institution, etc.) must also be taken into account in determining market location.

If there is a strong desire to 'modernize' the perishables marketing system, the first rule of thumb is: don't go the route of the parastatal. Perishable products normally demand entrepreneurial price decisions as market conditions change by the hour. It is likely that no parastatal can be managed to provide that type of response to market conditions and therefore the introduction of parastatals will increase inefficiencies in the market system.

Interventions in the perishables markets must take into account the system of social linkages existing between farmers, traders, wholesale and retail traders. Transport cost reductions, improved market information and improved alternative employment (to removed marginal/social welfare operators) would be effective in increasing market efficiency. However, only increased competition will ensure that cost savings are passed back to the supplier and forward to the consumer. Introduction of marketing credit, even forward contracting by producer cooperatives, might increase competitiveness. However, an effective method of increasing competition in these traditionally resiliant systems has not yet been found.

17

IMPROVING PHYSICAL MARKETING INFRASTRUCTURE IN AFRICA THROUGH MORE SELF-HELP

H.J. Mittendorf

A major cause of marketing problems often lies in defective infrastructure, particularly roads and transport services. It has been estimated that more than half the higher marketing costs in Africa in comparison with those in Asia are due to inadequate marketing infrastructure (Ahmed & Rustagi, 1987). The purpose of this paper is to discuss strategies to improve the physical marketing infrastructure in Africa south of the Sahara. It will address, in particular, strategies to improve roads from farm to market, transport facilities, communications, as well as storage, processing and market facilities. Particular attention will be given to policy issues affecting decisions on establishing marketing infrastructure, to the scope for self-help, and to management.

There has been some progress in the strengthening of marketing infrastructure in African countries during the last three decades. Considerable efforts have been made by African governments and donor agencies to build up and maintain road, telephone and rail services in addition to storage, processing and market centres. The private sector has complemented the public investments in roads with investment in trucks. This has made a considerable contribution to the better marketing of farm produce and to the strengthening of competition at the farm level. In spite of such progress, there are a number of critical issues to be addressed:

1. *Roads.* Many roads have been inadequately maintained (World Bank, 1987) with the result that transport costs have increased. Many governments have been unable to provide the necessary maintenance. Feeder roads have deteriorated and truckers are refusing to go into rural areas because of the high cost involved. This is aggravated by the shortage of foreign exchange for spare parts and petrol. Farmers, therefore, encounter increasing difficulty in finding transport services.

Greater emphasis has been given to tarred roads that serve large urban centres; comparatively fewer public resources have been allocated to the building of simple rural roads suitable for ox carts and bicycles. This has hindered more decentralized economic development (Lipton, 1987).

2. *Marketing facilities.* Governments have invested in physical marketing facilities, such as storage, processing, market centres and packaging facilities. Policies have, however, been oriented too much towards direct government investment, with often doubtful results, and much less towards encouraging investment by private traders, farmer groups and local communities. Mobilizing broad popular participation in investment in marketing facilities, in particular by the trade, has been seriously neglected.

3. *Communications.* Post, telephone and telegraphic services are important for an efficient marketing system. Often, they are inadequate or operate unsatisfactorily. A major problem is not only financial resources but lack of organizational skills.

To overcome such constraints, strategies have to be carefully reviewed, particularly with regard to policies and institutional aspects, and reoriented towards sustainability and the stronger participation of the beneficiaries.

Policy areas requiring consideration include:

1. commodity marketing policies because of their effect on marketing infrastructure;
2. rural finance policies and investment in marketing infrastructure;
3. decentralization of government policymaking because of its impact on local initiative.

COMMODITY MARKETING POLICIES

In many African countries in the 1960s and 1970s, marketing policies followed a centralized approach. Parastatal commodity marketing boards purchased crops from farmers and sold them on local and export markets. Little encouragement was given to the private trade to participate actively in marketing. Pan-territorial pricing provided little incentive for farmers, cooperatives and private traders either to invest in or to maintain marketing infrastructure. Cooperatives were often promoted by governments from 'above' ignoring or excluding an alternative 'bottom-up' approach and resulting commonly in high overhead costs and non-viability. It was expected that central government would provide the physical marketing infrastructure. Experience gained during the past decade has shown, however, that central governments are unable to enlarge the marketing infrastructure and, in many instances, have not even maintained the existing infrastructure, especially the transport network.

Structural adjustment programmes in the 1980s led to changes in agricultural marketing and price policies, focusing on stronger participation of the private and cooperative trade, and a reduction in the role of parastatal marketing organizations. The intention was to encourage the private sector to take greater initiative in the rehabilitation and maintenance of marketing infrastructure. Stronger private participation requires a more flexible government agricultural price policy. Seasonal and spatial price differentiation on a sustained basis is a prerequisite to private investments in storage and transport. Government-fixed transport tariffs have to be relaxed to reflect better the wide variation in real transport costs. Grain marketing margins have to be adequate to cover costs, including capital costs of investment. It is crucial that these new policies be accelerated and maintained for a number of years so that the private trade gains confidence, and is thus encouraged to make the necessary investments (Reusse, 1987).

Particular attention has to be given to the development of indigenous African trade at import/export, wholesale, assembly and retail levels. It has to be technically supported through training and advisory services to become competitive with ethnic minorities. In the interest of equitable development, there may even be scope for temporary protective measures, e.g. licensing until the indigenous African traders become competitive.

Investment in marketing infrastructure by the private and rural community sectors also depends upon the price level and margins allowed by governments in the marketing of grain and other products. Many African governments have had a consumer bias, resulting in producer price levels that provided inadequate incentives for production (FAO, 1987). Price stabilization policies have an important bearing on investments in marketing infrastructure since they reduce risk, although it has to be recognized that stabilization can be achieved only within a rather broad range of price targets through bufferstock operations in arid land areas, as in the Sahel countries, where rainfall fluctuates significantly from year to year. Governments cannot afford to operate a large-scale bufferstock operation because of the costs involved. Government price stabilization policies are less costly to implement if full use is made of lower cost storage at farm and village levels.

FINANCIAL POLICIES

The lack of capital for investment in rural infrastructure is closely related to inadequate policies for developing rural financial markets. The rural development strategies of the 1960s and 1970s were based to a large extent upon cheap external financing by multi- and bilateral aid agencies. Mobilization of local savings, was neglected. It was much easier for the government to borrow external funds than to mobilize small amounts of savings from large numbers of people. This has contributed to indebtedness and the present financial crisis of the developing countries, particularly in Africa. It was also behind the lack of initiative in establishing viable rural banks in Africa to which small farmers, rural people, traders and cooperatives would have convenient access. A heavy top-down approach by governments, and intervention that kept interest rates low constituted further obstacles to the development of effective rural financial systems (Adams and Vogel, 1985).

There is now a growing awareness of the need to reorient rural financial market strategies towards institution building on a viable basis (Braverman and Guasch, 1986); this includes the establishment of viable village banks, the mobilization of personal savings, flexible interest rates, efforts to reduce financial transaction costs and developing sounder credit and investment policies. Considerable efforts are needed to promote rural banking networks in Africa through technical assistance and training of personnel. Much can be learned from experience in Asia and certain African countries (Burundi, Ghana, Rwanda, Kenya, Cameroon) where emphasis has been given to the establishment of viable rural banks.

With decentralization of finance policymaking and an emphasis on savings, there may be better discipline in accepting loans and repaying them. Greater emphasis should also be given to developing financial instruments to promote medium- and long-term investments in marketing infrastructure.

Future finance policies should facilitate access to financial services by cooperative and private rural entrepreneurs. They can play a role only if they are able to borrow money to invest in physical marketing infrastructure, such as storage, processing and transport facilities. Reforms of rural financial markets, with emphasis on the establishment of viable rural banks, is a precondition to promoting self-financing of marketing infrastructure.

DECENTRALIZATION OF GOVERNMENT SERVICES AND PROGRAMMES AND THE PROMOTION OF SELF-HELP PROJECTS

The development and maintenance of marketing infrastructure in rural areas have been based primarily on central government programmes. Central governments were expected to finance storage, marketing and processing, as well as road building. Participation of local communities, farmer and trader groups was often neglected. Aid agencies have often not considered local participation in the planning, implementation and maintenance of local infrastructure. This has resulted in expensive projects, ineffective implementation and lack of interest of the supposed beneficiaries in using and maintaining the marketing facilities.

Following a review of the institutional aspects of rural road projects (Cook *et al.*, 1985), the Bank proposed a new programme for Africa on transport development in which particular attention would be given to institutions and human resource developments.

Involvement of local organizations, agencies and groups can contribute to a reduction in costs because of their local knowledge and by encouraging local self-help and finance. Some Eastern Asian countries, e.g. Korea, have allocated only small amounts of funds for roads, mainly for building materials and some technical advice, while the beneficiary communities have undertaken the major work on their own. Under the Korean Rural Infrastructure Project, for example, many small roads were built successfully by villagers. These roads were economically viable because they responded to a felt need of farmers. Kenya succeeded in promoting village self-help projects, which included marketing facilities.

Another form of cooperation which has helped to build, rehabilitate or maintain rural roads has been through close cooperation between local community administrations, in particular market centres, the trading sector and the farmers. All three groups are usually interested in improving rural roads. Farmers can provide labour during their off-season, traders can provide transport and local communities can assist in mobilizing funds, for instance through charging a fee on produce sold at the local market. The Government Agricultural Marketing Department in Zaire developed an active rural road rehabilitation programme in close collaboration with the rural transporters and commodity traders in the first half of the eighties to ensure the timely evacuation of agricultural produce after the harvest period. In India, rural markets in the northern part of the country charged 2% on grain sales in the market to finance the building and maintenance of farmer-to-market roads. Rural market centres play an important part in Africa and could serve as focal points in mobilizing resources to improve marketing infrastructure. Also, NGOs working in rural areas can make an important contribution to the mobilization of human and financial resources for rural road building.

The promotion of arrangements to provide adequate incentives to mobilize local resources is important in extending rural road networks. The experience in the USA at the beginning of the 1800s, when many private turnpike companies were established, based on local resources may be relevant. Private initiative made a considerable contribution to the initial road building. Around the year 1800 there were about 72 turnpike companies registered in the states between New Hampshire and Virginia.

One approach to selecting sites for rural roads would be for them to be based on demand measured by the willingness of beneficiaries to contribute in labour for construction and maintenance. Low population density and fiscal constraints necessitate inexpensive approaches. Low-standard, all-weather roads can be built to be upgraded only when demand manifests itself through a minimum level of traffic. Such low-standard roads would encourage the greater use of non-motorized vehicles, e.g. bicycles, oxen carts, donkeys, for short distance exchanges.

The maintenance of infrastructure is a general concern throughout the developing world and is particularly acute for rural roads. It is estimated by the World Bank (1987) that one third of the value of road networks in Sub-Saharan Africa has been lost due to neglect of maintenance, representing a capital loss of $4–5 billion. Local communities must take major responsibility for rural road maintenance. Some countries (e.g. Gambia, Kenya, Burundi, Dominican Republic) have adopted a form of 'lengthman' system, where individuals under contract assume responsibility for the routine maintenance of short stretches of road adjacent to their agricultural lands. In some areas of Mali it is the practice for two villages to be collectively responsible for the part of the road closest to them. Other forms of local participation include maintenance by small-scale contractors who mobilize labour and meet logistic support requirements. Increased use of private contractors would provide competition for government agencies, which should be given greater autonomy and held accountable for their performance.

Storage facilities can be built at farm level, controlled by individual farmers and at village level controlled by farmer groups, e.g. cereal banks in Sahel countries (FAO, 1985), by farmer cooperatives or by private traders. In countries which have liberalized marketing and introduced competitive systems it is expected that those agencies providing the best services to their customers at competitive cost will be the most successful in the long run. The same refers to processing initiatives in rural areas. There is ample evidence to suggest that rural people will respond to opportunities for investing in storage and processing facilities if they are offered adequate incentives.

Most rural markets have developed over time and may be owned by communal, private or sometimes by cooperative agencies. Investment in infrastructure has to be kept low for low-cost marketing. The active participation of users is essential to plan, implement and maintain rural market centres. There are many successful examples in West and East Africa where private entrepreneurs or groups of farmers have built simple sheds with no, or minimal, outside help. There are examples where private entrepreneurs have financed and built food markets and sold market stalls to traders in a similar way as shopping centres are planned, financed and established in developed countries. There are also cases where rural market centres have been planned by central governments in collaboration with foreign aid agencies without the participation of the beneficiaries. These have often been too expensive and not used adequately or at all because the proposed users were not convinced of the benefits to be derived from them.

A major constraint on rural transport is scarcity of spare parts, tyres, new trucks, petrol and road building equipment. Here, governments have to give higher priority to the allocation of foreign exchange for the purchase of these goods. They may also establish a special fund at the central bank into which foreign aid donations could be paid. Foreign exchange auction systems can facilitate access to funds by rural investors.

MICRO-LEVEL ACTION

The policy and strategy outlined above for central government have to be supported at field level through technical advice and training. The major objective has to be the promotion of variable, self-sustaining marketing systems and institutions through the encouragement of self-help. This would mean, in many instances, a drastic change from past and present practices, which have focused more on building marketing infrastructure without taking the trouble to involve the beneficiaries in its planning, implementation and maintenance. Technical advice and guidance should in future be based on a careful analysis of projects which should aim at establishing sustainable market infrastructure with a maximum contribution of rural people.

Many village storage and processing projects have been carried out by individual farmers or marketing entrepreneurs. There is less knowledge on how group action can best be organized and promoted for building marketing infrastructure, in particular feeder road and market centre building and maintenance. More case histories analysing the factors affecting the success of group action in this field seem needed. One way is for trader and farmer associations, as well as contractors, to build market infrastructure with the maximum support of rural people. An important point is to ensure that each group is convinced of the benefits. As far as facilities, such as storage, processing and market centres, are concerned, it can be expected that the private and cooperative sector will take the major initiative, if incentives are not removed by government regulations or support for monopolies. National trader, cooperative and local authority associations can assist in the necessary training and technical support as in a number of middle income countries.

Particular attention has to be given to the promotion of viable arrangements at the grass roots level. Here, a great deal of experience has been gained in the promotion of small informal, self-help groups, which can develop into effective change agents if proper leadership can be developed. National non-government support institutions to promote self-help groups can play an important role. On the other hand, one has to be much more careful with the promotion of formal farmers' cooperatives, which will be viable only if they are fully supported by their members.

CONCLUSIONS

The rehabilitation of marketing infrastructure requires high priority in government strategies. It has to be supported by policy changes which encourage the decentralization of government services, a stronger participation of the private sector in the planning, financing and maintenance of marketing infrastructure, and the promotion of self-help programmes at grass roots level.

Institutional arrangements should be promoted at national, communal and grass roots levels to support the development of self-help and the mobilization of local resources.

Aid agencies can support the development of viable marketing infrastructure through self-help by adjusting their programmes and projects accordingly.

National technical support institutes, fully accountable and oriented to well-defined objectives, deserve particular attention by aid agencies.

REFERENCES

Adams, D.W. and Vogel, R.C. (1985) Rural financial markets in low-income countries: recent controversies and lessons. *World Development* 14(4), 477–487.
Ahmed, R. and Rustagi, M. (1987) Marketing and price incentives in African and Asian countries: a comparison. In: Elz, D. (ed.) *Agricultural Marketing Strategy and Pricing Policy.* World Bank, Washington, DC, pp. 104–118.
Braverman, A. and Guasch, J.L. (1986) Rural credit markets and institutions in developing countries, lessons for policy analysis from practice and modern theory. *World Development* 14(10/11), 1253–1267.
Cook, C.C., Beenhakker, H.L. and Hartwig, R.E. (1985) Institutional Considerations in Rural Roads Projects. World Bank Staff Working Paper No. 748, Washington, DC.
FAO (1985) *Manual on the Establishment, Operation and Management of Cereal Banks.* Rome.
FAO (1987) *Agricultural Price Policies, Issues and Proposals.* Rome.
Lipton, M. (1987) Agriculture and central physical grid infrastructure. In: Mellor, J.W., Delgado, Ch.L. and Blackie, M.J. (eds) *Accelerating Food Production in Sub-Saharan Africa.* Baltimore, pp. 210–226.
Reusse, E. (1987) Liberalization and agricultural marketing. *Food Policy* 4 Nov.
Roth, G. (1987) *The Private Provision of Public Services in Developing Countries.* EDI Series in Economic Development, Washington, DC.
World Bank (1987) *Road Deterioration in Developing Countries.* Washington, DC.

18

A MARKET-ORIENTED APPROACH TO POSTHARVEST MANAGEMENT

A.W. Shepherd

Many past interventions in the postharvest sector have failed because, whilst being technically correct, they have been planned without reference to the market's needs and its ability or willingness to pay for the supposed improvement. The aim of this paper is to emphasize the need to place postharvest activities, particularly, but not exclusively, loss prevention activities, within a market context, so providing pointers for planners and technologists active in the postharvest sector.

Emphasis by donors and technical-assistance agencies on the postharvest sector expanded as a response to the increases in production which resulted from the 'Green Revolution' experienced by India and other Asian countries from the 1960s. Production increases led to even greater increases in the marketed surplus, causing the supply of food to the marketing system often to exceed the capacity of the system to handle it. Problems with losses were initially perceived to be problems with marketed grain, not with the processing and storage of grain held for own-consumption. During the late 1960s and early 1970s postharvest specialists tended to concentrate on the quantification of losses and, as Bourne (1977) points out, there was often a temptation to cite 'worst case' figures to dramatize the problem. Extrapolations from limited samples to produce country-wide figures were used and may have exaggerated the true picture. This reinforced concern and, no doubt, the willingness of donors to support further activities. However, it seems that researchers rarely attempted to quantify the extent to which the losses could be economically avoided.

With hindsight, there may have been an over-investment in loss-assessment methodologies. However, the approach did gradually change to one in which greater attention was paid to studying the postharvest system and to identifying its problems and bottlenecks. Even then, however, improvements were often seen in terms of what was technically possible

---

Source: AGSM Occasional paper No. 5, FAO, Rome, 1991: condensed.

rather than what was economically justified. Losses could be reduced by building stores – so stores were built. Consideration of alternative approaches to the problem, which involved examining a range of possible solutions, was often not done.

PRODUCING FOR THE MARKET

Food losses stem both from poor postharvest handling and from overproduction. In order to avoid wasteful overproduction, postharvest loss reduction activities should begin even before the crop is planted. If there is limited consumer demand for a product (whether sold at the market price or at an official government price) then production should only be undertaken if the market has been clearly identified. This, of course, refers primarily to horticultural crops, although there have been cases of governments promoting excessive production of grains, beyond the ability of the marketing system to store and market them. On the occasions when total harvest and marketing costs are likely to exceed the market returns the best thing to do is plough the crop back into the ground or, where possible, use it for animal feed. This is a fact of life which has been faced by farmers for generations (other than those benefiting from generous subsidies!) but is, nevertheless, difficult for both farmers and governments to accept. In these, fortunately infrequent, circumstances a food loss may be preferable to a financial loss.

Consideration of location can be important in extending produce availability and reducing gluts. A country with diverse climates, such as Colombia, has sufficient variation to ensure potato harvest throughout the year. Thailand can grow onions in three different areas within an altitude range of 800m. This provides three different harvest times which, when combined with simple types of postharvest storage, permits onion availability for nine months of the year.

ECONOMICALLY VIABLE LOSS REDUCTION

The application of known technology and infrastructure could, theoretically, reduce losses in the post harvest system to practically nothing. Unfortunately, it does not follow that such technology should be applied or that such infrastructure should be built. In certain cases there may be social benefit (e.g. food security) which can be used to justify uneconomic postharvest interventions. However, the guiding principle of all loss-reduction activities should really be that the assumed benefits through reduced losses, higher quality or higher prices must exceed the costs of the proposed improvements by a factor sufficient to justify the risk. As actors in the postharvest system, whether farmers or traders, are usually economically rational, any attempt to maximize loss reduction without reference to economic criteria will be doomed to failure.

Calculation of losses, and hence loss reduction potential, is fraught with difficulties. Many figures for losses are estimates rather than actual measurements and are often based on extrapolations from small samples. With grains, a survey of on-farm or trader stores immediately before the next harvest may well show high levels of infestation. However, as infestation builds up over a season, the *average* levels are much lower (Bourne, 1977). Grain drying improvements, if not accompanied by improved on-farm storage, may result in more grain reaching the market soon after harvest, so depressing prices (Cardino, 1982). Higher returns which may result from longer grain storage durations must be set against the cost both of the store and of the capital tied up in stocks, as well as the cost of possible quality losses. Even if calculations show a positive return, certain other aspects need to be taken into account. A farmer, for example, may have immediate cash needs such as school fees and government taxes, and may not be able to consider long-term storage. A trader may be unable to raise sufficient finance to tie up part of his operating capital in stocks. To address some of these problems, several Asian countries have recently introduced variations of a 'Paddy Pledging Scheme' which permit farmers and traders alike access to credit, with security provided by crops deposited in bonded warehouses.

An analysis of seasonal price patterns should be used to identify the advantages of medium to long-term storage in terms of higher prices. However, the impact on seasonal price variations if a large number of farmers or traders start storing for longer periods should be considered.

In theory, a portharvest improvement can be introduced, regardless of cost, as long as that cost can be recovered from the market and no other, more cost-effective, solution is available. Where possible, improvements should be relatively simple and low cost, ensuring that farmers, rather than technology suppliers, receive a significant proportion of the consumer's expenditure. Simple improvements to make existing structures proof against rats and vermin and to facilitate the application of insecticides may be more appropriate than the construction of new, more sophisticated stores.

INCENTIVES

A particular problem facing those trying to improve postharvest handling by small farmers is that those farmers often see no correlation between improved handling and market returns. An individual, small farmer practising improved techniques will receive no benefit if his produce is going to be bulked up into a larger consignment with the crops of others, particularly if the trader is not applying any form of quality control. In some countries an approach has been adopted which involves promoting downstream marketing ventures by small farmers. If, for example, a farmer retains his interest in a crop during processing (e.g. rice milling), both the relationship between handling and quality and the potential financial benefits become clearer.

Benefits of improved postharvest management must be capable of being demonstrated to those who are actually doing the marketing. For horticultural produce, the quality benefits of improved postharvest handling techniques often only show up at the retail stage, or when the consumer gets the produce home. If consumers suddenly find they can keep fruits for 3-4 days instead of the previous 1-2 days they may, over time, be prepared to pay more for those fruits. However, this price response will not be immediate and, meanwhile, it is unlikely that the marketing system will reward farmers and traders for improved handling. Again, improvements which are introduced at the top end of a consumer market are more likely to achieve success in the short term and, thus, demonstrable benefits to the marketing system.

Many modern postharvest techniques for horticultural produce are expensive, requiring a high initial investment, often in imported equipment. They also require highly trained staff and managers and immediate access to spare parts and skilled technicians. Thus, while there is a clear need to develop improved technologies such technologies should not, as a general rule, be significantly more advanced than the general level of technology in a society. As an example, cool chains require specialized refrigerated stores close to the production areas to remove the crops' field heat, as well as refrigerated vehicles. Produce, once stored in refrigerated containers, should then be refrigerated all the way to, and in, the retail shop. The cost can usually only be justified when an integrated chain is established, when there is a highly developed infrastructure (good roads, reliable electricity), when there is a skilled workforce and, most importantly, when there are consumers prepared to pay a high price.

There may be occasions when the social benefits from postharvest improvements may outweigh purely economic calculations. At the subsistence level, for example, standard cost/benefit analyses may not always be applicable. A family's response to high food losses may be to eat less; on the same basis reduced losses may lead to increased consumption, with positive nutritional benefits, rather than increased sales to the market which produce economic returns. At the national level improved food storage may offset precarious food supplies by reducing losses. However, the foreign exchange cost of building the storage facilities can often exceed the foreign exchange savings on food supplies which no longer have to be imported. Such calculations take no account of the lead time required to import food but they do illustrate the need to consider carefully the relative options before decisions to invest in expensive postharvest infrastructure are taken.

Storage of all crops needs to be considered in the light of possible alternatives. For example, an FAO consultant who visited the former South Yemen to advise on long-term onion storage concluded that storage was not the problem. The problem stemmed from the government's practice of offering farmers the same price for onions throughout the year, thus providing no incentive for off-season production. The pricing policy was reviewed and farmers started to produce onions year round, significantly reducing the need for storage.

With increasing liberalization in the marketing of basic foodgrains, particularly in Africa, many of the larger stores constructed by or for government marketing parastatals are now inappropriate. While some of these will be used to house food security reserves, others may, at least temporarily, go unused. Most traders are presently relatively small-scale and are unable to utilize efficiently large stores or silos built for marketing board operations. Few traders have the capital to undertake long-term storage or to contemplate the necessary investment in good-quality storage facilities. The implications of grain marketing liberalization for food loss prevention are therefore that there may need to be an increased emphasis on farm-level storage, to enable farmers to take advantage of seasonal price changes, together with increased attention to the needs of traders for small-scale stores.

On-farm storage can play an important role in national food security but government policies often discourage this. The application of pan-seasonal prices for grains, i.e. the same prices throughout the season, runs counter to market principles and encourages farmers to sell all their crop at harvest rather than invest in improved storage and drying techniques and wait for later price rises. There is also no incentive for traders to store. Having to buy a high proportion of the crop within a short period places great pressure on marketing boards and may be seen as a factor contributing to high losses. Pan-territorial prices, i.e. the same price throughout a country, introduced for egalitarian reasons, encourage production in remote areas. This can increase marketing costs and lead to losses because of the difficulty of evacuating large quantities from such areas.

PRODUCE STANDARDS

Except for grain drying, efficient postharvest handling cannot usually compensate for poor initial produce quality. The control of produce quality before it enters the marketing system is therefore vital. One way to encourage farmers to improve the quality of their production is through the enforcement of buying standards. Legally enforceable purchase standards can normally only be imposed by state buying agencies. Most such agencies operate standards but experience great difficulty in enforcing them at the buying-depot level. Lack of trained staff and a shortage of grading equipment often combine with an emphasis on maximizing quantity purchased and a lack of any incentive for depot managers to be too fussy over the qualities they accept, to defeat the best-drafted standards.

With the gradual reduction in the role of grain marketing boards and the increased role for private traders, it becomes increasingly unfeasible for official controls to be implemented at the point of first sale. However, in countries where grain trading has been in the hands of the private sector for many years, traders have often developed extremely sophisticated unofficial standards which govern the prices they are prepared to pay to farmers. Postharvest improvement activities should therefore aim to create an awareness amongst traders of the consequences of buying poor initial quality produce and of the need to impose more rigorous buying standards.

CONCLUSIONS

Planning for improved postharvest management and loss reduction necessitates a full awareness of and willingness to research the food system. Factors such as demand, the role of marketing agents, and their profit orientation are important elements to be considered in any food system analysis.

Postharvest improvements commence at the pre-production stage. Available market information can be used to plan which crops and varieties to grow, when to grow, when to harvest and in what quantities. Farmer support in the form of information and marketing extension services is therefore vital. Various techniques to expand seasons and hence reduce seasonal 'gluts' can be employed. Microclimates can be used effectively to provide almost year-round availability of some crops, although care must be taken to ensure that production in remote areas is supported by appropriate marketing services.

Farmers and others operating in the marketing/postharvest system are unlikely to accept new postharvest techniques unless the benefits can be shown to exceed the costs by a factor sufficient to justify the risk involved. Optimistic assessments of potential returns are to be avoided if sustainable development is to be achieved. An awareness of the impact of postharvest changes at one stage of the marketing chain on the efficient operations of other stages is important. Consumer requirements and the ability of the market to pay for improved quality must be borne in mind at all times. Simple improvements to the post harvest chain are often more cost-effective than sophisticated technologies.

REFERENCES

Bourne, M.C. (1977) *Post Harvest Food Losses – The Neglected Dimension in Increasing the World Food Supply.* Department of Food Science and Technology, Cornell University.

Cardino, A. (1982) Market needs for grain drying in the Philippines. In: Young, R.H. and MacCormac, C.W. (eds) *Market Research Needs for Food Products and Processes in Developing Countries.* International Development Research Centre, Ottawa.

IV
INSTITUTIONS AND POLICIES FOR MARKETING

There has been a continuing flow of economic policy books on marketing for governments of developing countries. Those of the 1960s favoured rapid industrialization with agriculture taxed through product prices to finance it. The 1980s saw a swing back to agriculture with the preoccupation of how to provide incentives for food production and exports.

M. Lipton propagated urban bias ('Urban bias and food policy in poor countries'. *Food Policy*, Nov. 1975) and 'urban bias revisited'. For a brief period, agricultural price policies acquired the status of burning issue of the day. Representative texts are C.P. Timmer's *Getting Prices Right: the Scope and Limits of Agricultural Price Policy* (Cornell University Press, 1986) and G.S. Tolley's *Agricultural Price Policies and the Developing Countries* (Johns Hopkins University Press, Baltimore, 1983). D. Ghai and L. Smith saw marketing deficiencies as the obstacle to production growth more than prices in their *Agricultural Prices, Policy and Equity in Sub-Saharan Africa* (Lynne Reiner, Boulder, 1987).

With a score of African countries accepting the IMF/World Bank call for structural adjustment and a greater role for private enterprise, research on its progress and impact became an engrossing interest. Papers presented at a World Bank symposium were brought together in D. Elz's (ed.) *Agricultural Marketing Strategy and Pricing Policy* (World Bank, 1987).ONCAD in Senegal was the first major marketing board to be abolished. The paper by Newman, Sow and Ndoye summarized here analyses the consequences. At the time of writing all the authors were at the Institut Sénégalais de Recherches Agricoles Dakar. J.M. Staatz, J. Dioné and N.N. Dembelé broaden their approach in 'Cereals market liberalization in Mali', *World Development* 17(5), 703–718, 1989. The *Report of the Nordic Workshop on Peasant Agricultural Marketing in Eastern Africa* (Swedish University of Agricultural Sciences, Uppsala, 1992) contains valuable analyses and conclusions.

J. Coulter and J.A.F. Compton summarize implementation issues in the liberalization of cereals marketing in Sub-Saharan Africa in the first volume of the Natural Resources Institute, Chatham UK, marketing series.

How far stabilization of the prices they receive is important to farmers and the extent to which governments should go in managing this has been a continuing theme of discussion. Knudsen and Nash's views presented here reflect the thinking of the early 1990s.

Structural adjustment has revived interest in government support services for marketing. Instead of intervening directly, setting prices, financing parastatals to buy, hold stocks and sell, a government should be setting ground rules for marketing operations by enterprises acting independently. It should be promoting competition between them by ensuring free entry of new firms and easy access to finance, information and training.

Writings on government support services for marketing include A.A. Schmid's *Legal Foundation of the Market: Implications for Formerly Socialist Countries of Eastern Europe and Africa* (Michigan State University staff paper, 1991) and the FAO papers summarized below. Recognition that farmers must take much more initiative in marketing and that small private traders have a role and should be helped led to interest in marketing extension. Often this had been left to cooperatives and the commodity boards, as noted in the World Bank's *Agricultural Extension in Africa* (1989), edited by N. Roberts. FAO's *Horticultural Marketing: A Resource and Training Manual for Extension Officers* came out in 1989. For the lead set by the Government of India in support to marketing, see S.C. Varma's *Agricultural Marketing in India* (Directorate of Marketing and Inspection, Ministry of Rural Reconstruction, Delhi, 1981). The sustainability issue – how to retain competent personnel for support services when official salaries are very low – is addressed in Abbott's paper in Part 1 (Chapter 6).

19

REGULATORY UNCERTAINTY, GOVERNMENT OBJECTIVES AND GRAIN MARKET ORGANIZATION AND PERFORMANCE: SENEGAL

M. Newman, P.A. Sow and O. Ndoye

GOVERNMENT OBJECTIVES

The Senegalese government announced a New Agricultural Policy (NPA) in April 1984. The NPA focuses on promotion of local grain production. This has been a recurrent theme in policy documents since the mid-1970s, including Senegal's Food Investment Strategy, and the Sixth Development Plan. The NPA aim of increased food self-reliance and gradual substitution of locally produced for imported grain focuses on incentives for production and consumption of local grain. It also focuses on the marketing system through which local cereals are exchanged.

'Assuring' market outlets for producers, assuring supplies to consumers and achieving more cost-effective pursuit of government goals through a gradual transfer of responsibilities to private sector intermediaries and producer organizations are all important elements of NPA objectives.

The incentives for marketing system participants, especially producers and intermediaries to supply locally produced and imported cereals to consumers are influenced by government regulatory policies *vis à vis* the marketing system. To evaluate such policies, for example options and potential impacts of specific rules and roles for public, parastatal, private and cooperative sectors, one must first understand the marketing systems for imported and local grain. The structural adjustment packages being adopted in many African economies emphasize increased reliance on the private sector to carry out agricultural marketing tasks previously handled by public and parastatal agencies. However, the empirical data upon which to base choices and strategies is often lacking (Mackintosh, 1985). The

---

Source: Paper presented at the XIX Conference of the International Association of Agricultural Economists, Malaga, 1985: condensed.

discussion that follows examines the rules and reality of the grain marketing system of Senegal. The discussion is based on surveys of market intermediaries conducted since 1983 in three regions of Senegal, but primarily in the central peanut basin.

PRIVATE AND PUBLIC CHANNELS

Senegal has a long tradition of state and parastatal involvement in the marketing of agricultural products. At the same time private traders have been important in the distribution of imported rice and sometimes the parallel assembly of grains, oilseeds and other agricultural products. In 1980, faced with high costs, considerable unreimbursed credit and a variety of other factors, ONCAD, the *de facto* national grain and oilseeds marketing board was abolished.

Since that time Senegal could be viewed as undergoing an experiment involving development of a legal private grain trade (except for paddy rice) and occasional forays into grain markets by a public agency, the Food Security Commissariat (CSA). Rice, sorghum and wheat imports are controlled by another public agency, the Price Equalization and Stabilization Board (CPSP).

An understanding of the process by which Senegal has remained extremely reliant on imported grain while local food production has been outstripped by population growth requires examination of both the rules governing the grain trade and the actual organization and performance of the marketing system.

THE RULES OF THE GAME

Regulations and their enforcement have a major impact on incentives for a marketing system to function in accordance with government policy objectives. Senegal's experience in this area provides a useful case study with broader implications.

The private sector

The rules governing assembly, transportation and storage of locally produced grain – millet, sorghum, corn and rice – have varied considerably since ONCAD was abolished. As in many countries where state or parastatal agencies have traditionally been involved in marketing of cash crops, the first stage in an attempt to 'assure' markets for food crops and provide incentives for their production is to specify who may participate in what marketing functions, when marketing transactions may take place and under what conditions, prices, etc.

At present, merchants who assemble millet, sorghum and corn must be licensed wholesalers. (In the case of paddy rice, assembly is still considered a government monopoly, so no licensing is provided for.) To be licensed, wholesalers are required to present evidence of a 3 million FCFA bank balance and certified storage facilities. The wholesaler must also certify that he keeps regular accounting records. In general, in addition to being licensed as traders it is necessary to be authorized (agréé) to handle a specific commodity on an annual basis. This requires certification of a 5 million FCFA bank balance (US$1 = 430 FCFA). The list of licensed traders authorized to handle millet has often been announced even later than the announcement of categories of legal participants in the grain trade, or not at all. During the 1983/84 and 1984/85 marketing seasons, specific authorization by commodity was not required, apparently, because the crops were considered by Ministry of Commerce officials to be too small.

The administrative regulations (décrêts) specifying that any licensed wholesale merchant could participate in the grain trade were signed in January 1984 for 1983/84 and December 1984 for 1984/85, while harvested grain began to move into markets in September–October of both 1983 and 1984. Similar situations have occurred in four out of the last six years (see Table 19.1). This situation obviously introduces considerable uncertainty for both private traders and producers. In the absence of annual regulations specifically permitting them to participate in the grain trade, traders must either subject themselves to risks of fines or seizure of commodities by conducting their business in quasi-legality or shift their activities to other sectors, leaving farmers without market outlets and themselves with foregone opportunities for profits.

Private wholesalers play a major role in the distribution of rice and sorghum imported by the CPSP. Quotas are assigned by a committee headed by the Ministry of Commerce and prices at wholesale to retail levels are more strictly regulated.

The public sector, parastatals and cooperatives

The rules governing the role of public and parastatal entities and cooperatives in the marketing system have also been variable over time and sometimes imprecise. When ONCAD was abolished, the public sector role in the assembly of local millet, sorghum and corn was passed to the Food Security Commissariat (CSA) which until 1984 was called the Food Aid Commissariat (CAA), due to its role as an aid distribution agency.

The public sector role in the assembly of locally produced hulled rice

Table 19.1. Selected official regulations of millet/sorghum marketing: 1979/80–1984/85

Event	1979/80	1980/81	1981/82	1982/83	1983/84	1984/85
Official opening dates of marketing season	Nov 19, 1979	Nov 19, 1980	Oct 1, 1981	Nov 15, 1982	Nov 2, 1983	Oct 15, 1984
Announcement of official prices	Nov 19, 1979	Nov 11, 1980	Oct 1, 1981	Dec 8, 1982	Nov 7, 1983	Oct 8, 1984
Signature of annual regulations specifying participants	Jan 4, 1980	Nov 11, 1980	Oct 2, 1981	Dec 8, 1982	Jan 23, 1983	Dec 21, 1984
Authorized purchasers/first handlers – producer level	Authorized licensed wholesalers (agréés)	Producer cooperatives	Producer[a] cooperatives	Producer cooperatives CAA Authorized licensed wholesalers	Licensed wholesalers Producer cooperatives	Licensed wholesalers CSA RDAS Producer cooperatives
Authorized purchasers from first handlers above	Authorized licensed wholesalers	CAA Processing industries Authorized licensed wholesalers	CAA CPSP Authorized licensed wholesalers	CAA Authorized licensed wholesalers	Licensed wholesalers CSA eventually	CSA Licensed wholesalers Rural development agencies

[a] In March, 1982, another decree was signed stating that, besides producer cooperatives, the CPSP, the CAS and authorized licensed wholesalers were allowed to purchase millet from producers.

was given to the Price Stabilization Board (CPSP) which is also responsible for commercial imports of rice and sorghum, and licensing of wheat imports. ONCAD's role *vis à vis* peanut assembly has gradually been shifted to the parastatal oil crushing firms. Rural development agencies have been given a role in corn assembly and an official monopoly on producer level assembly of paddy rice.

Cooperatives in Senegal have traditionally been state directed organizations used to distribute production inputs, assemble cash crops and recover debts. A 1983 reorganization of the cooperative system is aimed at broadening the scope of cooperative activities and increasing the degree of producer autonomy in their management. In grain marketing, a portion of the government purchasing target through official channels was set aside for cooperatives during the 1984/85 crop year: A preferential official price was also established to encourage cooperatives to use their own capital to assemble grain. Survey results indicate that this met with little success due to the overall level of official prices, timing of funds, availability and other factors discussed below.

During the current marketing year, uncertainty has also influenced the role of public agencies in the assembly of grain. On October 8, 1984, the Ministerial Council announced that the official marketing season would begin on October 15. The official regulation (decree) specifying that the Food Security Commissariat (CSA) could purchase grain at the producer level was not signed until December 21, 1984. Thus, the legal basis for direct producer-level purchases by the CSA was unclear (since for 1983/84 the CSA was to purchase from private traders) and the legality of producer-level purchases by private traders was also unclear. Nonetheless, both groups proceeded to assemble grain, though at prices that often differed considerably.

Prices

Prices are another area where regulatory uncertainty exists. As in many other countries, Senegal establishes 'official' prices at the producer, wholesale and retail level for locally produced grain, in this case millet, sorghum and corn. Producer prices for paddy rice and official wholesale and retail prices for imported cereals are also set. Official prices of locally produced grain do not vary over space or time, except for a 1 FCFA larger margin permitted in Dakar. Imported rice prices vary only by a subsidized differential for transportation from Dakar. This provides an obvious disincentive for private traders to move grain from surplus to deficit areas or to store grain, if official prices are respected.

For imported grain, government enforcement efforts make it clear that official prices are intended to fix prices to consumers and margins permitted to intermediaries. With locally produced grain, whether producer prices are intended as a floor or a fixed price is not clear. During 1983/84, the official price was treated as a floor price, and when market prices were above official prices, the CSA simply withdrew from the market (CSA officials report that this was in part because funds were not available). During the 1984/85 season, a number of cases of seizures by administrative authorities of grain traded at higher than official prices were reported during the early months of the 'official' marketing season. This calls into question what government wants 'official prices' to actually mean.

According to survey data collected in 36 rural markets, mean producer level prices for millet paid by private traders in the peanut basin stayed above the official producer price (55 francs CFA/kg until October 1984 and 60 francs CFA/kg after that date) for the entire July 1984 – June 1985 period. The lowest observed price offered fell below the official price during the harvest period only in the areas of highest production.

Thus, while the official price may have functioned as a floor, it obviously did not act as a fixed price at which all cereals were exchanged. In terms of presenting incentives to producers, this presents no problem. But imprecision as to the legal objective in setting official prices has resulted in other undesirable effects in addition to the seizures mentioned above. Foremost among these is that since legally permissible margins are established on the basis of official prices, private wholesalers often maintain accounting systems based on official prices at wholesale and retail levels even though no trades occur at those prices. This lets them appear to be in compliance with regulations and also trade grain. At best, this defeats the whole purpose of requiring or maintaining an accounting system. It also points out the importance of understanding participants in the marketing system and the functions they perform in order to evaluate policy options.

HOW THE MARKET REALLY WORKS

The private sector grain trade in Senegal's peanut basin begins with a system of assemblers who purchase small (less than 50 kg) and larger 50–100 kg) quantities at the village and periodic market level. Bagged grain (80–130 kg) is generally assembled by licensed and unlicensed wholesalers in periodic markets and handled in break bulk to move it to major regional centres where it is consumed, stored or redistributed to grain deficit areas. Some bulking of grain also takes place in villages.

Survey data indicate that despite the climate of regulatory uncertainty, the private grain trade is very active. The 1400 market intermediaries identified in 40 of the most important assembly markets can be divided into two general categories – small assemblers and wholesalers.

Small assemblers

1. *Day traders* who, with an extremely limited capital case (often 10000 FCFA or less) purchase small quantities of grain (3–5 kg) at a time. Often these traders resell one sack of grain before beginning to assemble another. During the period immediately following the harvest, volumes of 300–400 kg per day are not uncommon in areas where production is greatest. Quantities collected are generally resold to wholesalers before the end of the day. Margins of 2.5–5 FCFA per kg purchased are sometimes increased by imprecise weighing techniques.

2. *Commission agents,* who assemble grain with money advanced by or borrowed from larger traders, generally wholesalers. Their operating procedures are similar to those of day traders, but remuneration is on a flat daily fee or per sack basis, rather than a buy sell margin.

3. *Food deficit producers and non-producers,* who assemble grain in order to store for their own consumption later in the year.

Preliminary results of surveys of merchant operating procedures indicate that 39% of assemblers surveyed store grain. This is generally for short periods necessary for stock turnover, except for those who store for home consumption and a few who speculate on interseasonal price variation. Transportation of grain was performed by 35% of those assemblers surveyed. Though precise results on distances transported are not yet available, they appear relatively short.

Financing of transactions by small assemblers is largely from personal funds or a mixture of personal funds and those from other merchants and relatives. Of the 25% of small assemblers receiving funds from other merchants, one-third were relatives of the merchants. Use of bank credit for direct financing of small assemblers is negligible, although some bank credit to wholesalers may eventually reach the small assembler level.

Wholesalers

Wholesalers include both licensed and unlicensed traders handling a variety of mixes of locally produced and imported grain and highly variable volumes. Preliminary results of multiple visit surveys of 63 major wholesalers in the peanut basin provide interesting insights into private sector commercial activities in the grain trade.

Of the wholesaler sample, 58% were licensed, e.g. they had a 'carte de grossiste'. Of the licensed wholesalers, 72% also had a quota for selling imported rice. These constituted 42% of the total sample. Preliminary indications are that wholesalers located in zones of greatest production specialize in locally produced grain, while wholesalers based in deficit regions are more likely to handle both local and imported grains. Wholesalers handling imported grains are most likely to be licensed. Unlicensed wholesalers sampled handled approximately 30% of the total millet and imported rice by volume. At the first handler level the percentage of millet handled by unlicensed traders is much higher.

The total estimated millet volume handled by our wholesaler sample during the six months Sept 1984 – March 1985 is 9337 tonnes, valued at 722 million FCFA. The rice volume estimated for the sample during the same period was 4686 tonnes. This compares with total national millet volume handled by the CSA of 1752 tonnes nationally including 981 tonnes in the peanut basin, and 29 758 tonnes of imported rice distributed to quota holders in the peanut basin by the CPSP during the period. (This latter figure under-estimates total rice volume in the peanut basin since some of the 95 000 tonnes distributed in the Cap-Vert made their way into the the peanut basin.)

The sample of wholesalers studied handled more than nine times the volume of millet flowing through official channels in the peanut basin, and the sample probably represented 10% of the total millet marketed nationally.

Most wholesalers (82%) attempt to turn grain over within one month of purchase, despite the availability of storage facilities with excess capacity. One apparently standard procedure has been to turn grain over rapidly soon after harvest, when volumes are high and then to store beginning 5 or 6 months after harvest when volumes are lower and the 'hungry season' or 'soudure' approaches. Some traders indicate, however, that uncertainty as to potential foreign aid distributions and their impacts on price increase the risk of storing for speculative gains during the 'hungry season'.

Rapid product turnover by wholesalers is in part a function of capital rationing, the high cost of borrowed capital from unofficial sources, and limited access to bank credit.

FINANCING

Survey results indicate that wholesalers are even more reliant on their own funds to finance their operations than are small assemblers. Eight percent of wholesalers get at least some funding from relatives, 10% get some funds from other merchants and 6% report some bank financing. Twenty-eight percent of wholesalers reported using some borrowed capital in their operations. Senegal is a predominantly moslem country, so the issue of payment of interest is extremely sensitive. In response to a question about the amount which a merchant would have to repay if he borrowed 100 000 francs CFA for 1 month, interest rates ranged from 0 to 25% per month, with a mean of 7.2% per month. This compares to an official bank prime interest rate of about 15% per annum, equivalent to 1.25% per month.

While an interest rate of 7.2% per month seems extremely high, preliminary analysis of data on other aspects of financing indicate that reimbursement rates are extremely variable, but often quite low. Wholesalers reported that the percentage of the number of loans made since harvest that had been reimbursed was in the 5–100% range, with a mean of 62%. This implies that while interest rates from private credit sources are quite high, given reimbursement problems, they may not necessarily lead to extremely high returns, when the need to recuperate funds lost through bad debts is considered.

Apparently Senegal's current economic recession places wholesalers in a situation where they cannot move grain or other products without providing credit, yet when they provide credit they have reimbursement problems. Survey results indicate that wholesalers are quite active in providing loans to other merchants, retailers, farmers and consumers. At the same time, wholesalers indicated that they have little recourse in the short run if loans are not reimbursed.

Overall the cost of capital with which to do business and the opportunity cost of capital invested in the business are very important factors influencing the cost of marketing grain and the choices as to where wholesalers invest their money. If the cost of capital is accurately reflected by the mean figure reported above, financing the purchase of millet at the producer level at the official price costs 4.3 FCFA/kg per month, 36% of the officially prescribed margin. At the prime bank interest rate, the cost of financing is 0.8 FCFA/kg per month, or only 7% of the officially specified margin. Thus the cost of capital can be a very important factor in determining marketing margins necessary for the wholesaler to cover his or her costs. It is important to note that the above calculations include no allowances for the cost of capital invested in weighing equipment, storage facilities, transportation equipment or office/sales space.

If liquidity is a constraint, regulatory policy may inadvertently contribute to it. One of the requirements for licensing as a wholesaler (carte de grossiste) is that the applicant on a one time basis provides certification of a bank balance of 3 million FCFA. The above discussion indicates that significant financial capacity is necessary in order to function as a wholesaler. None the less, the certification process does not serve as a guarantee that the wholesaler will be able to meet his financial obligations. It does, however, serve as a barrier to entry into the grain trade. Regulations for the 1984/85 marketing season require that merchants making farm level purchases be licensed wholesalers, necessitating the 3 million FCFA attestation even though many small assemblers purchase daily quantities of 100 kg or less, with a maximum retail value at official prices of 7300 FCFA.

Transport

A similar regulatory constraint is imposed on movement of grain between surplus and deficit areas. For transportation of 200 kg or more of local grain across regional boundaries, a wholesaler's or retailer's license is required, even though the value at official prices is 12 000–14 600 FCFA. Such barriers, if enforced effectively, would further limit the available liquidity in the system and potential market outlets for farmers.

Government regulates transportation rates by setting a standard or 'barême' that government agencies pay for transportation. Surveys of actual costs indicate that the barême is perceived by wholesalers as relatively unimportant. Wholesaler data indicate that for hauls less than 100 km, transportation rates are considerably higher than the officially prescribed rates, but above 100 km actual costs per kilo tonne are lower than official rates. This can be in part explained by the fixed costs involved in immobilizing a vehicle for loading or unloading. None the less, the establishment of official rate regulations fails to take into account the economies of volume and distance that market forces are apparently incorporating.

Regulation

In evaluating the choice of tasks in the marketing process to confer on the private sector and those to retain for itself, Senegal and other governments seek to assure that marketing functions are performed efficiently or cost effectively without significantly compromising government objectives *vis à vis* producers, consumers and the balance of payments. If marketing operations are performed more efficiently, it may be possible to achieve higher returns to producers and lower costs to consumers. At the same time, the state may be unwilling to count on the unregulated private sector to share returns resulting from economies in operations with producers and consumers. For this, and other reasons, the state maintains a regulatory system and watches over the marketing system.

As noted above, frequent changes in regulations and relatively late announcements of the 'rules of the game' contribute to uncertainty in the grain marketing system. Overall, government regulation takes the broad form of specifying who may participate in grain marketing, and under what conditions, when transactions may take place, prices at which transactions may take place, etc. Regulations specify quality in broad terms, but no specific grades for local grains are used in local markets.

Government regulation also takes the form of checks of wholesaler licenses, permits for moving grain or other products when required, accuracy of weighing equipment, pesticides used, etc. Compliance with regulations constitutes an important transactions cost for wholesalers in terms of time involved in finding out what regulations are and being checked for compliance in addition to the costs of permits, licenses, and potential fines. Working out 'understandings' with regulatory officials is also a cost.

Survey results indicate that the radio and other merchants are the most important sources of wholesaler information on what the rules actually are. Regulatory enforcement officials, local chambers of commerce and newspapers were also listed as important information sources by some wholesalers.

The wholesalers indicated that they have considerable contact with regulatory enforcement officials, being checked on average of 4.2 times per month by a combination of economic control officials, customs agents, state and local police officials, natural resource officials and others.

There was not much difference between licensed and unlicensed merchants in terms of payment of 'fines' to regulatory compliance officials. 54% of the wholesalers reporting paying 'fines' during the 4–6 months after the 1984/85 harvest were licensed. In 35% of the cases reported, receipts were received for payments of 'fines'. In 65% of the cases an 'arrangement' was worked out. The total amount reportedly paid by 60 of the wholesalers interviewed intensively was 460000 FCFA, a mean per 'fine' or arrangement of about 5000 FCFA. Further analysis of the data will permit a more precise evaluation of the reasons for fines and other costs of regulatory compliance.

Regulatory uncertainty contributes to a climate conducive to ignoring regulations and/or corruption of enforcement officials in order to keep the system running. The fines reported above represent only 0.25% of the official margin, or 0.03 FCFA per kg of grain handled. None the less, this does not take into account the time spent on regulatory compliance by traders, and the marketing activities that do not take place because of fear of regulatory enforcement. For the state, regulatory enforcement represents a cost, and fines, if they enter the government treasury, help to offset the costs of enforcement.

In the final analysis, the key question must be whether the regulatory enforcement process is supportive of the government objectives that are its *raison d'être*. Some individuals are obviously able to work out 'arrangements' and conduct business as they wish. At the same time, working out such 'arrangements' may constitute a barrier to entry for traders who wish to begin or expand their activities.

REFERENCE

Mackintosh, M. (1985) Economic tactics in commercial policy and socialization of African agriculture. *World Development* **13**(1), 77–96.

20

DOMESTIC PRICE STABILIZATION SCHEMES IN DEVELOPING COUNTRIES

O. Knudsen and J. Nash

Preoccupation with the presumed adverse effects of price instability has led developing countries to establish a wide variety of mechanisms to stabilize domestic prices, not only of internationally traded goods but of nontraded goods as well. However, despite the ubiquity of such schemes, and the obvious importance with which price stabilization is regarded, the development literature is virtually mute on their characteristics and how they have performed.

In this article, we discuss the topic of domestic price stabilization programmes in a comparative cross-country context. We do not examine here the question of whether instability has adverse or positive economic effects, as a burgeoning literature continues to debate this issue. Rather, we take as given that stabilization is a desideratum of virtually all governments in the developing world. Our goals are to describe some of the characteristics of the different schemes used to stabilize prices, to examine how they have performed in meeting their objectives, and to consider some of the factors that determine their cost-effectiveness. Through such an examination, lessons can be learned that are of value in evaluating and designing such schemes.

THE MECHANISMS USED FOR PRICE STABILIZATION

Almost all developing countries have some kind of domestic price stabilization scheme. However, the mechanisms used vary widely and cannot be easily distinguished from other schemes that have other purposes, for instance, taxation of producers or consumers. One useful way of taxonomizing stabilization schemes is to divide them into two categories – those that physically handle stocks of the commodity and those that do not. In the former category are included buffer stocks and marketing boards, and in the latter are various kinds of trade taxes and restrictions.

In buffer stock schemes, the government agency responsible for price stabilization (frequently a nominally independent corporation with majority government ownership, a 'parastatal') imports commodities or procures domestic production at set prices, holds stocks, and distributes the stocks at trigger prices to the wholesale markets or at fixed prices through shops that are either government-owned or -regulated. The stocks are almost invariably basic consumption items – rice in the Philippines, South Korea, and Bangladesh; wheat and rice in India; corn and wheat in Mexico. Schemes that rely primarily on domestic procurement are common in South Asian and some Latin American countries. In other countries – for example, Indonesia – buffer stocks are maintained primarily through imports supplemented by domestic procurement. In all buffer stock schemes, government control of trade is an integral part of the mechanism either through quantitative controls on exports and imports or through parastatal trading monopolies.

The second broad category of price stabilization agencies consists of export marketing boards, which are most common in African countries, although they are used also in other countries, primarily for traditional tropical exports. Like buffer stocks, these boards buy and sell the physical commodity and usually enjoy a monopsonistic position in the market. Marketing boards, however, seldom hold stocks for very long, because they do not deal in commodities that are major consumables in the producing countries. While they usually severely tax producers, in some countries, such as Ecuador and Malaysia, these tax revenues are partially earmarked for reinvestment in support services for the export crop.

In the final category, trade taxes or quantitative restrictions are used to regulate domestic prices. The main distinguishing characteristic of this type of scheme is that the government need never handle the commodity to control its price. In Chile, for example, a variable import tariff is used to maintain a price band. When international prices are low, the import tariff is high, thus maintaining a minimum price to producers; when international prices rise, the tariff becomes progressively smaller, finally becoming zero at prices above a 'maximum' price. Thus consumers are offered some protection from 'high' prices, although internal prices are never lower than international prices. One variation of the variable tariff is the buffer (or stabilization) fund. When the world price of a commodity is high, a tariff is levied and the proceeds are put in a fund to be paid out as a subsidy when the world price is low. This kind of fund is used in Papua New Guinea (coffee, cocoa, copra, and palm oil), Côte d'Ivoire (coffee, cocoa, cotton), and South Korea (wheat).

THE PURPOSES OF PRICE STABILIZATION

Among the many countries that do have some kind of price stabilization programme, there is great diversity in the principal purposes of the stabilization. The reasons for the universal concern with price stabilization differ not only among countries but also for different crops in the same country. In some cases the implicit reason is political: schemes that are set up with the explicit goal of stabilizing prices actually function to change the average level of prices and transfer income from one group to another or to the government. For example, providing a minimum producer price (but no price ceiling) redistributes income on average from consumers to producers. More commonly, many marketing boards and parastatals have depressed the average price received by growers, thereby transferring resources from producers to consumers and the government. While the consumer beneficiaries of such programmes are frequently wealthier than the producers who are taxed, the government's goal is to support urban workers and encourage industrial development, or (in the case of traditional export crops) to extract the 'surplus' from inframarginal producers.

Apart from such distributive goals, price stabilization is generally believed to provide several economic benefits. To the extent that stabilization increases the certainty of producers about the price they will receive for their crop when it is harvested, it allows them to make more rational and efficient planting decisions regarding what crops to plant, how much to plant, and how intensively to cultivate. In addition, if price stabilization decreases the riskiness of income that producers face in planting a crop (and it is by no means certain that it will have this effect), it will induce more resources to flow into production of that crop if the supply of producers – potential and actual – is risk averse. While this may represent either efficiency gains or losses, depending on what other distortions are present in the market, policymakers use price stabilization to achieve this objective, implicitly viewing it as beneficial.

Finally, stabilization of the prices of tradable goods may function to insulate the domestic macro economy from external shocks. If changes in the international price of a country's major export, for example, were fully reflected in changes in domestic producer prices, then domestic production adjustment would tend to reinforce external price changes to generate greater instability in export earnings. (That is, when prices increase, production – and export volume – would also increase; conversely, when prices decrease, export volume would fall.) This instability in export earnings would create uncertainty in budgeting, as well as causing fluctuations in macroeconomic variables, if not sterilized by the monetary authority.

How Stable are Domestic Prices and Revenues?

The extent to which price stabilization schemes meet their goals depends on the degree to which prices and/or revenues are actually stabilized. In this section, we compare the stability of actual producer prices and producer revenue to the stability of these variables if domestic prices had been allowed to reflect changes in international prices, appropriately adjusted for exchange rates.

Given the popularity of price stabilization schemes, it should be expected that domestic prices and incomes are much more stable than their international counterparts. To examine whether domestic stability has been greater than it would have been had international equivalent prices prevailed, instability indexes for 15 crops across 37 developing countries with various kinds of stabilization programmes have been calculated, most of them covering the period 1967–1981. The index used is the standard error from a simple time-trend regression, using real producer prices as the dependent and time as the independent variable, divided by the mean of the series. (It should be noted that much of the instability in real domestic producer prices comes from the fact that nominal prices are typically held constant for long periods of time while inflation erodes their real value, with infrequent adjustments.) Producer prices are official prices, derived from a consistent time series of data constructed by the Food and Agriculture Organization of the United Nations. Border equivalent prices are constructed by multiplying the border prices by the official exchange rates

Table 20.1. Summary statistics for differences in instability of variables A and B

	A: producer price; B: border equivalent price	A: producer revenue; B: border equivalent producer revenue
Grain:		
Mean	−15.0	−12.0
Standard deviation	12.7	13.1
Beverage:		
Mean	−6.9	−5.4
Standard deviation	13.4	12.6
Fibre:		
Mean	−3.9	−2.5
Standard deviation	11.3	12.8

Source: Instability indexes calculated on basis of Food and Agriculture Organization (FAO) price data.

and dividing the result by the domestic consumer price indices. Producer revenue and border equivalent producer revenue are calculated as the appropriate price multiplied by annual production.

In Table 20.1, we report some summary statistics based on these indexes, grouped according to commodity type. Each cell in the table shows the average and standard deviation of the difference between the instability index calculated for the first variable at the top of the column and that calculated for the second. For example, the 'north-west' cell shows that across all grains in all countries in the sample, the average difference between the instability of producer price and border equivalent price is — 15.0 with a standard deviation of 12.7. This means that for these commodities, domestic prices were more stable than border equivalent prices by an average of 15 percentage points, but with a standard deviation in the sample of 12.7 percentage points.

Several interesting observations can be made based on the statistics in Table 20.1. First, regarding the central question of how successful these schemes are in stabilizing prices and incomes, the answer varies greatly depending on the type of commodity. For grains, producer prices on average have been significantly more stable than they would have been if they had followed border equivalent prices. However, it should be remembered that these are official prices. Where parastatal companies have segmented the market, prices in the parallel market may be destabilized. Consequently, these results may overstate the degree of domestic stability achieved. For beverages, the average reduction in instability is much smaller, especially compared to the standard deviation. For fibres, these schemes were on average remarkably unsuccessful in making domestic producer prices more stable than border equivalents. Why was domestic price stabilization so much less successful in beverages and fibres than in grains? The answer must be conjectural, but this evidence fits well with a general pattern of neglecting exportables (most beverages and fibres are exported from the countries where they are grown), while favouring import substitutes and food staples that are politically sensitive in urban areas. (Exportable crops are also often unprotected relative to importable food crops, as measured by nominal protection rates.) For revenue, the pattern is the same across the three commodities. It is, however, also of interest to note that for all commodities, the schemes were on average more successful in reducing price instability than in reducing revenue instability.

Finally, it should be emphasized that the figures for the mean reduction of instability are relatively small in comparison to the standard deviation across countries. Except for the cell comparing producer price to border equivalent price instability for grains, the standard deviations are larger than the means, in most cases much larger. The implication is that in a number of cases the 'stabilized' domestic variables were actually more

Table 20.2. Percentage of cases where instability of variable A is greater than instability of variable B

	A: producer price; B: border equivalent price	A: producer revenue; B: border equivalent producer revenue
Grain	9	15
Beverage	31	31
Fibre	35	38

Source: Instability indexes calculated on basis of FAO price data.

unstable than if they had followed border equivalent prices. This is even clearer in Table 20.2 which indicates that the pattern of greater domestic instability occurred in more than 30% of all cases for beverages and fibres, for both prices and incomes.

THE OPERATION OF DOMESTIC PRICE STABILIZATION MECHANISMS

In this section, we will examine in more detail several of the kinds of mechanisms used to stabilize domestic prices. For each mechanism we discuss considerations that bear on the costliness and efficiency of the scheme and then illustrate how the programmes work in practice, using some representative examples.

Marketing agencies

Schemes that stabilize prices by buying and selling the physical commodity are generally intended either to be self-supporting or – if one of the goals is to subsidize consumption for the poor – to operate at a modest loss. In reality, many have become major budgetary burdens for their governments, and of those that are not great burdens, most are ineffective in stabilizing prices. The reason for this is not so much bureaucratic incompetence, as is commonly believed, but rather the kind of pricing policies followed by the agencies, either by their own choice or by that of other government bodies. Governments have frequently been pressured by rural interests to keep procurement prices high, while urban interests press to keep retail prices low for staple commodities. As long as the subsidies could be financed through debt accumulation, with the adverse effects hidden, governments were free to ignore basic economic considerations and to follow the schizophrenic policy of high producer and low consumer prices.

Many marketing agencies (frequently organized as 'parastatal' companies) are given exclusive rights to deal in particular commodities. Since they exercise monopoly power, it is obvious that any pricing policies that generate even a small loss per unit sold will cause enormous losses over the whole market. But even those agencies that are intended to coexist with private traders, carrying out only marginal interventions, often incur heavy losses because in reality they have found that they are forced to carry out increasingly massive interventions in the market. The reason for this is that their pricing policies have generally made it difficult or impossible for private traders to coexist with them. Alternatively, if the agency does not have sufficient funding to purchase the entire harvest, private traders may coexist, but the market will be segmented into 'official' and 'black' markets.

Below, we describe the operation of two marketing systems that are intended to stabilize prices. The first, operating in Peru, has been able to maintain effective control over prices. The second, the system of parastatals in Tanzania, has not. These examples are quite typical of the way that such mechanisms operate and demonstrate the budgetary and economic costs they entail.

Case study: stabilization of rice prices in Peru

The rice marketing programme in Peru is similar to the stabilization schemes of many countries, especially in Latin America. The programme is run by a parastatal organization called Empresa Comercializadora de Arroz, S.A. (ECASA), which is not under the direct control of the government but has certain government-enforced rights, such as the sole right to buy, import, and sell rice at the wholesale level. As is typical of many such marketing agencies, ECASA's pricing policies – which are set partially by ECASA and partially by the government – have a number of goals, including price stabilization, subsidization of rice consumption by the poor, and enhancing development of underdeveloped areas by promoting the growing of rice there.

In most respects, ECASA's policies deviate substantially from the measures that would minimize its cost of operation. For one thing, ECASA has a legal monopoly in this market, so it must purchase, handle, and sell the entire crop. Second, its pricing policies are such that the spread between the average price paid to producers and the average prices charged at the wholesale and retail levels are completely inadequate to cover handling and processing costs. Third, to stabilize prices during the year, the price is not allowed to rise sufficiently to cover storage costs during the period when marginal consumption must come from stored grain; that is, there is 'pan-seasonal' pricing, thereby forcing out the development of private storage. Fourth, producers in the most remote regions are actually paid higher prices than producers in areas close to consumption centres, thereby effectively penalizing production in areas where it is most economical. Finally, ECASA does not pay producers differential prices according to the quality of the rice they deliver, as would buyers in a private market. Predictably, much of the rice delivered is of low quality.

The policies of pan-seasonal pricing and heavy subsidization of consumption may have had particularly adverse effects in Peru. The climate of the minor rice-growing area in Peru (the jungle) is dramatically different from that of the major growing area (the coast), and as a consequence several parts of the jungle reach peak production in seasons when production in the coastal area is at its lowest, and marginal consumption is coming from storage or imports. In a situation such as this, pan-seasonal pricing not only encourages rice consumption when it is most expensive to provide, but also suppresses off-season production in the jungle, which would otherwise reduce the amount that would need to be stored or imported to satisfy demand in this period. The heavy consumer subsidy is intended to improve the diet of the poor. However, it has the unfortunate consequence of encouraging consumption of a high-valued crop, which Peru could otherwise export to finance the purchase of more nutrition for the poor in the form of foods that are cheaper per nutrition unit. It is estimated that doing so would produce a net gain of 35% in calories and 115% in grams of protein per capita. The subsidy has also proven extremely hard to target to the needy; one study concluded that 36% of the subsidy benefited the middle and upper classes.

The net effect of all these policies is that ECASA is a severe drain on government resources. In 1983, ECASA's deficit was US$85 million, or about 0.5% of gross national product. The government is currently in the process of reviewing the role and structure of ECASA with an eye to the reform of some of its policies.

Case study: price stabilization in Tanzania

Tanzania relies on an extensive system of parastatals to purchase, process (in some cases), and sell virtually all crops, both in the domestic and export markets. The National Milling Corporation (NMC) handles most food crops, while export crops are handled by a variety of other parastatals. The NMC sets retail prices too low to cover costs of procuring (at a relatively high price) and processing the crop, incurring large losses. The loss per unit, as a percentage of the sales price, has been estimated to exceed 50% for all crops except wheat (15%) and finger millet (42%), with the loss for cassava and bullrush millet surpassing 90%.

With the margin between producer and consumer price set far too small to cover costs, one would expect that the private sector would be completely crowded out and that the parastatal would handle all purchase, processing, storage, and distribution of these crops. In fact, just the opposite has occurred in Tanzania; not only does NMC coexist with a thriving (though technically illegal) parallel market, but the parallel market has been of increasing importance in the market. The reason for this is that the official uniform (pan-territorial) producer prices are far below the prices that would clear the market in areas with low transport cost, that is, areas near consumption centers. These are the regions in which private traders operate, and producers who are located in these regions do not generally sell to NMC. In remote regions where high transportation costs would cause the market-clearing price to be lower than NMC's offering price, private traders have been squeezed out, giving NMC a virtual monopoly in the high-cost market. Both retail and producer prices in the parallel market are much higher than the official prices. While data are scarce on the exact magnitude of the differences, in the 2 years covered by one study, the parallel producer price was up to five times the official price. Consumer prices in the capital are commonly four to five times, and sometimes up to ten times, the official price. This means, of course, that consumer access to the official market must be rationed.

In the context of price stabilization, NMC's pricing policies have two important consequences. First, this pricing policy effectively divides the market into an official market, composed of those producers in regions where NMC provides the only source of transportation and those consumers that have privileged access to NMC sales, and a parallel market of all other producers and consumers. By making one large market into two smaller ones, the NMC actually increases the variance of production, consumption, prices, and producer income in both markets. The second consequence of the pricing policies is that NMC runs large deficits. In 1980–1981, its losses amounted to 31% of its total sales, or over 8% of the government's recurrent expenditures. Since these losses are financed by loans from the government, which are directly monetized, these deficits contribute significantly to inflation.

The parastatals that handle the export crops in Tanzania generally follow pricing policies that depress producer prices below export parity. Prices tend to be maintained constant, in nominal terms (which is to say, steadily lowered by the rate of inflation, in real terms) for 2 or more years, and then adjusted to bring them closer in line with export prices. The fact that these prices are in general below export prices, together with the relatively high prices on the parallel market for food crops, has meant that many producers have switched from production of export to food crops. In addition, because of their high operating costs, these parastatals have had substantial net fiscal costs, losing over US$80 million in 1980–1981.

Variable tariffs/subsidies and buffer funds

A second mechanism sometimes used to reduce the variability of domestic prices of tradables is a variable tax or subsidy scheme for exports or imports. An export tax, for example, means that the price to producers (and domestic consumers) is lower than the world price converted to domestic currency at the prevailing exchange rate, and the larger the tax is, the greater the difference between world and domestic prices. Therefore, if the tax rate is 'progressive', that is, high when the world price of the commodity is high, the domestic price does not increase proportionally and variations are smoothed out. Import taxes, of course, work in the opposite direction from export taxes, raising the domestic price over the world price. So, for the tax scheme to have the effect of decreasing the variability of prices, taxes must be lowered when world prices are high.

Progressive tax schemes are used by a number of developing countries for their exports. (This kind of scheme is less commonly used – or at least less commonly made explicit – for imports in developing countries, though it is an integral part of the European Community's Common Agricultural Policy.) In some countries, progressivity is not an explicit and systematic characteristic of the export tax regime but, rather, is established by year-to-year adjustments in the tax rates on a somewhat ad hoc basis. Papua New Guinea's standard export tax on many commodities (other than the buffer fund commodities discussed below) is simply waived when world prices fall to low levels, giving the system a progressive nature to some degree. Colombia also has very effectively used a progressive tax schedule on coffee exports to cushion the impact of the price fluctuations in that market even though the rate is not always predetermined by a fixed schedule. One study estimated that the magnitude of internal coffee price fluctuations is on average only about half that of world price fluctuations. Other countries, such as Argentina, have become so dependent on export taxes that their variability reflects changes in government revenue needs, as well as intentions to achieve increased stability.

Here, we discuss two particular variations of the variable tariff/subsidy idea: the progressive export tax system of Malaysia, which uses a predetermined rate schedule, and the buffer fund scheme of Papua New Guinea, which combines a progressive export tax (when world prices are high) with a progressive export subsidy (when prices are low).

Case study: progressive export taxes in Malaysia

Malaysia sets *ad valorem* rates on exports of certain commodities according to a progressive schedule based on the amount by which the actual export price exceeds a 'trigger price'. (This trigger has been changed from time to time.) Table 20.3 shows how the export tax system has worked for the two major agricultural exports, rubber and palm oil. As can be seen, the system has produced some degree of stability. For example, when the export price of rubber fell by 39% between 1980 and 1982, the producer prices fell by only about 23%. However, the effective tax rates are not entirely consistent from year to year. For example, between 1979 and 1980, when the price of rubber rose, the tax rate actually declined slightly. This kind of inconsistency, coming to some extent from changes in the trigger price from year to year, introduces an additional element of uncertainty into the prediction of prices by producers. Also, it is clear from the table that a progressive tax scheme tends to make government receipts from the tax quite unstable. Receipts from the rubber tax in 1979 were more than ten times their level in 1982. (The tax was virtually phased out in 1985.) This instability may make planning difficult, and, because there is a tendency to expand the budget when receipts are high but not to contract when receipts are low, this instability also tends to 'ratchet up' expenditures, which are then financed in years of low receipts by inflationary deficits.

Case study: the buffer funds of Papua New Guinea

The experience of Papua New Guinea in the field of commodity price stabilization is instructive because of that country's long history of involvement with such schemes and its reasons for selecting the unique manner – known as buffer funds – in which the programmes currently operate. Price stabilization schemes are used in Papua New Guinea for four export commodities: copra, coffee, palm oil, and cocoa. Here we describe in some detail the system used for cocoa, which is similar to systems used for the other commodities.

The industry board for cocoa was originally established with the goal of stabilizing the producer price by using a variable levy in combination with a support price scheme, based on the cost of production, plus a margin. At prices over the minimum price necessary to give a return equal to the minimum rural wage plus a small margin, a fixed progressive variable levy schedule was established, with the marginal rate increasing with the world price, up to a maximum of 50%. The advantage of this scheme, as opposed to the *ad hoc* export taxation scheme used in many countries, is that production prices can be predicted better, since the schedule is predetermined. No additional uncertainty is created by the inconsistency of the taxation rate from year to year.

While the system of stabilizing prices based on cost of production did succeed in stabilizing producer prices and reducing the uncertainty in their prediction, in 1977 the principles on which the system was based were reconsidered for several reasons. First, because world prices had remained

Table 20.3. Malaysia: export taxes

	1979	1980	1981	1982	1983	1984	1985	1986
Rubber:								
Export price (US/¢/kg)	100	136	111	83	106	93	77	80
Export tax rate (%)	24.9	23.7	14.0	4.1	7.5	4.4	0.1	0.0
Total tax receipts	511	493	231	47	117	66	1	0.0
Palm oil:								
Export price (US/$MT)	594	518	505	423	435	629	505	266
Export tax rate (%)	9.6	6.4	5.0	2.6	3.9	4.2	2.3	0.5
Total tax receipts	108	75	63	31.6	50	80	38	7

Source: Unpublished data from World Bank.
Note: Figures for prices from 1984 and later are not strictly comparable with earlier years, since series are spliced.

high, it became apparent that the scheme was taxing cocoa producers and significantly depressing the average price they receive. (By the end of September 1976, the fund had accumulated US$2.3 million.) Second, it was recognized that cost of production was a dubious criterion for setting the support price, since the cost of production per ton depends on many factors and varies widely across growers. Third, it was felt that if the support price was set too high or too low, relative to the long-term world price, significant costs would be imposed by the programme. If the floor price was set too low, the fund would seldom or never pay bounties but, rather, would act as a pure taxation scheme, as had been the case up to that point in time. But, if the floor price was set too high, the fund would tend to run out of money and, furthermore, would support an industry of an uneconomically large size.

For all these reasons, it was decided in 1977 to stop basing the target stabilization price on the cost of production and, instead to base it on the long-run world price, as proxied by a 10-year moving average of past world prices, suitably adjusted for inflation. Furthermore, the judgment was made that it would be infeasible to completely stabilize the price at the exact level of this moving average. Instead, the programme was set up so as to pay a subsidy or levy a tax equal to half the difference between the 10-year average in a given year and the world price in that year. Thus, the fluctuation in producer prices each year should be approximately half the difference between the change in world prices and the change in the moving average.

This scheme has reduced instability of cocoa prices by an estimated 46%, and has had three important concomitant advantages, compared to stabilization programmes tried in other countries. First, since the board regulates the price only by taxing or subsidizing the export price, it avoids the costs and problems of many programmes that require a marketing board to become directly involved in the purchase and transport of the crop. In order to insure that bounties as well as levies are passed on to producers through the private traders who are the initial payees or payers, price information is widely disseminated to growers even in remote areas by radio and other means to improve their bargaining positions *vis-à-vis* the traders. Second, this kind of system avoids some of the undesirable macroeconomic side effects of other schemes. The tax-subsidy system is self-financing, so it does not destabilize the government's current budget. Furthermore, at least 60% of the fund is deposited with the central bank, which can then easily sterilize any undesirable fluctuations in foreign exchange to neutralize inflationary or deflationary effects. Finally, the programme does not require any physical stockpiling of the commodity, thereby avoiding the kinds of costs associated with buffer stocks in other countries.

Lessons from Domestic Stabilization Schemes

The experience of the domestic price stabilization programmes of these and other countries provides some valuable lessons to be used in the design of such programmes.

Avoid handling the commodity when possible

While there have been a few exceptions, the experiences of most countries have been similar to those of Peru and Tanzania. Attempts to stabilize prices using governmental or parastatal marketing agencies have been extremely costly, in terms of both the government's budget and the efficient operation of the economy. Once established, these agencies frequently adopt (or have imposed on them) pricing policies that embody many goals other than price stabilization, causing various distortions of economic incentives. Given their importance in the economy, these agencies are the natural focus for pressure groups seeking to manipulate pricing policies for political ends. They also require truly massive bureaucracies to carry out their assigned task. The experiences of most countries indicate that for stabilization of the price of traded goods, variable tariffs and subsidies represent effective and generally less costly alternatives to marketing agencies.

Fears that changes in tariffs and subsidies on trade will not generate corresponding changes in domestic prices (and thus will be less effective in stabilizing prices than would a system of direct intervention) have proven unfounded. A detailed study of the responsiveness of producer prices to wholesale prices, and wholesale prices to export prices, in Thai agriculture found that the responsiveness on both levels is quite high. The experience in other countries, such as Malaysia and Papua New Guinea, confirms this. Likewise, fears that staples cannot be imported quickly enough to meet harvest shortfalls may have been overblown, since private storage (if not suppressed) can help buffer such shocks. A study of wheat storage policy in Pakistan found that interannual stockholdings could not be justified except under extremely low international prices (Pinckney and Valdés, 1992). Modest interseasonal storage could be justified because of the lead time required to import wheat, but the study noted that private storage had been quite sensitive to expected price changes.

Prices should be only partially stabilized

When it is necessary to physically handle the commodity – as, for example, in operating a buffer stock – it should be recognized that it is impossible to stabilize the price completely and that the more the price is stabilized, the more costly it becomes. The reason for this is that the greater the degree of stability desired, the greater the storage capacity that must be maintained and the greater the funding the agency will require. Since production each year is random, there will eventually come a time when many years of excess (or deficit) production will occur sequentially, defeating efforts to closely control prices through storage schemes and potentially bankrupting the agency. Simulation experiments have shown that as the degree of stabilization is increased, the costs of operating a stocking scheme rise much more rapidly than do the benefits. Bigman and Reutlinger (1979) for example, estimate that, for a hypothetical grain importer with a restrictive trade policy, an annual outlay of US$45 million would reduce the probability of a severe shortfall from 8% to 5%, but an additional expenditure of the same magnitude would further reduce this probability only an additional 1.1%.

Because of these considerations, the most successful buffer stock schemes will not try to stabilize prices a great deal but, rather, will allow them to deviate substantially from their long-run target levels with little or no intervention. This kind of policy is known as a 'price band'. Ideally, to the extent it can be determined what would be the storage each year that would be undertaken by private traders in the absence of whatever distortions might justify government stabilization efforts in the first place, this should be the storage undertaken by the agency. In some cases when there are no identifiable distortions, governments may none the less wish to avoid extremely high or low prices. Even in such cases, the aim of the buffer stock should only be to provide marginal storage – incremental capacity that would only be profitable under extreme price movements – and should otherwise allow prices to fluctuate, thereby encouraging private storage.

Price should approximate market prices

Pricing policies should attempt to mimic the market to the extent this is consistent with the agency's goals. First, the target stabilization price should approximate the long-run average market-determined price. If some other price is chosen for a buffer stock for a nontradable commodity, the agency will either continually accumulate reserves and spend its funds (if the target price is too high) or frequently run out of reserves (if it is too low). If a price different from the long-run world price is chosen for a tradable good, the agency can still stabilize the price around this target, but it will be costly to the economy, since it will cause too many or too few resources to be devoted to production of this crop. Setting the price too low may also have a severe adverse impact on the government's budget, since the stabilization agency must continually subsidize large quantities of imports.

Of course, choosing the right long-run market price is easier said than done. It is impossible to tell how much of any given price shift is due to transient factors, and how much is due to permanent changes in the fundamentals of the market. The policy of Papua New Guinea – using as a target a moving average of past world prices – is a reasonable approach to this problem, since it at least assures that the target price will not deviate substantially from the appropriate level, even if market fundamentals do change. While many countries do base their stabilization efforts on world prices, a surprisingly large number use some other basis, such as 'cost of production', or have no coherent basis. (In a small sample of 10 countries whose basis for the target price could be determined, only three used world prices.) Unfortunately, as a basis for price stabilization, the cost of production is virtually meaningless, since the marginal producer will always have a cost of production equal to whatever price is set.

Second, the short-term pricing policies of the stabilization agency should be such that private processors, storers, and marketers are not squeezed out of the market. If the agency is given a legal monopoly or if the private agents are squeezed out so that the stabilization agency must assume all of these roles for the entire crop, the costs of the programme are likely to be greatly multiplied as compared to a policy of marginal interventions. Pricing policies must recognize that prices between harvests must rise substantially to cover storage costs; that producer prices in outlying areas must be lower than those in areas close to consumption centres by a margin sufficient to cover transportation costs; and that prices of processed food must be higher than producer prices by a margin sufficient to cover processing costs.

Market forces can be used to stabilize domestic prices

Price stabilization schemes in general are costly and offer no certainty that domestic prices will actually be more stable than international prices. Consequently, the final lesson is that because of the cost involved in government-operated stabilization schemes, the market should be relied on to the maximum extent possible. And, indeed, there are a number of mechanisms inherent in the operation of economic markets that act to stabilize prices.

One such mechanism occurs through changes in the real exchange rate in response to fluctuations in the international prices of a country's exportables. When these world prices fall and export revenues decline, pressure is put on the exchange rate to depreciate, tending to partially offset the decrease in world prices. Thus, the price received by producers tends not to fall as dramatically as the international price. Of course, the devaluation of the exchange rate means that imports become more expensive and the real income of consumers of imports falls accordingly. Thus, the depreciation of the real exchange rate cannot really ameliorate the fall in the economy's real income implied by a fall in the international price of its exportables. But it does spread this loss around the economy, rather than concentrating its effects on the export sector. To the extent that the objective is to make more stable and predictable the price and/or income of this sector, exchange rate movements can thus contribute to these goals.

Other mechanisms are even more useful. One of the most obvious of these is private storage. An efficiently operating storage system transfers supplies from years of high production to years of low production, thereby stabilizing prices and consumption. Unfortunately, private storage facilities are in chronically short supply in many developing countries, due at least in part to the interseasonal pricing policies of official marketing agencies and to the uncertainty attached to government price policy from year to year. Not so obvious, perhaps, is the stabilizing effect on producer prices of the margin between the consumer and producer prices. A study of the rubber market in Malaysia has indicated that in the absence of regulations, this margin tends to be positively correlated with the prices; that is, when prices are falling, these profits also fall. As a result, producer prices fall and rise less than they otherwise would. As the study notes, the factors that cause this margin to increase and decrease with the price – such as the flexibility to rapidly cut cost and overhead when necessary – are not characteristic of government marketing agencies.

One other private mechanism that can be quite effective in reducing the uncertainty over prices faced by producers and intermediate consumers (such as agroindustry) is the use of futures markets. Hedging ameliorates short-term price risk (uncertainty) for economic agents, though it does not reduce the price instability. However, neither do futures markets increase instability; the evidence indicates strongly that the often-expressed fear that speculative activities in futures markets act to destabilize prices is not well founded.

Even international futures markets are not widely used in developing countries, and domestic futures markets are virtually nonexistent. These shortcomings seem to have a number of causes. A detailed study of this problem in Colombia found that the use of international markets was extremely rare for the following reasons. (i) Most important, exchange controls – especially prior licensing requirements – severely limited potential hedgers' flexibility. (ii) Import controls made hedging less useful. (iii) Unpredictability of government trade policy created risks to hedging. (iv) The advantage and mechanics of futures markets participation were not widely understood by potential participants. Colombia, since the late 1970s, has had an institution that could serve as the precursor of a domestic futures market, but its development along this line is impeded by its domination by one participant, the official marketing agency.

SUMMARY AND CONCLUSION

Governments throughout the developing world have established mechanisms to insulate domestic agricultural markets from the vagaries of international price movements. This has been done out of concern over the effect of price risk on producers and to promote macroeconomic stability. In some cases, rhetoric on price stabilization has seemed to mask another purpose of the schemes: to transfer income from producers to consumers or (less often) vice versa. What we found in examining the benefits of these policies is that they have not been universally successful in stabilizing prices and incomes and have sometimes had just the opposite effect. For a few countries in the case of grains, and in a significant number for beverage and fiber crops, domestic real producer prices and incomes were less stable than if they had followed border prices.

The cost imposed by stabilization schemes appears to depend substantially on the mechanisms through which they operate. To illustrate certain characteristics that affect the operational cost, both financial and economic, we drew from studies of a sample that is small but reasonably representative. We believe that the evidence argues strongly for a 'minimalist' approach to price stabilization: rely on market mechanisms when possible; avoid schemes that require physical handling of the commodity; do not try to stabilize prices too much; and try to mimic prices that would be established in a freely functioning market. By following these principles, governments could minimize the cost of mitigating adverse effects of price instability.

REFERENCES

Bigman, D. and Reutlinger, S. (1979) Food price and supply stabilization: national buffer stocks and trade policies. *American Journal of Agricultural Economics* **61** (November), 657–667.

Pinckney, T.C. and Valdés, A. (1992) Short-run supply management and food security: results from Pakistan and Kenya. *World Development* (in press).

21

TRAINING PROGRAMMES FOR HUMAN RESOURCE DEVELOPMENT IN AGRICULTURAL MARKETING

L.A. De Andrade and A. Scherer

In all developing countries considerable efforts and investments are under way to improve the educational level of the population as a basic precondition for social and economic development. Educational policies have not only to be concerned with formal training and the formation of new professionals. Equal attention has to be paid to the professional development of those persons who, through their day-to-day work, determine economic progress and who have not yet had the chance to receive an adequate orientation about the professional tasks they perform.

The agricultural and food marketing sector in Latin America presents a characteristic example. This segment often serves as an employment opportunity for persons who, due to their lack of educational or other professional qualifications, enter into marketing as a means of making a living. Consequently, this sector contains significant numbers of commercial operators who have never had the possibility of acquiring a basic professional background. However, their activities determine, to a large extent, marketing costs and efficiency. Often their professional limitations are the cause of shortcomings or restrict further marketing developments.

IN-SERVICE TRAINING

In-service training is the most difficult area of training because most instructors are unable to assess accurately what is needed.

To attract busy traders and their hard-working staff, a training programme should:

- be interesting and valuable;
- suit their time schedule;

Source: Paper presented at the FAO Marketing Workshop, Brasilia, 1982.

- gain their professional confidence;
- use a language they all understand.

In general, traders and their staff are not interested in lengthy, theoretical or general lectures by officers who have no practical experience. They want to be confronted with know-how and skills which, if applied in their work, will help them, for example, to:

- gain more money;
- do the same job with less effort.

Anyone attempting such in-service training must first know the people to be trained, which involves finding out their educational and professional backgrounds, working conditions and expectations, and the time they have available. He or she should involve them in planning the course from the start. Of such a group working in wholesale markets in Brazil, 15% had no schooling, 49% had primary schooling only, 33% had secondary or technical schooling and 3% had been to a university. The topics which interested them most were:

- technical marketing and product knowledge;
- grading and packing of produce;
- business organization and administration;
- market information;
- market equipment and its operation.

They were prepared to participate in training sessions of one to four hours on days when they were not very busy. The programme was organized by the Companhia Brasileira de Alimentos (COBAL). It had set up, with the support of the Federal Government, a network of local and regional markets and promoted the SOMAR voluntary chain with over 4000 affiliated food retailers.

TRAINERS

The development of successful practice-orientated training with operational target groups depends on the recruitment of training personnel qualified to execute such tasks. Experience proved that training personnel with the following qualifications achieved the best results.

1. The necessary social attitude, interest and enthusiasm, as well as the natural ability to transfer know-how to others.
2. Sufficient practical technical and commercial experience to be able to understand the thinking, language and expectations of their clients and to be able to demonstrate in practice what is taught in training sessions.
3. Willingness to learn, listen and adapt to new situations and to work hard

and systematically in a field where not much prestige can be earned.

For the execution of training projects, training teams, consisting of:

1 training specialist (psychologist, pedagogue),
1 organization professional (administrator, economist) and
2 or 3 technicians specialized in the area of work of the target group,

proved to be a workable and effective solution. This guarantees that all training aspects are adequately represented in any programme formulation and stimulates the desirable education process within the team, forcing each member to absorb and respect the professional knowledge and viewpoint of the others.

For transmitting special technical skills or operation procedures, the inclusion of skilled workers or operators for demonstrating such activities to their colleagues, etc., and correcting them during practical exercises until they become well acquainted with the task is also well accepted.

TRAINING PROGRAMME ORGANIZATION

Professional, technical and commercial training programmes have to be tailored to the requirements and priorities of the target groups. A decentralized training system, flexible enough to adapt itself to local standards, problems and priorities, is needed. Incorporating the various countrywide marketing facilities operated or supervised by COBAL, the following training set-up was established:

1. A central coordinating training unit at COBAL headquarters (CDRH) to:

    (a) determine training policy and priorities,
    (b) develop training material and programme content,
    (c) introduce and demonstrate training activities,
    (d) guide and supervise local staff in conducting training tasks,
    (e) undertake adequate training evaluation,
    (f) maintain finance and coordination control, and
    (g) organize training activities at headquarters.

2. A decentralized training programme implementation system, utilizing local marketing units and technicians.

Such a strategy allows training content to be adapted to local priorities and standards, and forces the involvement of local managements and technicians in training tasks, results and follow-up work. Only if each operational marketing unit acquires the capacity and understanding needed to apply training as a resource for achieving better results can the training inputs yield the anticipated dividend and take root as a permanent activity.

TRAINING EQUIPMENT

Training programmes for operational target groups often weak in educational background, demand strong support by visual aid material and written explanations in a simple language, complemented by adequate illustrations. Examples should, whenever possible, be based on local conditions, facilitating acceptance and understanding by the clients. They must be able to transmit the technical or operational details clearly, which implies that their preparation has to be orientated and guided by professionally qualified personnel involving a specialist in visual-aid material preparation and an experienced technician capable of determining the technical aspects the material has to contain.

Equipment which allows operation under varying conditions should be chosen. With the rapid development of the video tape equipment sector, the range of opportunities has expended even more. For practical demonstrations, the selection of suitable places and the availability of the required materials is essential.

It has to be stressed that visual-aid material fulfils only a supporting function in practice-orientated training programmes. It cannot be a substitute for the services of a qualified instructor or replace practical demonstrations where each participant has to show that he understood the training content.

ASSESSMENT OF TRAINING NEEDS

On the basis of the COBAL experience, a detailed assessment of the target group, their needs and expectations is essential for the success of any training programme with operational personnel. Thus, training in the horticultural marketing sector was preceded by collecting detailed evidence under all regional conditions of Brazil on:

1. training demand, need, priority, wishes and suggestions;
2. prevailing working conditions and constraints;
3. experience, age, opinion and improvement suggestions of the persons engaged in horticultural marketing.

If properly organized and conducted, such exercises enable, at the same time, training personnel to get acquainted with the situation.

The programme for market workers and porters, handling and transporting fruit and vegetables, was as follows:

1. *Objectives:*
Introduce better product handling and transport methods to reduce losses and increase efficiency.

2. *Duration:*
Ten hours on two different days when there was little market movement.
3. *Place:*
At the wholesale market, using wholesalers' stalls for practical demonstrations.
4. *Course content:*
Two hours – group introduction, definition of activities, communication games.
One hour – discussion of existing market regulations and improvement opportunities.
Two hours – slides, films on correct and incorrect ways of handling produce, etc., followed by discussion.
Five hours – practical skill training, during which groups of four or five perform all operations under supervision of instructor and experienced wholesalers.
5. *Services:*
Refreshments and lunch.
6. *Acknowledgement:*
Certificate presented to participants who complete the course successfully.
7. *Follow-up:*
Once a marketing community realized the advantages of professional training, it made its own requests. Periodic visits by a training organizer will help canalize these requests to the best advantage.

Considering the important role retailers can play as change agents, opinion leaders, innovators and educators of consumers, the training of retailers and their sales staff represents a decisive element for all agricultural and foodstuff marketing development strategies.

As this target group, through constant public contact and inquiries, is already aware of the advantages of professional know-how and since any training input can yield immediate marketing improvements and better sales returns, training projects in this sector are relatively easy to implement and are able to achieve a positive impact within a short time.

The programme for *horticultural retailers* was:

1. Theoretical (4–5 hours)
 (a) Product knowledge and basic nutritive values.
 (b) Functioning of the horticultural marketing and supply system.
 (c) Responsibility of the retailing sector and its personnel.
 (d) Sales orientation and hygienic aspects.
 (e) Proper treatment of customers.
 (f) Technical guidelines and loss-prevention measures.
 (g) Efficient purchase and sales control procedures.

2. Practical (6–7 hours)
 (a) Correct product handling and storage.

(b) Quality control and product classification.
(c) Pre-packing activities.
(d) Product presentation and replenishment, utilizing the various methods for different commodities.
(e) Promotion, advertisement and pricing activities.
(f) Correct use and maintenance of retail market equipment.

1. *Training period:*
during market days and hours with little movement, as indicated by the clients.
2. *Participants:*
from 20 to 30 retailers, divided into sub-groups of 4 or 5 during practical exercises.
3. *Supporting material:*
distribution of illustrated pamphlets, offering detailed operational descriptions and professional information.
From 30 to 40 slides demonstrating specific technical and operational aspects.
Demonstration material for practical exercises.
4. *Other incentives:*
awarding of diploma after full participation. Provision of refreshments and lunch, where funds were available.
5. *Arrangements:*
made together with local market administration, associations of retailers, etc.
inclusion of experienced retailers and nutritionists as instructors.
6. *Evaluation:*
observation of operational improvements, sometimes with visual documentation by making photos before and after training. Accompanying of sales developments.

Training programmes for *wholesalers* have to consider that they have considerable practical experience, are heterogenous in regard to theoretical knowledge and educational standards and that interests may differ according to products marketed, company size and area of operation. Professionally capable instructors, carefully verified training interest and the availability of suitable training materials are considered to be prerequisites for entering into training activities with wholesalers in order to avoid disappointments, jeopardizing the training idea as a whole.

Experience shows that gatherings of all types of wholesalers usually end in discussions of administrative and general subjects which everybody already understands, while meetings with specific groups handling the same product, using similar equipment, etc., yield more detailed professional exchanges.

For initiating and organizing training activities, the collaboration with available associations, etc., represents a recommendable approach. Training period, time and place should be decided upon by the participants themselves. Examples of successful training themes for wholesalers have been:

1. Introduction of regular meetings between wholesalers and market administration to discuss actual market operations, thus strengthening the relationship and interchange of available experience.
2. Arrangement of specialized courses covering specific products, equipment, marketing operations, administrative or organization questions, etc.
3. Organization of seminars promoting the discussion of marketing aspects with other members of the market chain (for example, producers, retailers, importers).
4. Information meetings providing a feedback to the wholesalers on the results of previous training activities, for example those with market workers. Apart from orienting them on the possibilities of training their own employees in a similar manner and explaining to them available training material, these seminars were also used to discuss specific complaints, shortcomings, etc., in order to arrive at solutions for improvement.
5. Organization of visits to other markets, invitation of experienced wholesalers from outside, creation of discussion circles.
6. Integrating experienced wholesalers in training programmes as instructors.

Training for *food retailers* affiliated with the SOMAR Voluntary Chain followed the working pattern shown below:

1. Preparation and testing of training material and procedures by a central training team of CDRH in conjunction with technicians of the operational commercial departments.
2. Orientation course with the supervisors of each wholesale distribution unit to prepare them for the training task including:
  (a) detailed orientation about the SOMAR operation, philosophy and procedures;
  (b) detailed discussions of each training content aspect covered by a pamphlet with illustrations;
  (c) practical exercises at retail shops until the supervisors were acquainted with the training task.
3. Retailers and sales staff training through the responsible supervisor of the area (20 retailers per supervisor) at the retailer's shop.

Supervisor and retailer determine beforehand the suitable training time and the priority in which the various training contents are discussed on the basis of the set of pamphlets the trainee receives.

The initial programme included the following:

For the retailer:
(a) list of necessary documents to establish a commercial entity and admit sales personnel.
(b) internal and external aspects of the store.
(c) layout and departmentalization.
(d) product presentation and promotional activities.
(e) product replenishment and pricing activities.
(f) production and use of posters.
(g) control of stocks, costs and accounts.
(h) use of accountancy systems.
(i) sales tax information.
(j) useful data on labour legislation.

Operational personnel:
(a) how to receive merchandise and stock it in the storeroom.
(b) useful hints on merchandising.
(c) replenishment of merchandise on the shelves.
(d) how to receive checks and other cashier activities.
(e) packers at checkout.
(f) personal parcel storage

4. Evaluation including visual, operational and economic development aspects conducted by the supervisor and his coordinator before and after training was carried out on the basis of checklists.

5. Incentives:
Awarding of diploma after full participation.
Selection of most successful retailers of an area, offering them a fully-paid round trip to the annual Rede SOMAR Convention, etc.

Other specific improvement programmes or the introduction of new sales items, for example horticultural products, are based on a training system where the supervisor, together with a specialist, visits the retailer at his shop, orientating and educating him and his staff within his working environment and determining on the spot the required improvement actions. It is the task of the supervisor to undertake the necessary follow-up training work until the objectives are reached.

22

AGRICULTURAL MARKETING EXTENSION

C.Y. Lee

Production-oriented agricultural extension is now available in most developing countries, although the number of field-level workers and their adequacy may be criticized. Field-oriented marketing extension is rarely there at all. For the effective implementation of marketing improvement programmes an efficient marketing extension network is essential. Marketing extension, especially that provided by the field-level worker, will vary by country, but will generally include the following.

Advice on product planning

The Marketing Extension Worker (MEW) should be able to advise farmers on what crop and variety to grow in the coming season and at what time. Even for small farmers, the concept of product planning, i.e. careful selection of the crops to be produced considering marketability, is important. The small farmer is often very slow in adapting to changing market situations. He keeps on producing the same traditional crops or handicraft products to which he is accustomed. He should take up lines of production for which the smaller grower has some advantage. He should select activities which require relatively less land but intensive cultivation with high labour inputs and with relatively higher market value. High-value cash crops, such as mushroom, vegetables – especially off-season vegetables – poultry and pig raising, fish, etc., may do much more to raise small farmers' incomes than cereals.

Source: Paper presented at FAO Regional Consultation of National Coordinators on Small Farmer Marketing Development. Seoul, Korea, 1984.

Market information

Small farmers need two types of market information: forecasts of market trends and current prices and market arrival information. Information on market trends and expected price movement is useful in planning production. An expected fall in grain prices, for example, would suggest that a small farmer grows the minimum to meet family requirements, and devotes the balance of his resources to other products. For this purpose, 'market forecast information systems' have been introduced in some countries to assist farmers in planning their production. A marketing extension service will provide such information and field-level workers will use it to advise farmers. In addition, the MEW will also provide current price and market arrival information for each locality on a daily, or at least weekly, basis to keep farmers informed of price trends and market arrivals in relevant rural and urban markets. The information must be that which has direct impact on the area where the field-level marketing extension worker is operating. The small farmer needs market information for the places where the price of his produce is determined, not just for the capital city, which may be far away.

Securing markets for farmers

The MEW will assist small farmers in finding markets for their produce. This can be done in several ways. For grains, the MEW can advise farmers on how to obtain government guaranteed minimum prices where applicable, and on other more attractive outlets that might be open to them.

For cash crops the MEW can help with production/marketing contracts between farmers and processing industries or wholesale traders. For fruits and vegetables the MEW can assist in making arrangements with wholesale markets for regular and continuous shipment from the farmers. An active MEW can be an important link or 'match-maker' between small-farmers – more effectively on a group basis – and wholesale traders or processing industries who buy their produce.

Advice on sales timing

In order not to 'glut' small local markets, it is important for farmers to stagger harvesting and shipments to the market. Planned shipments are especially important for fruit and vegetables, livestock and other perishable produce. Such staggered marketing can be arranged only through coordination among the farmers in the local market area. Here the MEW can play an important role. He can advise farmers on a group basis on which group should harvest which crops on which day for marketing. Under the guidance and coordination of a MEW, such staggered marketing arrangements can help stabilize local market prices and raise farmers' incomes.

Improved marketing practices

The MEW can advise farmers and/or train them in improved harvesting methods, grading and standardization, improved packing, handling, and storing methods, etc. The MEW may organize training programmes for farmers and/or can provide direct advice on an individual basis. For this purpose the MEW can use farmers' fields for demonstration and training, or local and wholesale markets.

Promote group marketing

Because of the 'smallness' of the marketable surplus of individual small farmers, group marketing can bring important advantages in bargaining over sales and organizing transport. Differences of interest between farmers, distrust of group leaders, lack of managerial skills and operating funds often prejudice group marketing in practice. Even so, the MEW can assist in organizing promising groups. He can help them make contracts with wholesale traders or processors, he can assist in arranging group transport of farmers' produce, in establishing and operating small packing houses at village level, and in obtaining operational funds for group marketing.

Advice on the establishment and operations of rural markets

In most developing countries there are periodic rural markets used especially by small farmers. Farmers with a large marketable surplus tend to bypass such markets. The small farmers are their main customers, both in selling their chickens, eggs, vegetables, fruits, a bag of paddy, a bag of maize to meet cash requirements, and in buying necessities such as clothing, salt, shoes and other consumer goods. In most cases these markets are owned by local governments and are managed directly by them or by a contracted party.

Most such rural markets have no storage; many do not even have shelters, just an open space. Various unregulated measures, dubious scales and weighing practices are used, tending to disadvantage the occasional user. Towards the end of the day, with farmer-sellers reluctant to take back their produce, it becomes a buyer's market. The 'market authority' is usually concerned only in collecting market fees. The MEW can advise and/or assist the local authority in planning rural markets, in designing their structures, and in managing and operating them more efficiently. He can assist in introducing improved weighing methods, and better packing and grading practices. At the same time he can use the rural market for collecting and disseminating local market and price information, and to train farmers and traders in improved marketing practices.

SOME EXAMPLES OF MARKETING EXTENSION

Marketing extension services are already in operation in some developing countries, though their quality varies and often needs strengthening.

In South Korea the National Agricultural Cooperatives System constitutes the core of marketing extension. Primary cooperative societies employing some 20 full-time staff have a 'Guidance Section' which advises on kinds of crop to produce, marketing time and market outlets, packing, grading, market information, group marketing, arrangement for marketing loans, etc. Wholesale markets are mostly operated by county or city level agricultural cooperatives and the links with primary level cooperatives are relatively easily established. Market information is provided through five information centres in the country, so that the information may have direct relevance to local markets.

In China, city or county-level fruit and vegetable corporations – government-owned procurement and distribution agencies – employ 'production personnel' whose main responsibility is to advise farmers on what crop to produce, how much, when to sell, and where to sell. Each city company employs 50–200 such marketing extension workers.

In Papua New Guinea livestock officers provide a marketing extension service to the small farmers who fatten 10–15 cows obtained through bank loans. They advise farmers on which animal is ready for market and when to sell. They also arrange transport to the slaughter-house where the animals are processed and payment is made directly into the farmer's account under a pre-arranged price agreement.

In Nepal, Food and Agricultural Marketing Offices are being established in eight agricultural areas, each with three or four people on marketing. They are under the supervision of the Food and Agricultural Marketing Services Department of the Ministry of Agriculture. The Marketing Offices will provide marketing advisory services to the crop production extension workers, marketing advice directly to small farmers and will assist local government in designing and operating rural markets. They will also collect and disseminate marketing and price information.

In Zambia, a Marketing Officer is designated in provincial level agricultural cooperative departments; he provides marketing information such as the price and stock position.

INSTITUTIONAL SET-UP FOR MARKETING EXTENSION

It is not necessary to have a marketing extension worker at every village. For practical purposes, it would be useful to have a marketing extension worker at the 'above-village level', such as district, or say for every 3000 – 5000 farm households. Or there may be a marketing extension worker at a special development area where most of the farmers specialize in a particular cash crop.

There are two ways to institutionalize a small farmer marketing extension service. One approach is to establish a new marketing extension service by employing and training new staff as marketing extension workers. In Kenya, marketing extension workers are being trained at Egerton College; it is the intention of the government to have a marketing officer at each district level. The other approach is to use existing staff. A number (say one out of every 10 – 20 production extension workers) may be trained to become the 'marketing specialist' in a locality. They would support production extension workers in the marketing aspects of the extension work. They would be part of the existing extension system, but specializing in marketing. Staff of primary cooperatives of government marketing board's field procurement points, or of local governments who are concerned with supervision and management of rural markets, can also be trained to become marketing extension workers at field level.

To support the MEWs there must be a well-qualified marketing officer at provincial level. The provincial level marketing extension office may be a separate institution from the existing production extension service because it involves cooperatives, government marketing boards and also local governments dealing with rural markets. Such marketing offices are being established in Nepal. Provincial level marketing officers have been designated in Uganda. The provincial marketing officer can also be a staff member of the existing provincial level agricultural extension office or agricultural development office. Similarly there may be a separate marketing extension unit in the marketing department of the central government, or it may be a division of an existing agricultural extension department.

TRAINING FOR MARKETING EXTENSION

Well-trained MEWs are the key to a successful service. Many will be high-school graduates. Therefore the training programme must be geared to a basic education.

A general agricultural marketing extension course could include:

- concept of marketing;

- planning for production;
- use of marketing and price information;
- appropriate harvesting methods;
- grading and standardization;
- improved packing and handling practices;
- simple storage practices;
- production and marketing contracts;
- group marketing arrangements;
- designing and operation of rural markets;
- training farmers in marketing.

The introduction course may be 3–4 weeks, especially for those who have no prior training in marketing. In addition, specialized courses may be offered, as on market information, rural markets, grading and quality control, field-level storage, etc., each lasting 3–4 weeks or more if necessary. Such training is best organized in a rural environment, at government experimental farms, in farmer's fields, at rural markets or primary cooperatives. Demonstration by instructors and trial and practice by the trainees are most important. In-service training with primary cooperatives, rural markets, or at commercial enterprises should follow.

23

AGRICULTURAL MARKET INFORMATION SERVICES

B. Schubert

Market information services have the function of collecting and processing market data systematically and continuously, and of making it available to market participants in a form relevant to their decision-making. Their purpose is to increase market transparency, which in this context is defined as the degree of information that primary suppliers, traders, final consumers and market control institutions have about parameters relevant to their marketing decisions. Adequate knowledge of prices, conditions of sale and qualities is, from the microeconomic point of view, indispensable for rational production, marketing and consumption decisions. From the macroeconomic point of view, such knowledge is one of the conditions for the functioning of stimulus and control mechanisms, which, by way of competition and/or state intervention, take over the coordination and direction of economic processes.

Important functions of market transparency in agricultural markets are:

1. Creation of production stimuli by indicating market opportunities.
2. Stimulation of competition among suppliers and among traders.
3. Promotion of the adaptation of supplies to the development of demand:
    (a) in the short term by means of spatial and temporal adjustment (transport and storage);
    (b) in the long term by means of production development oriented towards the preference structure of the consumers and towards comparative location advantages.
4. Reduction of seasonal and erratic price variations and their associated marketing risks.
5. Provision of data as a precondition for the planning and control of agricultural market policy interventions.

Source: Passages from *Agricultural Services Bulletin* No 67, FAO, Rome, 1983.

TRANSPARENCY DEFECTS

The level of information in agricultural markets in developing countries is, in general, low and at the same time unbalanced. Obstacles in the procurement of information are as follows:

1. Lack of standardization of weights, packaging, qualities and contract conditions.
2. Supplies are particulated into units which are unsorted, small, discontinuous and spatially widely separated.
3. Lacking or unsatisfactory infrastructural prerequisites (roads, markets, communication facilities).
4. Opaque trading practices, e.g. covered bidding.

Under these conditions, especially in the case of small farmers, only inadequate information opportunities are available. To a very large extent they are dependent on information obtained from other farmers and from the purchasing traders. Their statements are, however, determined by their own interests, and can hardly be checked. On the other hand, bigger traders who have extensive contacts as well as a telephone at their disposal have a considerable information advantage. These traders often run their own information service through which they secure competitive advantages.

In the face of these transparency defects the task of an agricultural marketing department is:

1. to raise the information level of all market participants with respect to the most important decision-related parameters; and
2. to ensure that the information needs of the weaker market participants are satisfied, in order to reduce the existing differences in the level of information.

PERFORMANCE OF EXISTING SERVICES

The criteria for the evaluation of the effectiveness of market information services are derived from the information needs of the market participants. In order to aid decision-making, market information must be:

1. relevant, i.e. its content must be related to the information needs of the target group;
2. meaningful, i.e. precisely specified with regard to location, time and other features and formulated in a way which can easily be understood;
3. reliable, i.e. accurately and regularly collected and transmitted;
4. promptly available, i.e. published within a few hours of being collected; and

5. easily accessible.

Investigations of existing market information services in developing countries indicate that these criteria are fulfilled only in very few cases. The most important cause of inadequate performance seems in many cases to be of a conceptual nature. The services are integrated in agriculture ministries or similar authorities which regard themselves rather as control and administration organs and less as service institutions. They regard the function of information collection, processing and distribution as the procurement of data for administrative and statistical purposes such as for example, the calculation of indices and the preparation of annual reports. Farmers and traders are not, or only secondarily, their target groups. Many of these services 'publish' their information in bulletins which are hardly attainable to the public. In addition, the data from these services, when published are usually so out of date and so little related to the information needs of the market participants that they are, in any case, not interesting for farmers and traders.

Even when there is conceptual orientation towards the target-group farmers and traders, government information services in developing countries have to struggle with a series of institutional difficulties which have a negative influence on their performance. An administration used to normal office hours and bureaucratic procedures can only with difficulty adapt to the requirements of the market rhythm. There is a tendency among the data collectors not to expose themselves to the daily grind of collecting data in often overcrowded and unhygienic markets or from widely scattered producers. Low pay, transport problems, informants' distrust and insufficient identification with their task have a negative effect on their motivation. In addition, there are management problems. The data collectors are often untrained, they do not know the purpose of their work and they are hardly ever checked. There are neither compulsion nor stimulus mechanisms for good work – the results are corresponding.

PRECONDITIONS FOR THE ESTABLISHMENT OF MARKET INFORMATION SERVICES

In the initial stages of economic development, agricultural market information services do not play an important role. In the first place, their establishment requires a certain stage of development of infrastructure, educational level and administration. Secondly, an unavoidable need for an information service first appears with a rising share of market production and falling significance of subsistence production as a result of increasing economic development. For instance, special efforts to establish or to improve their information services were undertaken in such countries as

Taiwan and South Korea some years ago and are to be observed at present in countries such as Malaysia and Indonesia.

Previous experiences indicate that the improvement of agricultural market information services in developing countries presupposes the determined and continuous support of the development administration; it is associated with considerable expenditure of time and resources and requires a very systematic approach. However, what we are concerned with is the establishment or expansion of government service institutions which have to be able to provide private producers and traders with market information relevant to decision making in a continuous, quick and reliable way.

Determined and continuous support from the development administration refers to creation of a capable institutional base, ensuring cooperation with other institutions (extension service, media) and the continued provision of personnel and budgets.

The planning of agricultural market information services should begin with empirical studies of the marketing systems involved, of the information needs of the target groups and of the performance of already existing information systems. The data obtained form the bases for decisions on target groups, regions, product groups and the kinds of information to be included in a pilot project. As a rule the pilot project should be limited to raising the information status of producers and traders with respect to current producer and wholesale prices for selected products in the main areas of cultivation and in the most important centres of consumption. After completion of the pilot project cycle – planning, implementation and evaluation – further stepwise horizontal (additional regions and products) and vertical (additional kinds of information) expansion of the service can be planned.

SPEED OF DISSEMINATION

The more rapidly information is made available, the more relevant it is. As far as possible, not more than a few hours should elapse between the collection and publication of data. Speed is a question of organization and of the means of communication. Since telephones, telex, teleprinters, radio and television are now available in almost every part of the world, information can be transferred much more rapidly than goods. Before lorries reach the urban wholesale markets with vegetables bought early in the morning in the production regions, wholesalers and retailers can have already obtained information on producer prices. When producers and buyers meet in the morning, the producers can already have found out the prices paid in urban wholesale markets the day before. In many cases this is possible without any great additional outlay for expensive technical gadgets, but by means of the rational exploitation of available systems of communication.

In speeding up data publication the scope for improving market information services is considerable. To achieve this bureaucratic procedures should be streamlined, mailing should be partially abandoned for other speedier channels of communication, and data collectors should have access to a telephone wherever possible. Preconditions for achieving these improvements are adequate budget provisions for current expenses (i.e. to cover telephone costs) and an organizational set-up, which ensures that market information service personnel are not restricted to office hours but can fully adjust their work schedules to the requirements of the market process.

STAFF CONTROL AND MOTIVATION

Many existing services lack an effective system of control and motivation. No notice is taken of whether the collector works conscientiously or whether he invents the data at home. If this is discovered, it does not lead to any action being taken. Services operating with such data are wasting their time.

In addition to inaccuracies due to negligence, intentional manipulations were observed in a pilot project in Thailand. These were justified by saying that the publication of falling prices would only have encouraged traders to pay less to the farmers.

Measures to improve such a situation include:

1. the provision of an independent data collecting staff for the market information service;
2. repeated in-service training for data collectors, with the aim of making comprehensible the purpose of market information, in addition to the command of data collecting techniques. To this end collectors should be involved in conducting surveys;
3. employing data controllers whose task it is to check regularly and systematically the way in which data collectors work and the quality of their data;
4. setting up a system of management which involves the data controllers in monitoring and evaluating the impact of the information service. In this way performance can be continuously checked and management can immediately take steps to correct any adverse tendencies.

24

THE MARKETING DEVELOPMENT BUREAU IN TANZANIA

E. Seidler

The Marketing Development Bureau (MDB) was established in the Ministry of Agriculture of the United Republic of Tanzania with FAO/UNDP assistance in 1970. It subsequently received World Bank funds. From 1961, when it became independent, Tanzania had adopted a policy of socialism and self-reliance which, in agricultural marketing, centred on parastatals and/or cooperatives. This placed a very heavy burden on the Ministry of Agriculture, which was responsible for marketing. This Ministry had a seat on the board of each marketing parastatal; it was also very much concerned with pricing policies and was responsible for the cooperatives. In manpower terms, it had a workload that was barely supportable.

The work programme of the Marketing Development Bureau included:

1. market research and export promotion;
2. training of marketing staff for government services at the Cooperative College, Moshi;
3. issuance of marketing intelligence bulletins;
4. provision of advice on pricing policies;
5. review of the operations of parastatal marketing bodies, and assistance in solving their problems.

In addition, the MDB was asked to monitor the food supply situation. An international programme of storage construction and provision of initial reserve stocks resulted from the food security studies it initiated.

As constraints and deficits multiplied in the parastatal sector, the problem-oriented work of the MDB increased both in scope and complexity. Accounting and financial management became crucial areas of attention, the more so as substantial investment resources were pumped

Source: Updated passage from *Marketing Improvement in the Developing World*, FAO, 1986.

into the crop authorities. The abolition of the cooperative unions in 1976, and the requirement that crop authorities purchase directly from the villages, imposed a further supervisory burden on the MDB.

In such an economy, a problem in promoting marketing efficiency is that it is not possible to measure norms for parastatals' performance by referring to norms of enterprises motivated by market forces for which survival equates success. In the absence of independent initiatives to generate new ideas, proposed courses of action are confined to modification of the existing system. Relative efficiency is mistaken for the only type of efficiency possible.

By 1976, the MDB was making a full annual review of agricultural prices. This analysed, in some depth, all factors relevant to government decision-making on producer prices. It was the first attempt to present price proposals in an overall balanced study, rather than disjointed decisions made around individual crops.

In its advice to the government on pricing decisions, the MDB brought in concern for the impact on the marketing organizations involved and a live awareness of the effect of inflation. This is illustrated in the following two excerpts:

> More serious is the Government's decision not to follow MDB's recommendation to differentiate between white sorghum, which is readily acceptable for human consumption and red sorghum which with its high tannin content and bitter taste is useful mainly as an ingredient in both traditional and modern sector brewing and also for stock feed. A price of 130 c/kg with a premium of 30 c/kg for Mwanza, Mara and Shinyanga was recommended for white sorghum while the price of red sorghum was to remain unchanged in money terms at 100 c/kg. Although the National Milling Corporation does not keep separate figures for white and red sorghum, it is thought that the majority of its purchases are of the higher yielding, lower risk red sorghum. It is probable that the 1982/83 announced price for sorghum will lead to purchases of red grain substantially in excess of domestic market requirements. NMC can expect to have storage capacity tied up by this product, and to incur heavy financial loss from deterioration in storage and loss-making exports, as was the case in 1978/79.

> It was recommended that the retail price of beans be increased from 350 to 480 c/kg (Grade I) and 275 to 380 c/kg (Grade II) in order to maintain real income from bean production at about the present level. It will be recalled that the present produce price for beans has remained unchanged since 1977/78, at which time the price was probably pitched unduly in favour of the producer. It is considered that the approximate 30 percent drop in real value which the

unchanged money price of beans implies will have a significant negative impact on NMC purchases. It is likely therefore that availability of beans for sale by NMC in 1982/83 will fall short of demand, which in 1980/81 resulted in NMC sales of 25 000t.

A new priority consideration in marketing around this time became the cost of fuel. More than half of the United Republic of Tanzania's export earnings went on imported energy. Export crop prices and quantities were largely static, or declining. Transport requirements became a major issue. The feasibility of continuing pan-territorial pricing in accordance with socialist principles was another aspect that was sharply under question.

In the mid-1980s Tanzania began a series of economic recovery programmes and structural adjustment in cooperation with the IMF and The World Bank. The MDB undertook a series of studies and provided policy and planning advice to the Government on the move to liberalized marketing of both domestic food crops and cash export crops. The MDB has in recent years taken on the role not only of assisting in planning marketing liberalization but in monitoring its progress and facilitating its implementation. A grain marketing information system covering the main urban markets in the country was established with market prices being reported over the radio two to three times a week for the benefit of private traders. This market price reporting scheme has facilitated market transparency and promoted interspatial arbitrage.

The MDB continues to receive international assistance which complements increasing local staff numbers. Whilst some of the original national project staff have left the project on promotion to senior positions in other departments in the Ministry, a number of the original Bureau staff still remain in senior level posts. The continued success of the Bureau in promoting meaningful policy advice has been in large part due to the Bureau's ability to retain its staff. The Assistant Commissioner in charge of the Bureau has been there since its inception 21 years ago. The provision of overseas training possibilities to national staff, the opportunity of working with international personnel and the provision of transport services and efficient secretarial support kept national staff committed to the Project, but declining real salaries in the public sector in recent years have seriously undermined this.

The success of the MDB can be regarded as being due to the long-term international support provided, stability in the national staff under the guidance of its head who has provided continuity throughout its life and recognition by the Government and external donors of the need to maintain the high level advisory services, data bank and policy analysis capability of the Bureau. This has been especially so in recent years with the transition in economy from centrally planned, government directed marketing to a more liberalized, free market based system.

V
PROVISION OF SEEDS AND FERTILIZERS

How to organize the distribution of agricultural inputs to large numbers of small farmers became important with the advent of high yielding varieties of rice and wheat and hybrid maize. The first studies looked at the prospective demand as a basis for investment in production facilities and infrastructure for distribution. Financing requirements (see the paper by Abbott) and subsidy policies then came to the fore. Meetings sponsored by the Government of India and the aid agencies stimulated a flow of policy papers. The Berg report for the World Bank, of 1983, which also covered output marketing, was the first step in its campaign for a general liberalization and a retraction of state enterprise in favour of independent competing initiatives.

FAO papers by H.J. Mittendorf and A.W. Shepherd have compared fertilizer marketing costs in African, Asian and Latin American countries. The shift to private wholesaling and retailing and reduction in subsidies in Bangladesh was studied by an IFPRI research team (*Fertilizer Policy and Foodgrain Production in Bangladesh*, Washington, 1983), and K.L. Moots *Evaluation of the Bangladesh New Marketing System* (IFDC Muscle Shoals, 1982). The IFPRI report by Pray and Ramaswami summarized here gives a balanced view of the issues in seed distribution.

'How to do it' books include the FAO guide *Fertilizer Marketing* by K. Wierer and J.C. Abbott, 1983, a *Handbook on Fertilizer Marketing* (Fertilizer Association of India, New Delhi, 1976), A. Fenwick Kelly's *Seed Planning and Policy for Agricultural Production* (Bellhaven Press, London, 1989), and *A Manual on Seed Marketing Management in developing Countries* (FAO, Rome, 1987). P.R. Mooney (*Seeds of the Earth: Private or Public Resource*, Inter Pares Ottawa, 1979) stresses the risks involved in the concentration of seed development and distribution in the hands of a few private companies.

25

FINANCING FERTILIZER DISTRIBUTION NETWORKS

J.C. Abbott

Promotion of fertilizer consumption depends on the availability of finance for distribution. Funds are required for investments in storage, mixing plants, local warehouses and trucks, management and technical know-how, and for operating capital; in addition, the fertilizer itself must be financed through distribution and, to a great extent, through purchase and use by the farmer. Unfortunately, the tendency in many countries has been to focus on investment in production and related physical facilities, while paying lip service to the need for operational credit and less to the financing of management and know-how. Experience shows, however, that the training of staff, the introduction of advanced management methods, and the availability of funds to hold or purchase the fertilizers are as important as the physical facilities.

DISTRIBUTION FINANCING

A fertilizer distribution system must have sales agencies with stocks of the types in regular use within easy reach of all farmers. Preferably there should be more than one agency, so that the farmer has an alternative source of supply and there is a check on slow service and petty exploitation. They must all have storage and staff. Central and regional storage is needed to serve them, also transport. Operating capital is needed all through the system to pay staff and transport and to finance stocks from their acquisition from the factory through wholesale and retail distribution until the farmer receives payment for the produce on which the fertilizer was used. Bags would be another direct capital outlay. Financing of outlays on fertilizer stocks and handling costs is required over the whole of the period in between receipt of fertilizer by the wholesaler and sale of the

---

Source: *Monthly Bulletin of Agricultural Economics and Statistics* **26**(5). FAO, Rome, May 1977.

resultant crop by the farmer, which in single crop economies is commonly six to nine months. The total amount of finance needed at one time is somewhat less where stocks do not have to be built up for one single season of use. If there are two clear cropping seasons, and if purchases are spread more or less equally over the year, operating capital need is reduced to about two thirds of the value of the year's sales. Part of it will have been repaid during the year by the farmer and will again be available to finance sales later.

Where fertilizers are imported another operating capital or credit element is involved, and another time dimension. If procurement and distribution are efficient, the time between ordering imports and their use on the farm may not exceed six months; if there were another six months until the crop was sold the total duration of financing would be one year. However, procurement and distribution take more than six months in some importing countries; in some of the land-locked countries of central Africa it has even taken a year.

To commit all the financial resources available for a development project to physical facilities and financing sales, leaving no money for the training of staff, is a great mistake, being one of the reasons why many fertilizer plants in developing countries operate far below capacity in their initial years. It has been estimated that 2–3% of the investment cost of a fertilizer manufacturing plant should be reserved to build up personnel. To staff adequately a fertilizer distribution system the percentage should be much higher; at the wholesale stage it may reach 3–5% and in retailing 5–10% of the investments in physical facilities. Where fertilizers are distributed through traditional village traders, the need for additional physical facilities may be small but the need for management training could be considerable, the cost of organizing a one-week practical training course for stockists in Kenya (including the provision of training materials) being US$50 and $100 per person in 1974. At the farm level the main expenditure on staff training is in providing practical advice on the benefits of fertilizer use and the techniques of application; $10 per ton of annual fertilizer consumption has often been taken as basic budget for this. The investment for fertilizer distribution ranged from $180 to $260 per ton of annual throughput in seven World Bank case studies; the operating cost averages $50 per ton. Thus the investment and operating costs of distribution are on a par with those of fertilizer production (Sheldrick, 1976).

A condition for the successful organization of fertilizer distribution is that it be effectively integrated with extension, agricultural marketing and other related rural services. An adequate basic infrastructure of ports, roads, bridges and communications is essential for moving fertilizers and necessary for the supply of other inputs, the marketing of farm products and other economic activities. In fact, fertilizer distribution is a function enmeshed with the rest of agricultural marketing. Trading contacts with farmers for buying their produce are a valuable basis for selling fertilizer; management training for the employees of a cooperative handling fertilizers should serve the cooperative in its total development; and advice to the farmer on fertilizer use should benefit his whole farming enterprise. The difficulty, however, 'of identifying and putting together well defined system and investment packages which can easily be superimposed on existing infrastructure and institutions' has been one of the obstacles to international financing.

SOURCES OF FINANCE

Fertilizer distributing firms and cooperatives use three basic sources of financing:

1. original capital of the owner, or shareholders, or as provided by the sponsoring government or other body;
2. other financial resources and savings of the enterprise; and
3. loans from banks or outside financing agencies.

Some west European fertilizer manufacturers choose as agents firms which can show that they have the means; then a discount is made on the price to cover the costs of receiving and holding stocks. Their measure of an enterprise's ability to operate on this basis would be based on a visit to the would-be agent's premises, evidence of his turnover of related business, and a banker's reference on his financial standing. The same approach was used in Pakistan when the market was opened to private enterprise, but there are places where very few suitable agencies are available. There may be firms which have the financial standing and the rural business contacts, but it is questionable whether fertilizer distribution would ever attract their first attention, and whether they would devote much effort to selling the manufacturer's product.

In many developing countries the resources of local traders are limited so that to start fertilizer distribution they frequently need outside financing. They can go to banks and seek operating capital there, but the question then is generally on what terms and against what security. For investments in storage and transport facilities distribution enterprises might need loans of three to perhaps ten years' duration; for working capital also three years or more must be envisaged if the firm is not forever to be squeezed by a rigid repayment schedule. In some cases government banks are reluctant to finance operations of private traders; on the other hand, there are commercial banks who would lend more to private traders rather than to cooperatives.

The possibility of obtaining capital by launching a joint stock company is still open in many countries. It depends on public appreciation of the commercial prospects of the enterprise, which often lags behind the awareness of the individuals closely concerned; unless they are outstanding promoters or are blessed with good contacts, mustering capital in this way can be a slow process. For this reason mixed companies were set up in Ecuador to market various agricultural products, their basic capital being provided by government institutions with private enterprise participating. Another possibility is for fertilizer wholesale companies to seek capital from foreign investors. Here the need to remit returns in foreign exchange will be an obstacle in some countries and the risk of expropriation a deterrent; on the other hand, they may get tax exemptions during the initial years and duty-free import of essential equipment and materials as an incentive.

The simplest way of meeting seasonal financing requirements is for the credit needed at the wholesale level to be gradually transferred by the wholesale distributors to the retailers and from them on to the farmers. This can be credit in kind, the fertilizer which the distributor is expected to sell, and such a direct passing of financial resources contributes greatly to marketing efficiency and eliminates the trouble, costs and uncertainties of an independent search for credit by each enterprise at successive stages of distribution and use. In addition, it facilitates sales at each stage; experience over a number of years in Mexico shows that the private distributors were able to sell most of their fertilizer to farmers on credit specifically because they received ample credit from the wholesale company. If credit requirements at wholesale, retail and farm level are treated independently and have to be met from various unrelated sources, this can increase the total financing requirement considerably. Furthermore, delays and uncertainties in obtaining such credit will slow down the process of marketing and add to its cost. Behind the Mexican wholesale supply company, however, was the Government; when European fertilizer manufacturers were asked whether they stood security for bank loans to their distributors the reply was 'No, we are manufacturers, not bankers!'.

Cooperatives dependent for their capital on farmers' subscriptions face the same difficulty. Farmers may see clearly the potential advantages of a new distribution channel, but those best able to contribute capital may already be fairly well served. The smaller farmers who stand to gain most may have little capital to contribute and a high reluctance to let it out of their own hands. In several countries where cooperatives are involved in fertilizer distribution a special cooperative development bank has been set up to provide the necessary operating capital. In practice such financing for cooperatives seems essential.

Fertilizers may be stocked at a retailer's store on a consignment basis, while the retailer – he may be a cooperative or a private dealer – receives a commission. Although this solves part of the credit problem of the retailer, the experience has often been that he does not make the same sales effort as under conditions where he owns the fertilizers.

For both private and public distribution systems it is important that they be able to accumulate some capital from their current operations; cutting profit margins to the point where they cover only the immediate costs means that no money will be available to maintain the system. Not only must fertilizer distributors be able to meet staff and other direct operating costs, they must earn enough to cover depreciation of plant and equipment and replace it when needed. They must also be able to maintain sufficient operating capital. Under conditions of inflation this means expanding the original amount in pace with the decline in the value of money.

For major new investments in fertilizer distribution in developing countries the burden of financing generally falls on the government's own resources in local and foreign currency, and loans and aid from multinational sources and from other governments. Smaller contributions may be expected from the suppliers of equipment and machinery, and from private companies allowed to participate.

Such investment projects generally include a foreign exchange component which may be as high as 50% and a local currency component. Stores for fertilizer distribution can generally be set up with local currency, while motor vehicles, handling, mixing and bagging equipment (and often the bags too) call for foreign exchange. The financing of management and know-how may call for local currency and foreign exchange, according to where staff are recruited and training takes place. Repayment of a foreign exchange loan involves generating not only the necessary cash flow over the designated period but also assurance that it can be converted into foreign exchange, which can only come via a government guarantee. Provision of the local currency component is the normal role of a development bank, again often backed up by the government.

Flexible policy attitudes on the part of both external and national financing agencies are essential. Too often loans have been limited to the so-called hardware component, or investments in physical facilities, the local cooperating agency being left to supply operating capital and finance investments in management, staff and know-how. The result of such a policy can be large physical facilities which are not used to full capacity. It is important, therefore, that loans cover a balance of fixed investment, operating capital and technical know-how.

A continuing constraint on the establishment of effective distribution systems in developing countries in parallel with the creation of national fertilizer production capacity has been the apparent ineligibility of small-scale distributors for public finance; nor do they figure in the international loan programmes of most developing countries, unless they are cooperatives. The problem is both administrative and political. There is difficulty in formulating programmes and supervising credits to many small enterprises, some of which may be engaged in various lines of business at the same time, and there is reluctance to make public loans to enterprises often thought too deeply engaged in mobilizing capital out of the small earnings of their rural customers. How to find suitable ways of channelling development finance into fertilizer distribution systems involving small-scale private enterprise is one of the challenges before the development bankers of today.

CREDIT FOR THE FARMER

There are some people who argue that credit to enable the farmer to buy fertilizer without having to pay cash at the time of delivery is not an issue. They say that if the economic return from applying fertilizer is sufficient, then the money to buy will always be found. Clearly there are degrees of profitability in using fertilizer. At a high level of profitability the farmer will try all ways to find the cash to pay for at least some fertilizer if he has no access to credit; at less obviously profitable levels, access to credit becomes a vital issue in his decisions on how much fertilizer to use. Moreover, for the farmer fertilizer credit requirements are mainly seasonal. If he diverts to a seasonal use his own capital, he will be unable to apply it to some longer term improvements and equipment that might be important for the enterprise as a whole, and for part of the year it will be unproductive.

In organizing fertilizer credit services for farmers two major practical considerations stand out:

1. How can the farmer obtain credit to buy fertilizers easily, promptly and without having to go through complicated procedures, so that he can be sure of having it when he needs it?
2. How can fertilizer credit be integrated with related services such as extension, water supply where irrigation is important, and with the marketing of farm produce, to best assure that the fertilizer credit is used productively and to minimize the risks of non-repayment?

The spectrum of developing countries shows a great variety of sources of credit to farmers to buy fertilizer: private retailers, cooperatives, commercial banks, specialized agricultural development banks, specialized marketing and processing enterprises or boards, and agrarian reform and area development institutions. In many cases the advantages of linking credit, fertilizer distribution, marketing and extension have been recognized, and these services are offered as a package. Too often, however, the coverage is limited and the service is poor. The bulk of the farmers, especially small producers of domestic food crops, have no access to credit to buy fertilizers, possibly lacking land titles or other security acceptable as collateral. A small loan costs as much to supervise as a larger one, so credit agencies find it too expensive to justify advances to small farmers on the basis of crop sales alone.

Another common complaint is that the procedures followed by official credit institutions in making fertilizer and other production loans are too complicated and time-consuming. There are months of delay in approving loans, with the result that the fertilizer arrives too late at the farm, or that the farmers end up being obliged to use credit from other sources on very unfavourable terms.

Private retailers

Where private retailers have a major role in the marketing of agricultural produce, they may also provide efficient supply services to farmers for fertilizer, including selling it on credit. Often they are well placed to judge the creditworthiness of farmers and to follow up the repayments, and usually they are more flexible than institutions in providing their credit quickly and without bureaucratic procedures. However, the extent to which a private distributor can finance his farmer customers depends on his own ability to obtain finance, traditionally done via a commercial bank. The quantity that can be obtained from this source reflects the trader's own volume of business and proved reliability in repayment, and the interest charged is likely to be the going rate for commercial advances. If the retailer is to carry the risk and cost of assuring repayment he will need some recompense which, in a competitive situation, may have to come from within the margin he obtains on selling the fertilizer. Where all the bargaining power is on the distributor's side, the farmer may have to commit to him the produce grown with the fertilizer and forego choice of an alternative outlet.

Over 60% of the credit for fertilizer sales in Kenya is provided by private distributors. Generally they find it easiest to do this if they are also buying the farmer's produce; fertilizer retailers dealing only in farm supplies have not this edge in securing repayment. In Europe the general practice is to allow credit for six weeks or so, while in some countries such as Brazil, although private fertilizer distributors do not provide credit themselves, they help farmers obtain it from a bank. Distributors can do more in providing credit (and in providing it at lower costs) when they themselves are financed directly by a major manufacturer or importer. Mexico has already been mentioned as a country where private fertilizer distributors finance about 70% of their sales, themselves receiving credit from the government-controlled manufacturing and wholesale enterprise.

Cooperatives

Cooperatives may be set up to distribute production supplies and make these available to members on credit. Most link this with selling members' produce, since in this way they can obtain direct repayment of the loan from the proceeds of selling the borrowers' crops. A notable example is the Federation of Coffee Growers in Colombia, which handles about 15% of the fertilizer used in that country. Practically all these fertilizers are sold on credit, which is recovered from coffee sales through the Federation. Likewise, a cooperative in southern Brazil set up to market members' fruit and vegetables supplies them with fertilizer on credit.

These are examples of cooperatives built up by homogeneous groups of specialized producers with a clear common interest. They are strong enough to supplement their own financial resources by large-scale borrowing from banks for which the federation is accepted as guarantor. Cooperative systems set up by governments to serve large numbers of small farmers rarely have much lending capacity of their own and depend on central financing for their credit operations. This is the case of Egypt, where fertilizers are provided on credit through a network of cooperatives drawing their funds from the government-sponsored General Organization for Agricultural and Cooperative Credit.

Experience indicates that credit should be provided by the same cooperative that handles the fertilizer and markets the crop. In Iran and a number of other countries, selling fertilizers through cooperatives and offering credit separately through an agricultural bank has been less successful.

Commercial and agricultural banks

Commercial banks are a potential source of credit for farmers for fertilizer purchases in most countries, but in practice their role has been limited. With various opportunities of employing their funds, lending to small-scale agriculture in developing countries has seemed too risky and troublesome.

The approach followed by some governments has been to set up banks committed to lending to agriculture and staffed accordingly. The Agricultural Bank of Colombia finances more than half the fertilizer sales in that country and is itself a direct distributor for about 25% of the fertilizers used. These and other inputs are sold on credit mainly to small-scale farmers through more than 400 local agencies throughout the country. While this bank does not itself provide technical advice to farmers, it has an agreement with the government extension agency to help all farmers buying on credit. In several other Latin American countries also public banks are engaged in fertilizer distribution.

Agricultural banks are also the main source of credit for fertilizer purchases in many developing countries of Africa and Asia, but less commonly are they direct distributors. The Tanzania Rural Development Bank has distributed fertilizer, but its clients were relatively large cooperatives. In China fertilizer is ordered for production units by the supply and marketing cooperatives of the commune. While the less well-off units can obtain it on credit, the emphasis is on payment at the time of collection.

Other private and governmental institutions

The problem in serving many small farmers is how to get the fertilizers into their hands at the right time and see that they are used efficiently. Providing fertilizer on credit as part of a package obtainable from one source and intended to cover all a farmer's requirements has been favoured in some countries. One model is the agricultural service centre of Sri Lanka. This combined offices for fertilizer distribution, extension, crop insurance and the Bank of Ceylon, with stores for fertilizers, seeds and equipment. In the BIMAS package for agricultural development in Indonesia, fertilizer accounts for about 60% of the total credit granted. In Iran and Morocco about half of the total fertilizer supply together with credit has gone through such special programmes. While the provision of credit and fertilizer through these package programmes has accelerated development, the overhead costs are often high and only specific sets of farmers are served. An important criterion of such a programme is whether it constitutes a model that can feasibly be extended over the whole country.

Some large enterprises engaged in the processing and marketing of specialized crops enter into fertilizer credit operations because they are concerned that the farmers supplying the raw material should meet certain quality standards or fit into a tight processing schedule. Commonly they provide fertilizer to their growers on credit along with seed, pesticides and advisory services under production and marketing contracts. In Venezuela, for instance, almost 15% of the fertilizer has been distributed on credit by the sugar processing plants; these credits are recovered when the farmers bring the sugarcane to the plant. Similar arrangements are usual for tobacco, oilseeds, fruit and vegetables and other crops for processing. Integration of extension, credit, input supply and marketing on a crop basis can, however, conflict with integration of these services on an area basis. This is an issue meriting careful consideration.

There are some African countries, such as Malawi, where a government marketing board distributes fertilizers on credit and recovers the loan from payments to the farmers for their crops. However, some of the boards long established in marketing are reluctant to engage in credit collection.

They see it as complicating their operations with the risk of inducing some farmers to sell their produce through other channels.

Credit terms

For the farmer it is important that the terms on which he obtains credit for fertilizer fit closely his production and marketing programme. The benefit from applying fertilizer is realized in the larger crop that results and the increased income obtained when it is sold. Taking into account this sequence of events, the date at which a fertilizer loan has to be repaid is often a more serious consideration for the farmer than the rate of interest charged. If to repay the loan he must sell his crop immediately after harvest when prices are normally at their lowest, then much of the income benefit of his higher productivity may be lost.

In Mexico the duration of credit from the fertilizer distributors ranges from 30 to 300 days according to the crop. In Brazil and Peru, some fertilizer credits are extended until 30 days after harvest. However, a post-harvest allowance of up to three months would do more to ease the pressure on small-scale farmers to sell their produce as soon as possible. Another solution would be if farmers could obtain marketing and storage loans from another source and use these to liquidate their fertilizer credits. Sometimes this is organized through cooperatives for members who deliver their produce to them for marketing later in the season.

There are also fertilizer treatments from which the benefits accrue over a number of years – phosphates and lime – and applications to perennial crops like coffee, tea, and oilseeds. These should be eligible for medium-term loans of corresponding duration.

Repayment of loans

Assurance of repayment is a major issue for all credit systems. Banks often require land title as security even for short-term credit, which greatly increases the complexity of borrowing procedure and automatically excludes those farmers who have no such title. Since fertilizer credits are intended to expand farm output, the crops and livestock which benefit should be the security.

Committing a farmer to sell his produce through a particular marketing enterprise which undertakes to deduct credit repayments is being pursued in many places. In East Africa this system is known as 'letter of instruction' by which the farmer instructs his marketing agent to direct the crop proceeds through the bank providing the credit.

Usually such 'stop order' arrangements relate to credit advanced for a particular crop. They are easiest to manage in a one-channel market controlled by a marketing board or state corporation and for crops such as tea, cotton, and sugarcane which must pass through a limited number of processing plants. Another device applicable under controlled marketing arrangements is the credit identity card. Holders must sell their produce to an official buyer, who is obligated to check the credit card and deduct repayments on loans indicated there.

For the much more difficult situation where farmers raise various crops which they can sell to a range of buyers, experiments are going on with joint pledging of repayment by a number of farmers. Credit for fertilizers from the Agricultural Bank in Afghanistan at one time required the participation of ten farmers. All had to agree to make up for the failure of any one of them to repay his loan. This approach has been used for some time with the small farmers in some parts of Mexico; 50% success is reported.

The toughest line in securing repayment of production credit is perhaps that of some irrigation districts in Ecuador and Costa Rica which have used access to water as security. Cutting off the water, or threatening to do so if a credit is not repaid, is a very effective control.

Streamlining of procedures

Whatever the problem of repayment, there is no excuse for some of the delays and arbitrariness of official institutions. Up to eight different documents have been required: a loan application form, copy of the registration card and the land title, a certificate that taxes have been paid, a copy of the last bank account, one or two letters of guarantee from other persons, and a detailed description of the farm.

There have also been cases where as many as six different sections in the same institution had to read, evaluate and approve the request – the sales office, the general credit department, the accounts section, the legal department, the office of the general manager, and the economic analysis department. Delegation of authority to branch level and systematic pursuit of such goals as 'ninety percent of loan requests approved or rejected within 15 days' are essential. Institutions with no such sense of urgency should not be in the business of credit for fertilizer distribution.

CONCLUSIONS

This analysis raises the following basic questions: Do fertilizer sales go faster if the credit is provided by the fertilizer distributor or by a separate institution? Can we say that the first holds good wherever farmers have become familiar with fertilizer use? Is a tripartite system, with the credit coming from a separate institution on the advice of extension officers and the distributor merely handing out fertilizer against chits, useful mainly for development phases, or can its continuation be justified on the grounds that all a farmer's credit should come from one source?

Existing marketing enterprises with storage and transport facilities, and established contact with farmers can usually be found, but there must be ways of putting more capital into their hands if they are to provide the sales initiative and service that are needed for fertilizer distribution. Fertilizer manufacturers, banks of various kinds and governments must all take a share in this.

REFERENCE

Sheldrick, W.F. (1976) *The Role of the World Bank in Helping to Meet the Fertilizer Requirements of Developing Countries.* The Fertilizer Society, London.

26

PROBLEMS OF MARKETING AND INPUT SUPPLY

World Bank

The central problem in marketing and input supply is the very general tendency to give too large a set of responsibilities to public sector institutions, and too few to other agents – individual traders, private companies, and farmers' cooperatives.

MARKETING AGENCIES

Export crops are almost everywhere in Africa marketed by state trading organizations; often they use 'licensed buying agents', private traders, to help in village-level purchases. Government monopolies also exist in many countries for food crop purchases, though these are generally less well organized than export-crop marketing and are in most cases unable to purchase more than a minor share of marketed output.

The performance of the export crop marketing agencies is of major importance in several respects. First, their degree of efficiency affects the share of export proceeds that can be paid to producers. Because of long distances and the frequently difficult problem of access, the cost of marketing tends to be high even under conditions of efficient marketing operations. In Kenya, for instance, charges for marketing, storage, transport, and administrative overheads averaged 34% of the f.o.b. border price for maize, 23% for wheat, and 48% for rice during the 1972–1979 period – and the agency in question is not regarded in Kenya as a particularly inefficient marketing institution.

Second, the crop marketing agencies are the major point of contact between peasants, the money economy, and the state bureaucracy. Unless the marketing transactions are done fairly and efficiently, there are high

---
Source: *Accelerated Development in Sub-Saharan Africa.* World Bank report, Washington, 1981.

risks of peasant disaffection from both the bureaucracy and the market economy.

Serious inefficiencies characterize the operation of most marketing agencies. Some of these arise from problems found in almost all parastatals – overmanning, inadequate non-salary budgets, and management scarcities. There are also inefficiencies peculiar to the export crop parastatals due to the lack of competition. And there are additional problems in these agencies when marketed volumes stagnate or decline: decreasing turnover is compensated by higher overhead per unit, the producer price being the residual. The result is an upward spiraling of costs and a parallel downward spiraling of exports. Classic examples of this are groundnuts in Mali and several export crops in Tanzania (Ellis, 1979).

In food-crop marketing, parallel marketing channels exist in many countries of the region; the legal and official marketing agency coexists with a semiclandestine private trading sector. This is most often the situation with respect to foodgrains. In these markets, attempts at controlling marketing and prices are most extensive but they are effective to varying degrees. In countries importing wheat, price controls are often fairly effective for flour and bread; for rice, the degree of control depends primarily on the share of paddy grown in government-controlled schemes. For domestic cereals, the share of official trade in marketed production may be as high as 25–50% (in some East African countries), or as low as 1–2%. Price and marketing controls are conspicuously absent for roots and vegetables, no doubt because of the problems and risks associated with the perishability of these crops. Most governments do not put much trust in the private sector's ability to cope with the task of providing stable supplies of food to the urban masses, although private traders handle the bulk of the trade almost everywhere. In most cases private traders are tolerated openly or tacitly as indispensable partners but are not allowed to work in an economic environment that would enable them to realize their full potential. The uncertainties associated with the ambiguous position of private trade and traders discourage full-time involvement in food marketing, investment in transport and storage, and a systematic approach to developing an adequate supply network.

Official marketing agencies are responsible for collection, transport, sometimes processing (as with rice), and distribution to the wholesale and sometimes even retail level. But producer prices and consumer prices are fixed by government with little regard to the actual cost of collection and distribution. Marketing agencies are also not always or not fully reimbursed for losses incurred in the process. Several of them have accordingly accumulated large deficits reflecting operational inefficiencies and the cost of government-imposed subsidization of consumers. In several cases, deficits have reached striking proportions, given the rather modest quantities of foodstuffs controlled by these agencies. Some agencies are passive, buying whatever small quantities are offered to them at official prices in postharvest periods when market prices are low, or during bumper crop years, buying all they can pay for. Others exercise varying degrees of compulsion, occasionally bordering on outright requisition.

INPUT SUPPLY

Input distribution agencies are another part of the rural marketing system that has contributed to the poor performance of agriculture. Input supply involves more than just the provision of fertilizer and seeds. It can include the supply of farm equipment, fencing and building materials, tractor hire services, and spare parts.

Unless farm inputs are made available to farmers on a regular basis and at the right time, there is little chance that agricultural production and productivity will move forward. Unfortunately, there are only a few countries in Africa where this important condition is fulfilled. Procurement and distribution of inputs is another field monopolized by governments or parastatal agencies. In more than 60% of African countries, governments reserve full control of the procurement and distribution of fertilizer, seeds, and most other services as well (see Table 26.1). The motives for entering this field are similar to those advanced for government involvement in food crop marketing: inputs are seen as vital commodities that should not be left to the care of the private sector, which is regarded as exploitative and unreliable. Policymakers also frequently perceive a need to subsidize the

Table 26.1. Relative frequency of government and private sector control in the procurement and distribution of agricultural inputs, 39 countries

Item	Percentage of countries			
	Fertilizer supply	Seed supply	Chemicals supply	Farm equipment supply
Government control[a]	64	61	47	42
Private sector control[a]	11	11	17	22
Mixed government and private sector involvement	25	28	36	36
Total	100	100	100	100

[a]Procurement and distribution activity is considered 'private' if more than 80% of it is in the hands of the private sector, and 'government' if more than 80% of it is in the hands of the public sector.
Source: World Bank data files.

service provided, which is a rationale for monopolizing its distribution.

Subsidies need not involve monopolization. The market mechanism can be used. But most officials do not believe that markets work well enough to be utilized this way. They believe, in this case, that subsidies granted to importer-wholesalers or to private traders would not be passed along to farmers.

Many officials also believe that only by public distribution will inputs be made available to the remote areas that private trade is assumed to neglect because of low profitability. While this may be true in some cases, it is mainly the policy of panterritorial pricing – fixing official prices of inputs uniformly for the entire territory without regard to actual transport costs – which impedes private trade from effectively competing in remote areas.

There is no a priori reason why government agencies should not be able to fulfil the input supply functions efficiently, but due to the structural problems besetting many public agencies – scarce management, lack of incentives, conflicting objectives, overstaffing, and lack of control – they have rarely succeeded in meeting the rigorous requirements of their clients – input delivery at the right time, at the right place, and in the right amounts.

Government agencies have failed to meet these needs because they have difficulties in adapting bureaucratic, financial, and administrative procedures to commercially oriented operations. For example, they fail to buy inputs on a phased basis because they are geared to the time of release of funds from the budget, and these are not necessarily the optimal times. Likewise, pay scales and hiring and promotion procedures tend to be similar to those in government. This leads to reduced individual initiative, unwillingness to make quick and independent decisions, and consequent efficiency losses.

The absence of competition in input supply also leads to a lack of innovation. Inputs are ordered in routine fashion without regard to location-specific requirements. A recent study in Senegal has revealed considerable scope for savings in fertilizer cost to farmers, notably by tailoring the nutrient content more closely to their needs, by eliminating ineffective elements, by reducing transport cost through higher concentration, and by determining optimal dosage and composition on economic rather than on technical grounds. Such adjustments, and the supply of inputs in variable package sizes convenient for farmers, would be introduced more readily and on a broader base in a system leaving more scope for a private sector participation in input supply.

The general problems outlined above are exacerbated by the common practice of government subsidization of inputs, in particular fertilizer. This has a number of negative consequences. First, in monopolistic input distribution systems the funds budgeted for input subsidies limit the total amount of fertilizer made available; under the budgetary constraint that many countries experience, the actual amount of inputs that can thus be purchased remains far below the quantity desired by farmers at the subsidized price. Therefore, rather than supplying more farmers than can be be served under private trade conditions, governments end up serving considerably fewer. Second, since the quantity delivered remains well below the level of demand, price is driven up and, in spite of the subsidy, some users may pay as much as – or conceivably more than – they would under free-market conditions. Third, even where the input distribution agency has a source of finance independent of the budget (such as bank credit or sufficient working capital as equity), subsidization ties its operation to the budget year, causing delays in procurement and untimely delivery to farmers.

REFORM OF PRICE, MARKETING AND INPUT SUPPLY POLICIES

While there is not much disagreement with the general propositions that higher producer prices would stimulate production and sales, or that marketing systems should become more efficient, pushing beyond these propositions is not easy because the problems are complex and involve broad aspects of development strategy. For example, the appropriate level of producer prices, the relationship between prices of export crops and food crops, and between prices of individual crops in each category are all a function of a government's development goals and social policy objectives. None the less policy changes are needed and the directions of change are discussed in the following paragraphs.

Export crops

Trade data for the 1970s and the level and trend of export taxation suggest that in many countries there is scope for increasing producer prices for export crops. The slow growth of world demand for many primary commodities is not a valid argument to the contrary as long as Africa does not even maintain its market share. Higher producer prices in real terms would stimulate production directly. It would also allow elimination of most of the subsidies on inputs, equipment, credit, or water that now hamper the distribution of these commodities and services and distort the allocation of resources.

Four objections to a high price export crop policy were noted earlier: the need for government revenues; the limited freedom for manoeuvre on producer prices due to overvalued exchange rates; high marketing costs; and the conflicts that develop with food self-sufficiency objectives.

With respect to government revenue, a number of observations are in order: first, revenue preservation should take second place to the need for maintaining or increasing the pace of export production; second, reduced taxes should raise export levels so that higher volumes would to some extent compensate for the reduced rates; finally, and even more significantly, higher producer prices should still leave scope for taxing some of the 'rent elements' prevailing for some crops – coffee, cocoa, tea, and even cotton.

The exchange rate issue has to be confronted directly or indirectly.

In many cases, lagging export growth and soaring food imports are attributable to this factor. An adjusted exchange rate, or revised tax and tariff policies with equivalent effects, would allow better incentives for export crop production and would, if the resulting increase in import prices is passed on to consumers, curtail demand for imported cereals and put food production programmes aimed at import substitution on a sounder economic basis. At the same time, a corrected price relationship would boost and stabilize demand for traditional staples, which usually can be grown at lower cost than wheat and rice. Although there are supply problems for traditional staples as well, these problems are not unrelated to the mismatch between the structure of cereals demand in African countries and the associated structure of domestic supply.

More will be said about marketing problems in later sections. Two points suffice here. First, the export crop monopolies suffer greater propensities to inefficiency than the food marketing agencies, because they are exposed to no market pressures pushing them to reduce costs. This means they risk becoming a steadily growing drain on export proceeds, with producers paying the price.

Second, the situation has evolved since the first years after independence, when many or most of the export crop monopolies took their present form. The export crop marketing monopolies arose because, except where there were public agencies, the export trade was almost entirely in the hands of foreign firms and immigrant merchants. Few societies, much less those newly independent, would accept that control over trade in vital commodity exports be so completely dominated by foreigners. Since nongovernmental alternatives were scarce, governments took control. It is now evident, however, from 20 years of experience in many different settings, that a high price is being paid to keep the export monopolies in place. Since new abilities, both private and organizational, have now developed, it would seem time to make export crop marketing more competitive. Cooperative marketing could be more widely encouraged, as among coffee growers in Cameroon (see Box 26.1) and private traders might be allowed fuller entry, perhaps for sale to a state export agency. In this, as in all proposals for structural change, it should be stressed that there are no ideal solutions. Rural markets function imperfectly in many respects, and there are risks that some farmers may suffer from unequal

> **Box 26.1. Arabica coffee marketing in Cameroon's Western Highlands**
>
> Since 1958, arabica coffee producers in Cameroon's Western Province have marketed their production through six cooperatives under the leadership of the Union of Arabica Coffee Cooperatives of the West (UCCAO). In recent years, about 100000 smallholders have used the cooperatives to market about 18000 tons of coffee and to purchase about 20000 tons of fertilizer annually.
>
> Although chartered by the Government, both the cooperatives and UCCAO fully control their own affairs including finances and terms of employment of staff.
>
> Members of each cooperative elect a delegate assembly which in turn elects directors of the cooperative who appoint an executive committee and a chairman to manage routine operations. These include the purchase of coffee at the collection centres throughout each district, where the coffee is weighed, graded, sorted, sacked and shipped to central warehouses to await sale abroad. The cooperatives also distribute fertilizer and equipment, and administer the seasonal credits which are used by members to finance the purchase of production inputs.
>
> UCCAO has a board of directors – the chairmen and selected executive directors of the cooperatives – and is run on a day-to-day basis by a director general and a central secretariat. Its primary function is to market coffee overseas. It also arranges for a line of credit to finance the purchase of the crop, and when coffee is delivered performs additional electronic sorting to improve quality. UCCAO also is the central buying agent for the fertilizer and agricultural material required by its members. In this, it arranges for delivery and distribution to cooperatives and helps defray the distribution costs from the 1% commission it takes on the value of the coffee sold.
>
> UCCAO also represents the interests of the growers with the Government, which sets coffee prices to producers and determines the amount to be paid from the price stabilization funds at the end of the crop year. UCCAO also maintains vehicles and equipment and keeps the central accounts. Finally, UCCAO manages the reserves accumulated from the difference between the f.o.b. price for coffee and the payments made to producers minus the operating costs of the Union. By law, about 20% of these reserves must be retained as long-term protection for coffee producers' incomes. Most of the remainder can be invested in developmental projects with Government approval.
>
> In 1978, UCCAO became the implementation agency for an integrated rural development project which was supported by IDA. As part of the project, the cooperatives of UCCAO also made additional investments in their infrastructure, and are diversifying their participation in the development of rural areas to include food crop promotion.

bargaining power. But the present arrangements have proved so generally inadequate and the costs are so high that new departures are justified.

There is, finally, the issue of export-crop/food-crop price policy interactions. If export crop prices rise, it is feared that food crop production will fall. This is, however, not necessarily so, and even if export crop output were to grow at the expense of food crop production, it is not necessarily bad.

Empirical evidence does not support the hypothesis that expanding export production leads to declines in food production. This may occur in some cases (the northern groundnut basin in Senegal may be one), especially in the short run. But the bulk of the evidence points the other way. Countries that have been doing well in cash crop production have also been among the most successful in expanding food production. This is confirmed by aggregate data as well as by examples at the level of individual countries. This complementarity is not surprising. First, export crops are the nucleus around which extension, input supply, and marketing

> **Box 26.2. Measuring comparative advantage**
>
> A country is said to have comparative advantage in a given commodity when it can produce that commodity relatively more efficiently than most other commodities. Comparative advantage can be assessed with the help of a measure called 'Domestic Resource Cost' (DRC); this measures the cost of domestic resources (labour, materials, etc.) used to save or earn a net unit of foreign exchange. The lower the DRC, the more efficient the activity.
>
> The table compares DRCs for food and export crops in 11 countries. It shows a pattern of strong comparative advantage in exports. These results are based on price relationships, input costs, and technologies of the mid-1970s, and can change over time. But projections for the 1980s do not predict changes in price relationships on world markets which would significantly alter these conclusions.
>
> **Domestic resource costs per net unit of foreign exchange for export crops and food crops in selected countries**[a]
>
> | Export crops | | | Food crops | | |
> |---|---|---|---|---|---|
> | Cocoa | | | Groundnuts | | |
> | Ghana | (1972) | 0.30 | Nigeria | (1979) | 1.40 |
> | Ivory Coast | (1972) | 0.36 | Zambia | (1974) | 0.50 |
> | Coffee | | | Zambia | (1977) | 0.94 |
> | Ivory Coast | (1972) | 0.51 | Maize | | |
> | Kenya | (1975) | 0.44 | Nigeria | (1979) | 1.76 |
> | Cotton | | | Zambia | (1974) | 0.58 |
> | Ivory Coast | (1972) | 1.12 | Zambia | (1977) | 1.16 |
> | Mali | (1972) | 0.21 | Millet | | |
> | Senegal | (1972) | 0.42 | Mali | (1972) | 0.62 |
> | Togo | (1977) | 0.37 | Nigeria | (1979) | 1.21 |
> | Zambia | (1974) | 0.53 | Senegal | (1972) | 0.62 |
> | Zambia | (1977) | 0.34 | Rice | | |
> | Groundnuts | | | Ivory Coast | (1972) | 1.50 |
> | Mali | (1972) | 0.23 | Ivory Coast | (1975) | 1.80 |
> | Senegal | (1972) | 0.36 | Mali | (1972) | 0.67 |
> | Palmoil | | | Mali | (1976) | 0.56 |
> | Ivory Coast | (1972) | 0.36 | Nigeria | (1979) | 2.55 |
> | Nigeria | (1979) | 0.39 | Senegal | (1972) | 1.02 |
> | Tea | | | Sorghum | | |
> | Kenya | (1975) | 0.67 | Mali | (1972) | 0.62 |
> | Tobacco | | | Nigeria | (1979) | 1.66 |
> | Zambia | (1974) | 0.54 | Senegal | (1972) | 0.62 |
> | Zambia | (1977) | 0.82 | | | |
>
> [a] At official exchange rates
> Source: World Bank data.

services are built; these also benefit food producers. Second, food production directly benefits from after-effects of fertilizer expended on the commercial lead crop. Third, the existence of a commercial crop facilitates the propagation of productivity-increasing equipment. Finally, where individual farmers undertake cash crops to such an extent that they develop a food deficit (which they usually do only if there is a reasonably well-developed local or regional food trade), cash crop production creates a local market for food crop producers that is often more secure and stable than distant urban markets. The general point is that the benefits of a changing, dynamic agriculture are not restricted to a single crop or sets of crops. When change accelerates, the productivity of the whole farming system also increases.

Even if it could be demonstrated that export crop increases have come at the expense of food production, the conclusion would not necessarily follow that a strategy of self-reliance requires a substitution of food production for exports. Most African countries have distinct comparative advantage in export crop production. An export-sacrificing policy of self-reliance would therefore have costs in terms of income (see Box 26.2). A policy aiming at food security at the price of lessened emphasis on exports has a further pitfall: most methods of intensification imply increased use of inputs such as fertilizers, insecticides, and fuel for pumping (in irrigation schemes), i.e. they rely heavily on imported inputs. Thus, agricultural production under these known methods of intensified cultivation becomes more vulnerable to external disequilibria. If the pursuit of food self-sufficiency diverts resources from export crops to food crops, declining export earnings may lead to balance-of-payments problems jeopardizing the self-sufficiency objective itself. Sudan and Tanzania are countries that have, in recent years, deliberately sacrificed export expansion for the sake of increasing food production. Their present balance-of-payments crises, as severe as those of some mineral exporters, are partly related to that policy.

Food crops and input supply

While many details regarding food price policies can only be assessed in the context of individual country situations, two principles are central. First, food imports should be subject to duties, so that the import price reflects at least the true cost of foreign exchange. Otherwise, low-price imports will continue to replace domestic production, with negative effects on rural income and growth. Second, there should be a gradual freeing of domestic food markets, to encourage greater competition. This would in most cases merely recognize the existing reality, which is that, whatever the legal or formal situation regarding public monopoly, the main part of the cereals trade goes through private channels and, consequently, the great majority of consumers already pay 'free market' prices.

With regard to food marketing and input supply, the proposal to allow a fuller degree of competition means encouraging cooperative actions by farmers and allowing private traders an increased role in these markets. Some observers object to proposals for more competitive marketing arrangements on the grounds that rural African markets function imperfectly, that traders would therefore exploit farmers, and that indigenous traders are still few in number in parts of the continent, so that trade in foodstuffs might once again come to be dominated by nonnationals. But many recent studies suggest that African food markets are reasonably competitive, that trader profits are rarely 'excessive' and that farmers are usually well protected against 'exploitation' by market information and the availability of alternative points of sale (Nicolas, 1972; Hayes, 1979; Mukui, 1979; Southworth *et al.*, 1979, IDET-CECOS, 1980). Even if this were not so, governments can more efficiently protect farmers by making markets more competitive through better information, roads, and marketing facilities than by acting as substitutes for traders.

In any case, it is important to recall that in a large number of African countries, food markets continue to operate, as they have in the past, without much public control. Performance is generally impressive; in Nigeria, private trade supplies two very large cities (Lagos and Ibadan) and many more towns of 100000–500000 people. In Mali, despite the uncertainties of public policy, private trade in the mid-1970s supplied two-thirds of the cereals consumed in the Sixth Region, the most remote part of the country.

Indeed, even the most casual visitor to a market town in Africa has to come away impressed by the range of goods and services available for sale, their variety and quality gradations, as well as by the evident dynamism and liveliness of the bargaining that characterizes the simplest transaction. What is particularly striking is that one finds in these markets almost everything *except* goods which are sold by monopoly suppliers of the public sector: fertilizer, seed for main crops (though one finds seeds for vegetables), and animal-drawn implements.

The keystone of any marketing reform, then, must be to capitalize on the indigenous trading system, a proven asset, and let it play a bigger role in the distribution system. The private sector, with its small-scale, decentralized and flexible structure, is particularly well suited for this task. Devolution of marketing functions to private enterprise may be more difficult in some parts of Africa, where the tradition of indigenous entrepreneurship is weak, but this should affect only the pace of change, not the objective.

In most instances, governments will be reluctant to allow food marketing to become exclusively a private sector activity. A variety of agents can, of course, coexist; indeed, this should be encouraged. Cooperatives can take on many activities in this area and the state role in food marketing would remain substantial even after considerable liberalization. Governments could improve market functioning, easing market access by both traders and farmers through greater emphasis on rural road development and maintenance, by providing better information on crop size and prices, via radio and otherwise, and by gradually introducing uniform weights and measures, a task that governments have neglected. State grain agencies would also continue to have other major functions: they could manage grain imports; they might buy and sell in the open market for special purposes (e.g. localized production crises); they might operate buffer stocks for seasonal price stabilization; they could do grain storage extension work, especially for new grains (e.g. maize in parts of West

Africa); they could constitute and operate a reserve stock of cereals as a first line defence in case of drought or other food emergencies; and they could provide for the needs of collective consuming units, such as the army. This is obviously a large set of tasks; to carry them out well would strain existing capacities in public sector food marketing organizations. But they cannot perform those tasks well while they are grappling with the intractable problems of trying to control trade in food grains.

The role of the private sector in input procurement and distribution should also be enlarged. The private sector should contribute to the distribution of inputs down to the farm level, and to their importation and wholesale distribution. While the latter is a task for larger commercial enterprises, distribution offers scope for small traders as well (see Box 26.3). The private sector is a major partner in input supply activities only in a few African countries, and a subsidiary partner in a few more. There are many reasons for this, besides frequent lack of encouragement by governments. The shortcomings of research and extension work have held back farmer demand for farm supplies, in countries experiencing foreign exchange shortage import quotas on raw materials have limited the possibility of local fabrication of simpler equipment. An important issue is the 'critical mass' factor. A private company is not prepared to establish, for example, research or demonstration plots for sales of a few thousand tons of fertilizer. Yet, in many parts of the world, private corporations' research efforts and demonstrations parallel, or even surpass, those of government institutions. Therefore, some African governments might find it useful and acceptable to create incentive schemes designed to attract private companies willing to provide these services. An excellent example of useful innovation by private firms in Africa is the introduction of the ultra low volume, hand-held sprayer by leading chemical companies. These sprayers have tremendous possibilities for insect and weed control.

A field with broad possibilities for private sector participation is production and distribution of quality seed; poor seed quality and irregular and late renewal of seed are important sources of agricultural stagnation, for instance in groundnut cultivation. In many countries, there has been a marked deterioration of seed quality through improper production and multiplication practices by parastatals (Mali, Niger, Senegal, and Tanzania are examples). The success of the Kenya Seed Company may be replicable elsewhere.

Before governments began to monopolize input distribution, rural traders handled this function in association with produce marketing and retailing of consumer goods. Input supply activities can be more attractive for private trade, and costs of distribution could go down in the process, if private traders could engage in marketing of food and export crops. The more functions private trade is encouraged to fulfil, the more scope there is for spreading transport costs and overheads, and this will reduce cost to

Box 26.3. Privatizing input supply systems: the Bangladesh experience with fertilizer distribution

The Government of Bangladesh has recently introduced reforms in fertilizer distribution arrangements which shift functions from government agencies to private traders and may have some lessons for Africa.

Due to heavy population pressure on land, growth in agricultural production in Bangladesh depends largely upon the increased use of modern inputs; the achievement of a 3.5–4% rate of growth of output depends on an annual increase of about 15% in fertilizer use. The Bangladesh authorities concluded in the mid-1970s that without reform of the fertilizer distribution system such increases were not likely.

Under the Old Marketing System (OMS), the Bangladesh Agricultural Development Corporation (BADC), an autonomous body under the Ministry of Agriculture, was responsible for the procurement and marketing of all agricultural inputs, including fertilizers, pesticides, seeds, irrigation pumps, and various types of agricultural machinery. BADC employed almost 7000 people in its fertilizer marketing operations alone, drawing supplies from three ports and three factories from which it moved fertilizer to 67 intermediate warehouses ('godowns') or directly to 423 thana-level warehouses. (The 'thana' is an administrative unit 100 to 150 square miles in size with an average population of about 22000 comprising about 10 unions or 150 villages.) The final distribution level was private dealers. Of the 32000 licensed dealers in 1978, 20000 were active, each selling an average of about 25 tons per year under a price structure which gave them too small a profit to provide a real incentive to promote sales. Dealers accounted for about 75% of total sales, cooperatives for the remaining 25%. Dealers were required to register at the thana warehouse of their area, to purchase from that warehouse, and to sell only within their union. BADC held all the storage space at transit points and intermediary warehouses, as well as in most thanas.

While the major problem under the Old Marketing System was inadequate supply of fertilizer, the fertilizer distribution system was also handicapped by inadequate planning and coordination and unavailability of sufficient transport and storage facilities. Consequently, there were frequent local or national shortages. In order to develop a system capable of effectively distributing greater quantities in the future, BADC and the US Agency for International Development (USAID) set up a small pilot distribution scheme in 1976. This was followed by a detailed study of fertilizer marketing and distribution, financed by the IDA. This study provided basic information concerning fertilizer use by season and district, storage utilization, seasonal and geographical flows, prices, and so forth, to help BADC plan its fertilizer marketing storage programme. This programme was developed into a USAID-financial project which included, among other things, construction of 350000 tons of additional storage capacity, and a phase programme to develop greater private sector involvement in fertilizer distribution.

In 1978, BADC and USAID undertook lengthy and detailed field studies to develop specific proposals for a New Marketing System (NMS). The NMS is designed to reduce restrictions on private traders and move toward a more open fertilizer distribution system. Under the NMS, BADC is gradually withdrawing from fertilizer sales. BADC sells mainly to wholesalers at 'primary distribution points' while retaining responsibility for sales to retailers in remote and inaccessible thanas. All private dealers and cooperatives are permitted to buy from all BADC warehouses. Private movement of fertilizer is unrestricted except in the five-mile border area. The Government agreed to develop a system whereby private dealers can obtain sufficient credit from commercial banks, although credit has not yet proved to be a problem for traders. USAID also financed consultants to help BADC set up and monitor the new private sector fertilizer marketing system, and to devise measures to reduce internal transport and storage problems.

In 1978 and 1979, BADC took the first steps to liberalize marketing. It increased official dealers' margins, permitted farmers to buy from any traders, whether or not in the farmer's own thana, and made it easier to become a trader. Backed by a USAID Fertilizer Distribution Improvement Grant, BADC introduced the NMS as a large-scale pilot operation in December 1978 in the Chittagong Division, which accounts for one-fourth of the area of the country and a third of total fertilizer consumption.

The marketing system introduced in the Chittagong Division enjoyed a reasonably successful start. Sales increased over the previous year and forty-five thana warehouses became redundant, leaving mainly those in remote thanas which did not attract wholesalers; retail prices dropped in areas around the primary distribution points and were below official prices, except in remote thanas. The new fertilizer wholesalers demonstrated their ability to move fertilizer cheaply and

> effectively from surplus to deficit areas, selling to both farmers and retailers.
> Based on the successful pilot in the Chittagong Division, the NMS was adopted and extended to the rest of the country. Major accomplishments of the NMS as of mid-1980 include the following:
>
> - BADC's fertilizer points of sale will be reduced by 55–60%; about one-third of the original 130 thana warehouses have been closed.
> - In the Chittagong Division, farmer access to fertilizer points of sale has greatly increased.
> - Prices paid by farmers for fertilizer under the NMS are lower than under OMS.
> - A new class of private wholesalers developed as intermediaries.
> - Despite the change in system and a local drought, fertilizer sales in the Chittagong Division, as a percentage of national sales, remained unchanged.
>
> There have been problems: the NMS has worked poorly in underdeveloped areas where transport and communications are inadequate and fertilizer sales low; whether or not dealers assume the distribution function has often depended on whether transport facilities are good. Commercial credit programmes for assisting dealer sales to farmers have not developed as expected. The BADC has not yet worked out a new staffing programme for staff rendered unnecessary by the NMS
> Bangladesh's experience shows the importance of careful preparation of a marketing reform and also how long a process it can be. It has taken five years of intensive effort to bring only the wholesale function successfully into the private sector. BADC is now transferring its seed and pump operations to the private sector through similar long-term programmes.

farmers. A broader variety of goods brought to rural areas is also an effective way of inducing farmers to produce a marketable surplus of produce on a more sustained basis.

But a marketing system based on competition of government and private sector is inconsistent with the principle of applying uniform producer prices throughout the territory (panterritorial pricing), a system currently applied almost universally in Africa. This system has been adopted mainly to help poorer regions. It does so imperfectly and at heavy cost. First, the pricing policy either absorbs scarce public resources (if the transport cost differential is covered by a subsidy) or it penalizes producers in more favourable locations (if the extra cost is covered by averaging the producer price). With transport costs sky-rocketing in recent years, the principle of panterritorial pricing has become more costly than ever. The same conclusion applies to the pricing of inputs such as fertilizer. In Zambia, for instance, comparison of costs of transport in an accessible and in a remote area showed that each hectare of maize grown on the farm costs the nation K28 in crop and input transport in the former location, and K128 in the latter. Net revenue to the nation is K36 per hectare in the first location, but there is a net loss of K188 per hectare on maize grown in the remote area.

Second, uniform producer prices, without regard to transport costs, are an impediment to regional specialization. Finally, the system of panterritorial pricing distorts competition between private and government trade both in crop marketing and in input supply: private traders occupy profitable markets and leave unprofitable ones to the state agency. The private traders buy in the most productive regions and sell where unit marketing costs are low, while the state agency is constrained to buy and sell everywhere, and at the uniform national prices. There is no way that the state agency could avoid deficits under these conditions.

It is also important to give special emphasis to transport policy in making distribution systems more competitive. After a decade of rural development projects the crucial role of transport, in particular feeder roads, is gaining enhanced appreciation as a decisive element in the chain of conditions linking farmers' motivation to produce to the existence of stable and permanent market outlets – which is not only a matter of prices and marketing institutions but also of physical access to markets.

In a number of countries, in particular in East and Central Africa, more feeder roads will not have much impact without more trunk roads, which makes road development a costlier proposition than in other countries where the basic network exists, or where distances are short. Moreover, feeder road development by itself is no remedy if import policies and foreign exchange allocations are not handled in such a way as to give priority and encouragement to the importation of trucks and spare parts; in many countries, feeder roads are presently underutilized because of a pervasive lack of spare parts, which has effectively reduced the number of operating vehicles. Nor will feeder road construction do much good unless the roads are maintained.

Since the early 1970s, investment in feeder roads has been intensified, often within the framework of rural development projects, but the initial impetus has been blunted by the lack of local financial and organizational resources needed to maintain them. An increasing share of resources spent on transport development is now devoted to maintenance and rehabilitation. Rural road development and maintenance should continue to hold prominent places in rural development. They are vital complements for the liberalization of marketing and input distribution advocated above. A complementary necessity is to help farmers equip themselves with means of transport (carts), increasing thereby their capacity to deliver produce to an accessible spot without inordinate expense in terms of labour. If a larger number of farmers owned, or had access to, animal-drawn transport, this would expand the zone of effective coverage along both sides of a feeder road. This process would also increase the economic rate of return to road development.

Donors have a major role to play in helping African governments move toward a restructuring of incentives in agriculture. Infrastructure – especially rural roads – is a high priority almost everywhere. In some countries, import credits for rehabilitation of the road network and the transport system are preconditions of renewed growth. Such credits can suitably be provided within the framework of a structural adjustment loan, which involves policy discussion between donor and recipient, as in the World Bank Export Rehabilitation Credit to Tanzania.

Changes in food policy everywhere pose especially sensitive problems.

Donors can help governments that need to make adjustments by providing technical advice and bridging finance – for example, to smooth efforts to align domestic prices for foodgrains more closely with world market prices.

Donors can also help by responding sympathetically to African concerns about food security. Without greater assurance on this score, indeed, governments may be reluctant to engage in restructuring incentives.

African governments, in their quest for food security, tend to emphasize buffer stocks of cereals. However, buffer stocks are an expensive and risky road to food security. Initial investment is high and annual costs (losses, interest, treatment, overheads) may amount to 15–20% of the investment. Stocks need to be rolled over every 2–3 years in order to avoid deterioration, which could disrupt the domestic grain market. Administration is demanding, and there are high risks of additional losses through inadequate management. They are, therefore, best limited to a bridging role, a first line of defence, until imports arrive. Donors should explore with African governments more cost-effective alternatives, including the possible use of future markets as insurance devices.

Adequate food security facilitates policy changes, but should be designed so as to minimize negative influences on domestic production. The most effective food security objective in Africa today is, after all, a reversal of the declining trend of production. If Africa had maintained a 1% annual growth rate in cereals yields from 1961 to 1979, cereals production would be 6 million tons higher; this is more than 1979 commercial imports and food aid combined. Real food security comes from a dynamic agriculture.

REFERENCES

Ellis, F. (1979) *A Preliminary Analysis of the Decline in Tanzania Cashewnut Production 1974–79: Cases, Possible Remedies and Lessons for Rural Development Policy.* Economic Research Bureau, University of Dar es Salaam.

Hayes, H. (1979) *Marketing and Storage of Food Grains in Nigeria.* Samaru Miscellaneous Paper no 50, Institute for Agricultural Research, Ahmodu Bello University, Samaru, Nigeria.

IDET-CEGOS (1979) *Pre-étude de la commercialisation des produits vivriers au Cameroun.* Ministére de l'Agriculture, République Unie du Cameroun.

Mukui, J.T. (ed.) (1979) *Price and Marketing Controls in Kenya.* Institute of Development Studies Occasional Paper no. 32, University of Nairobi, Nairobi, Kenya.

Nicolas, G. (1972) Processus d'approvisionnement vivrier d'une ville de savane: Maradi (Niger). *Travaux et documents de géographie tropicale,* no. 7.

Southworth, V.R.W., Jones, O. and Pearson, S.R. (1979) Food crop marketing in Atebubu district, Ghana. In: *Food Research Institute Studies,* vol. 17, no. 2. Palo Alto.

27

A FRAMEWORK FOR SEED POLICY ANALYSIS IN DEVELOPING COUNTRIES

C.E. Pray and B. Ramaswami

The seed industry is crucial to agricultural development. Seed was perhaps the most important form in which the technology of the green revolution was transferred to farmers. As the new biotechnology moves from the laboratory to the field and seeds incorporate functions such as the ability to resist pests and generate nutrients that were previously supplied by other industries, the seed industry may become even more important.

After several decades in which seed policies were relatively non-controversial and largely unacknowledged in the literature on agricultural development, the seed industry became the centre of a number of debates in the 1980s. In the international arena the debate centred on how to preserve the genetic resources of plants, who owned these resources, and how the access of poor countries to such resources could be ensured. At the national level declining government budgets, pressure from donors and the local agribusinesses, and the failure of some government seed corporations are forcing policymakers to privatize the seed sector. At the same time concern is growing about the ability of the local private sector to supply adequate seed to farmers.

KEY ISSUES FOR POLICYMAKERS IN DEVELOPING COUNTRIES

How does a country build a seed industry? Inefficient or weak seed industries are often cited as major constraints to spreading new crop varieties in developing countries. Policymakers must decide whether the seed industry is an important constraint or whether the so-called improved varieties are actually inferior to the local farmers' varieties.

If they determine that the seed industry is inadequate, is government seed supply necessary? Policymakers must decide whether private firms

---

Source: International Food Policy Research Institute report. Washington, 1991. (Condensed.)

would rapidly increase the supply of improved varieties of seed or whether the government should supply commercial seed. Policymakers may have other options for increasing the supply of improved seeds such as subsidizing seeds or providing technical assistance to private seed firms.

How much privatization is optimal in more developed seed industries? In countries where the seed industry is more advanced, governments are being pressed by the local private sector and by foreign donors to privatize the production of seeds, the system of distributing seeds, and the breeding of plants. Government officials must decide what, if anything, should be privatized.

What is the relation among genetic diversity, yields, and crop failure? As modern varieties replace traditional landraces, scientists and environmentalists have become concerned about declining genetic diversity in farmers' fields. They fear that reducing the genetic diversity will produce disastrous disease or insect epidemics that could devastate poor farmers in developing countries. They are also concerned that we will lose genes that could increase yields in the future. What is the relationship between seed industry policy, genetic diversity, yields, and crop failures? Can policies on seeds reduce the probability of crop failures and the loss of genes that might increase yield?

Would granting plant breeders' rights, a form of patenting, improve the welfare of farmers? Such rights are being promoted in developing countries by private seed firms and the governments of the United States and some European countries as a means of encouraging more private research and transfers of technology. Its opponents view such legislation as a way for private companies – particularly European and US multinational companies – to extract money and genetic resources from farmers in developing countries. The opponents of breeders' rights have proposed strengthening farmers' rights instead. They propose that developed countries compensate farmers in developing countries for the use of their genetic resources in the past by financing the research and conservation of genetic resources in those countries. Policymakers must decide whether supporting the rights of plant breeders or strengthening their patent laws will increase or decrease the amount of technology available to farmers. This decision may be particularly important as biotechnology and stronger property rights in developed countries make more agricultural technology proprietary.

Does restricting seed imports and foreign seed companies help farmers? Many countries wish to be self-sufficient in the production and research of seeds because seeds are an essential ingredient in agriculture. This goal is reinforced by shortages of foreign exchange. As a result developing countries restrict seed imports and foreign investments in the seed industry. Donors and some local interest groups are now pressing these governments to liberalize their policies on trade and foreign investment.

Table 27.1. Public sector involvement in maize seed industries, by region, 1986

		Number of countries with			
Region	Number of countries reporting	Seed regulating agency	Seed quality agency	Public seed enterprises	Private seed enterprises
Africa	12	9	9	10	9
Asia	12	11	11	11	8
Latin America	13	9	11	11	12
All developing countries	37	29	31	32	29
Eastern Europe and USSR	4	4	4	4	0
Developed market economies	7	6	6	0	7
All countries	48	39	41	36	36

Source: CIMMYT (1987).

Will farmers gain from liberalization? Will society as a whole gain? These questions must be answered before a policy on seed imports and multinational seed companies can be formulated.

The development of the seed industry varies widely by country and crop. FAO (1985) rated seed industries on the basis of three activities: the improvement of varieties, quality control, and production and distribution. The seed industry in South America has reached an advanced level in food and industrial crops. About half the countries in Asia and Central America have also reached advanced levels in their ability to improve varieties and control the quality of seeds in food crops. Central America is behind Asia in production and distribution of seed for food crops. Asia and Central America are less developed than South America in industrial crops. In Africa, the majority of countries have pilot activities in food crops, but only a few have attained advanced status. The situation is worse in industrial crops: many African countries report no efforts to control the quality, production, or distribution of seeds. All regions report low levels of development in vegetables and pasture crops.

The industrial structure of the seed industry in developing countries is a mix of organizational forms including public research institutions, public sector seed corporations, private local firms, farmer associations, multinational companies, and nongovernmental development agencies. Research to develop new varieties, particularly in rice and wheat, is mainly performed by public institutions and international research centres. In

Table 27.2. The private sector's share of maize seed sales in noncentrally planned countries, 1986

		Share of total seed sales		
Region	Number of countries reporting	Improved varieties	Hybrid	All commercial
Africa				
Eastern and Southern Africa	4	45	99	92
West Africa	6	9	77	61
North Africa	1	73	100	78
Total	11	57	95	83
Asia and the Middle East				
Middle East	2	0	45	38
South Asia	3	38	63	54
Southeast Asia and the Pacific	4	69	99	73
East Asia, excluding China	2	69	38	39
Total	11	62	62	62
Latin America				
Mexico, Central America, and the Caribbean	6	66	71	68
Andean region	4	65	91	86
Southern cone of South America	3	81	98	97
Total	13	70	96	92
Developing countries	35	65	92	85
Developed market economies	7	100	100	100
All noncentrally planned countries	42	65	98	94

Source: CIMMYT (1987).

some countries, local- and foreign-owned firms are conducting research on hybrid crops and vegetables.

Many developing countries invest in public seed corporations that multiply and distribute seed. One survey of the maize seed industry found that the majority of developing countries produce at least some seed in the public sector (Table 27.1). As the same survey shows, private seed firms are also important in supplying seed and in fact supply the bulk of commercial maize seed (including open-pollinated varieties; see Table 27.2). The private sector (including cooperatives and farmer associations) also plays an important role in the production of seed for crops like rice and wheat that do not have commercial hybrids, e.g. in Mexico, Thailand, and India.

In most developing countries imports are not an important source of commercial seed. The United States and the countries of the European

Community are the major exporters of seed to developing countries, but this trade is limited by differences in agroclimatic conditions, disease and pests, and government trade barriers. Trade among developing countries is limited by the underdeveloped seed industry and trade barriers. The United States and the European Community shared an estimated 90% of the US$1.2 billion global trade in commercial seeds at the beginning of the 1980s (Groosman *et al.*, 1988). Developing countries import 47% of the United States' seed exports and 27% of those from the European Community. They import principally seeds of hybrid maize and sorghum, vegetables, and potatoes from the European Community.

Multinational firms participate in the seed industry of developing countries to varying degrees. In Argentina they control about 80% of the maize and almost all of the sorghum seed market, but local companies control most of the wheat seed market. In contrast India kept almost all multinational corporations out of the seed industry until 1987. In all countries they concentrate on hybrid seeds of major field crops and on hybrid vegetable seeds. They export commercial seed to developing countries. Most large multinational seed companies prefer, however, to grow part of their commercial seed in the country where it is sold and to import part. This strategy saves transportation costs, takes advantage of inexpensive labour to produce labour-intensive seeds, and diversifies production risk. Multinational corporations may sell the parent seed in return for a royalty on the commercial sales. In other instances, they may decide to invest in a local partnership or a wholly owned subsidiary (Kania and Goldberg, 1982).

Despite a large number of acquisitions and mergers, seed companies still compete fiercely for world markets. Table 27.3 describes some of the major seed companies. Despite the publicity accompanying the acquisition of seed companies by chemical companies, many of the largest companies are primarily seed companies or are owned by food companies. These three types of companies have different goals, and their competition is based on differences in technology, price, and services.

In most countries, governments intervene to control quality, seed imports, seed prices, and ownership of varieties. Quality can be controlled at different stages of the production process. At an early stage, varieties may be tested in trials and identified as superior because they possess desirable characteristics. These varieties may be subjected to a system of field trials and seed inspection to maintain genetic purity known as seed certification. This certification process may be voluntary or mandatory. At the marketing stage, certified and uncertified seed may be tested for purity and germination quality according to standards established by law.

Almost all countries have some restrictions on seed imports. Many forbid commercial imports of certain crops. Most countries restrict the importation of parent seed less than commercial seed, although some

Table 27.3. The principal international seed groups, 1987

Group	Country	Primary activity	Sales (US$ million)
Pioneer	United States	Seeds	692
Sandoz	Switzerland	Chemicals	382
Limagrain	France	Seeds	234
Cargill	United States	Food processing	200–250
Upjohn	United States	Chemicals	217
Provendor (Volvo)	Sweden	Food processing	213
Ciba-Geigy	Switzerland	Chemicals	213
DeKalb-Pfizer	United States	Seeds	154
R-D Shell	Great Britain–Netherlands	Petroleum	150–200
Orsan (Lafarge)	France	Biochemicals	139
KWS	Federal Republic of Germany	Seeds	127
ICI	Great Britain	Chemicals	98
Lubrizol	United States	Chemicals	83

Source: Pierre-Benoit Joly, 'Should Seeds be Patentable? Elements of an Economic Analysis', in *Patenting Life Forms in Europe: Proceedings* (International Coalition for Development Action, Brussels, 1989).

countries even ban them. Imported seeds are usually subject to quarantine regulations and other quality-control laws applicable to local seeds.

Seed prices of maize are subject to government control in 22 of the 34 noncentrally planned developing countries that replied to the 1986 survey conducted by Mexico (CIMMYT, 1987). Seed prices are subsidized in 11 of the 34 countries, but the subsidy is often subject to the vagaries of state budgets. Patenting of varieties is a controversial issue in developing countries. A weak form of patents known as plant breeders' rights have been in existence in Western Europe and the United States for some time. Except for Chile and Argentina, however, developing countries either have not passed legislation endorsing plant breeders' rights or do not enforce it (Siebeck *et al.*, 1990).

The yield improvements from new varieties have been extensively documented in the United States, Europe, and developing countries. Long-term gains in yield of 1.0% a year in wheat and maize are quite common in countries with well-established research programmes like the United States and Australia. About half of that increase can be attributed to genetic improvement, the rest to increased inputs. In countries recently affected by the 'green revolution', the gains have been greater. Wheat yields grew 2.5% annually in Pakistani Punjab and 5.1% in the province of Parana, Brazil. The contribution of genetic improvement to wheat yields has been

0.5% or less in dry areas and around 2% in irrigated areas.

Qualities other than yield also affect the decisions that farmers make. The importance of improved eating quality in rice in Asia and varieties of cotton with higher-quality lint in Pakistan have been documented (Unnevehr, 1986). Early-maturing varieties are favoured in many regions because they allow a second crop to be grown or produce higher yields of the second crop.

Price of seed

Because improved varieties and hybrids cost more to produce, their seeds are also more expensive. Does that prevent farmers from adopting them? It has been demonstrated many times during the past 25 years that truly superior seed will almost 'sell' themselves. Marketing difficulties, however, are encountered when the seed represents a solid, demonstrable, but only modest improvement over the seed presently planted by cultivators (e.g. yield advantage of less than 20%). The implication is that the demand for improved seed is very elastic if the substitutes (farmer-saved seed) are close in yield and quality characteristics. In other words, a small increase in the price of improved seed will lead to a large decline in the quantity purchased. The demand is less elastic if the seed is truly superior. For example, Indian farmers are willing to pay 30 times the price of conventional cotton seed to acquire some hybrid cotton varieties.

Price is more of a constraint for poor farmers in crops that require a large amount of seed such as groundnuts and potatoes. In this case, seed is a major component of the costs of production, and farmers are quite sensitive to seed prices.

Many governments regulate seed prices and, at least implicitly, believe that the price of seed is an important determinant of adoption behaviour (CIMMYT, 1987). No one has actually estimated the price elasticity of demand for improved seed. Limits are placed on subsidies, however, because most seeds can be eaten. If the subsidies on seed push prices too low, people will buy seed to eat rather than to plant.

Price of other inputs

The use of improved seeds is often associated with the use of other purchased inputs such as irrigation, fertilizers, and pesticides. Since seeds cost less than these other inputs, the demand for seed is likely to be affected by their price or availability (as is the case when inputs are rationed). The role of subsidized credit in facilitating the adoption of new technology has been well documented in the literature (Feder *et al.*, 1985).

Relative price of crops

The demand for seed depends on the output plans of the producer. The demand for sorghum seed, for example, depends on the market incentives for crops competing with sorghum for the same land. In Kenya, the demand for maize seed fluctuates as the price of maize (relative to other crops) changes according to government policy (Lynch and Tasch, 1983).

Farmers' forecast of weather conditions and prices

The demand of farmers for seed is conditional on their forecast of weather and prices at the time of planting, so the demand for seed varies year to year. This variation is important to seed enterprises, which need to project demand accurately in order to avoid being saddled with unsold stocks or being unable to meet demand. Some uncertainty is unavoidable, however, since firms make their seed production decisions many months before farmers decide on their demand for seed.

Costs of reaching distribution outlets

In developing countries, the density of retail outlets is sometimes so low that farmers incur significant transportation costs if they wish to obtain seeds from the outlets. This can substantially reduce the demand for seed sold through an organized distribution network, as illustrated by Pakistan's experience with diffusing improved wheat seed. In their survey of farmers using improved varieties, Tetley *et al.* (1988) found that when farmers change varieties, their most important source of seed is other farmers. Retail seed outlets are important for the initial introduction of seeds, but not for diffusion, which takes place largely through transfers of seed from farmer to farmer. Seed transfers are highly localized and therefore probably reflect the high transportation costs of contacting the formal seed system.

FACTORS AFFECTING SUPPLY

The basic input in seed production is the breeder seed, which embodies the improved genetic characteristics. The costs of producing the breeder seed are essentially the costs of research and development plus the cost of multiplying enough breeder seed to distribute to seed companies. The key inputs are scientists; germplasm; the stock of scientific knowledge; laboratories; and land, labour, and other farm inputs. Private sector seed industries in the United States spend 3–4% of the value of their sales on research and development. In India private companies spend about 4%, while in Argentina they spend about 5% of sales.

The costs of research and development exhibit the following characteristics:

1. Since an improved variety cannot be produced without breeder seed, but once this is developed it can be multiplied indefinitely, research and development costs are essentially fixed.

2. Research and development involve years of testing and selection. The gestation lag from the initial expenditure on a plant breeding project until a variety is ready for the market can be more than 12 years.

3. Improved varieties developed by research are often superior to local varieties only in a limited agroclimatic region.

4. Expenditures made for research and development do not assure that improved varieties will be developed.

The economic consequences of these characteristics are profound. Owing to the fixity of costs and gestation lags in research, the natural barriers to entering the seed industry may be formidable. Although plant breeding is more easily divisible than some other types of research that require large laboratories, large investments may be required to develop a competitive programme of research and development for major markets. Smaller firms often depend on public research and development to provide them with new varieties or basic research to overcome their disadvantage. Since research is also crop- and region-specific, smaller firms sometimes survive by breeding for market niches. Agroclimatic specificity limits the transfer of finished varieties from the advanced seed industries of the developed countries to those of the developing countries. However, advanced breeding lines and other genetic material from developed and other developing countries can greatly reduce the cost of producing new varieties. In addition, because basic biological science is much less location specific than applied technology, basic scientific advances in developed countries can also reduce the cost of developing new varieties.

The lags and the uncertain payoffs mean that firms will probably require a substantial risk premium on their returns to investment in research and development. More important, once the breeder seed of a variety of a self-pollinated crop is available, anyone can copy it. Therefore, companies cannot capture all the potential economic benefits from breeding a new variety, and these benefits are passed on to farmers and consumers. Because investing in research and development is risky and companies cannot capture all the benefits, private firms will invest less than the socially optimal amount in plant breeding research.

Multiplication costs

Several rounds of seed multiplication are involved once the initial quantity of seed leaves the breeder. At least one stage intervenes between the production of breeder seed and commercial seed. The output of the intervening stage is referred to as foundation seed or basic seed. Multiplication costs frequently account for the largest share of the total cost of production. In the United States they make up 30% of the price of hybrid maize. In India between 40 and 60% of the price of pearl millet and sorghum seed is accounted for by multiplication costs.

Seed enterprises usually subcontract farmers to multiply foundation seed into commercial seed. The price of grain sets the minimum floor that contract farmers receive for multiplied seed. The actual payment is usually higher than the minimum, however, because growers must take more care and observe special precautions when growing seed. Significantly more labour and supervision is involved in producing hybrid seed, and so labour costs are one determinant of how much of a premium over the price of grain is needed to induce farmers to grow seed instead of grain. Besides grain prices and wage rates, the costs of producing commercial seed also depend on its yield or multiplication ratio. Seed yield depends on a number of factors, including management practices and environmental conditions. Seed yield also depends on the yield potential of the seed parent. It is lowest for single-cross hybrids because the parent lines are inbred. This adds to the cost of producing single-cross hybrids. Seed yields can be increased and costs lowered by producing three-way crosses and double crosses, where one or both of the parent lines are single crosses. Even then, the cost of producing a double-cross hybrid is greater than the cost of growing an open-pollinated variety of seed.

Postharvest cost

After the seed is harvested it is dried, cleaned, chemically treated, packaged, and stored until it is distributed. These operations account for about 15% of the price of hybrid maize seed in the United States and for between 8 and 18% of the price of hybrids and between 20 and 30% of the price of varieties in India. Many postharvest activities require a reasonably well-developed infrastructure to operate efficiently. For instance, electric dryers make severe demands on the electrical system in developing countries with poor infrastructure. However, traditional floor-drying methods are cumbersome and inefficient. Since harvesting, processing, and storage occur at different locations, transportation is always an essential element of cost. In developing countries, however, inadequate infrastructure makes transportation costs even more important. Transportation costs as a proportion of processing costs are four times more important in developing countries than in the United States. Inventories need to be held because of the lag between harvesting and selling the seed. Inventory costs can be quite high in developing countries, however, because credit tends to be expensive.

Distribution and marketing costs

Distribution and marketing costs are largely determined by transportation costs, publicity costs, dealer margins, and overheads. Together they account for more than 10% of the price of hybrids in the United States. In India, they vary from 8 to 20%.

As mentioned earlier, transportation costs tend to be high in developing countries. Transportation bottlenecks often seem to pose a severe problem in getting seed to the distribution (retail) outlets from the multiplication and processing centres.

In many developing countries, distribution is in the hands of government seed corporations. The Pakistan Seed Corporation has been criticized for its ineffective distribution and marketing network. Part of the distribution failure is due to artificially low dealer margins (below 10%), which reduce the incentive for private dealers to distribute the corporation's seed. Where private seed companies compete with public sector seed corporations, their dealer margins are consistently higher (above 10%) than the dealer margins of state seed corporations. The state-run Kenya Seed Corporation is an exception: its success in distributing hybrid maize has been attributed to its policy of offering attractive margins to retailers (Lynch and Tasch, 1983).

Public sector seed corporations are not always tightly managed, and in the absence of competitive pressure they tend to run up high overhead costs. The Tanzanian Seed Corporation (TANSEED) earns a huge margin on its seed operations, but runs a loss year after year because of the high overhead at its headquarters. In many countries, the government absorbs most of the overhead costs without passing them on to the consumer (farmer).

In order to increase their market, firms and their dealers spend a substantial amount of money on demonstration plots in farmers' fields, field days for farmers, and advertising on radio and television and in print. Companies that breed their own varieties may have a team of technicians that sets up demonstrations, trains dealers and farmers, and deals with complaints about the seed.

THE DEVELOPMENT OF THE SEED INDUSTRY

The development of a seed industry proceeds through four stages. In stage 1, no seed industry exists because no improved varieties exist. In stage 2, farmers begin to use varieties developed by formal research and development. Most seed is still produced by the farmers themselves, but the new varieties are produced and distributed by government or commercial seed companies. In stage 3, the use of improved varieties spreads as the private sector begins to be a significant source of new technology. Both public and private enterprises produce and market seed. In stage 4, most of the varieties planted by farmers are bred in private research programmes and all of the commercial seed is produced and marketed by private firms.

In the first stage, formal research and development has not produced new varieties that have had an impact on farmers, but farmers themselves have gradually improved their varieties over time by selecting the best grain in the fields, trying off-types found in their own fields, and experimenting with varieties brought from other regions. Many villages or regions have households that specialize in selecting, producing, and storing seed. No demand exists for commercial seeds because they are not superior to farmers' seeds.

In the second stage, superior varieties are developed by local research or introduced from outside the region. The seeds of these varieties have to be multiplied and distributed if they are to increase production or improve crop quality. Institutions are established to produce and distribute seed.

In most cases, the force behind the movement from stage 1 to stage 2 is government research. The government's motivation to invest in agricultural research is a combination of broad political pressure to increase food production and the specific interests of groups of merchants, processors of agricultural products, and large farmers who are organized around a commercial commodity. In a few cases, the commercial interests may themselves be organized into a coalition large enough to undertake direct research activities in commercial crops. Due to the long gestation lags and inherently risky nature of research, formal research and development can only be initiated by an organization that has access to finance and can bear the risk. In many developing countries the government is the main institution that can do this.

Once public research produces improved varieties – particularly varieties with clearly superior yields like the green revolution wheat varieties – the demand for improved seeds increases significantly, which provides opportunities for private firms to supply seed. Large farmers and seed traders are the most likely to sell improved varieties developed by the government or imported from other countries. The government may train new seed firms to produce seed and provide future seed firms with markets for their crops. Governments may also respond to the demand for improved seeds by directly undertaking the production and distribution of seeds.

In the third stage, private companies, through their research and development, successfully develop commercial crop varieties. The private varieties are almost always hybrids. Public research maintains a dominant role in the development of improved open-pollinated varieties. Private companies produce foundation seed of both private and public varieties. The demand for improved seeds in general, and private hybrids in particular, increases. The availability of complementary inputs like fertilizers and pesticides is less of a constraint. Since more of the improved varieties are hybrids, farmers cannot produce them as easily as they can produce improved varieties. Therefore the demand for commercial seed increases. Between 5 and 40% of the seed planted is commercial seed produced by government agencies or private companies. If the main crops of the country are self-pollinated, the percentage of commercial crops (that is, the proportion of seed planted that is purchased from the seed industry) will be lower. If they are cross-pollinated, the commercial percentage will be higher. With the growth of private companies that have proprietary hybrids, private advertizing and technical services increase as a means of improving their competitive position. The widespread use of hybrids increases the average cost of seed. The popularity of improved varieties may also have environmental consequences. Planting genetically similar varieties over large areas and using higher doses of fertilizers and more irrigation may lead to more disease and pest problems than in the past and require farmers to replace their varieties much more often.

The dynamic factor that moves countries from stage 2 to stage 3 is the profit that firms make producing and distributing seed. Credit markets are imperfect everywhere, particularly in developing countries. The ability of a firm to generate an investable surplus is therefore critical in allowing it to take on risky projects with long gestation periods. If in stage 2 government policies allow seed firms to make sufficient profits, and if there is scope for producing hybrids of the major crops, some firms will start to invest in plant breeding programmes. The success of private research and development in producing new hybrids and the expansion of private seed production and distribution mark the beginning of stage 3. At the same time, continued public investment in extension and rural infrastructure contributes to increasing the demand for improved seeds.

The fourth stage is distinguished from the third in that private companies do most of the research on plant breeding, produce most of the breeder seed, and produce virtually all of the foundation and commercial seed. In Western Europe, most governments leave the breeding of finished varieties to the private sector. In the United States, the Department of Agriculture recently stopped producing finished varieties, but state agricultural universities continue to be important sources of new varieties. The public research institutions that continue to do research arrange to have the seed multiplied and distributed by groups of seed companies or farmer cooperatives. The public sector puts greater emphasis on basic science and other activities that do not interest the private sector (for example, open-pollinated varieties or minor crops).

Once again, profits from the activities of the earlier stage are the dynamic force that moves the private seed industry to venture into new areas. Private firms become more active in developing pure-line varieties, especially if, as a result of their lobbying, the government enacts a plant variety protection act. In a mature seed industry, the investment in research is large enough to produce a rapid turnover of varieties. In fact, maintaining a technological edge by quickly developing new varieties becomes a major basis of competition.

GOVERNMENT SEED PRODUCTION AND DISTRIBUTION

Seed production and distribution do not appear to possess the characteristics of public goods or externalities that justify public intervention. In stages 1 and 2 of seed industry development, however, private firms may not have accurate information about the risks and benefits of multiplying and distributing seed. This market failure would lead private companies to underinvest in seed production and distribution. Therefore government investment in pilot seed production programmes and market development may have positive benefit to cost ratios. Alternatives such as technical assistance and training on seed technology, coupled with subsidized inputs, especially capital, might be equally effective in moving a seed industry from stage 1 to stage 2.

Some government seed programmes, such as the Thai rice seed programme, seem to have been quite effective in producing and distributing seed. Some of the Indian state seed corporations, like the Andhra Pradesh State Seed Corporation and the Maharashtra State Seed Corporation, also seem to be quite effective. Indonesia has developed an effective rice seed production programme, and the Kenya Seed Company seems to do a good job. Unfortunately, a larger number of government seed programmes have been less successful. Many farmers complain about the quality of seed sold by PRONASE in Mexico. The National Seed Corporation in India annually incur massive operating deficits that must be financed by the government.

There are enough examples of rapid distribution of high-yielding varieties in countries with no government production and distribution of commercial seed to put into question the necessity of government production. A variety named Pajam, which was developed in the Philippines, was introduced into Bangladesh (then East Pakistan) around 1965 by the

Academy for Rural Development in Comilla. The government would not officially approve the variety so that the Agricultural Development Corporation did not multiply Pajam and extension agents were supposed to discourage its use. Yet because of its high yield and relatively good eating quality, it was the most popular improved rice variety in Bangladesh in 1978. The farmers of the Indian Punjab replace their wheat varieties more often than most farmers in the world (except Mexican farmers in the Yaqui Valley), yet their state seed corporation is the smallest in India, and they buy the smallest percentage of their seed from the state seed corporation. The secret of their success seems to be that the university distributes good foundation seed through farmer fairs and demonstrations on farmers' fields, and then farmers rapidly multiply their own seed.

In Argentina private companies and large farmers started producing seed of government-developed varieties and selling them to other farmers in the 1920s and 1930s. In the 1950s hybrid maize was introduced and popularized by private companies like Morgan and Cargill. Hybrid sorghum was introduced by DeKalb. The semidwarf wheat varieties adapted to Argentina were developed simultaneously by the government breeding programme and private companies. The government encouraged the multiplication and distribution of its varieties through an association of farmers called Productores de Semillas Selectas Coop. Ltd.

In stages 2 and 3 government production of seed cannot be justified on grounds that private firms have insufficient information. It may be justified, however, as a means of preventing a monopoly, although there may be better ways of fighting a monopoly than by government seed supply. A stronger case may be made for producing public seed on grounds of equity rather than efficiency. Selling seeds to larger commercial farmers growing hybrids, vegetables, and cash crops is easier than selling them to small farmers growing subsistence crops in marginal rainfed areas. The latter do not represent a high-value, high-growth market and will likely be ignored by the private sector. Private firms will breed subsistence crops if hybrid varieties can be developed, but in most subsistence crops hybrids have not been developed. Therefore government involvement in the production and distribution of seeds may be necessary to meet income distribution goals. Recent work at CIAT and in several national programmes indicates that increased attention to developing small farmer seed production and sales can help meet farmers' needs.

The actual consequences of government action may not, however, improve the distribution of income. For instance, the state seed corporation in Tanzania, which has a monopoly on the production and distribution of seed, is so inefficient that farmers simply do not have enough hybrid maize seed to plant. In such a situation, small farmers have the least access to seed. Since state agencies are also political institutions, they, like the private sector, tend to direct their activity toward agents with large endowments.

Governments must recognize the limitations of private firms. The most efficient mix of public and private activities in the seed industry varies among countries and stages of development. The optimal mix has not been established at any stage. There are, however, limits to what developing countries can expect the private sector to do. Private companies will conduct research on hybrid crops like maize, sorghum, and sunflowers that have large markets, but will rarely breed self-pollinated crops like wheat and rice unless plant breeders' rights can be enforced. Large private companies will not invest much in the production and distribution of most self-pollinated crops because they cannot compete with farmers. Nor will large commercial companies invest in the production and distribution of hybrids for small markets.

Therefore at a minimum the governments of developing countries must continue to breed self-pollinated crops and produce enough foundation seed to ensure that farmers can spread new varieties. Some type of relationship will have to be worked out between government research and private seed producers so that breeder seed and foundation seed are kept pure and available to private companies for multiplication and sale to farmers.

Farmers lose if technology transfer is restricted. Restrictions on seed imports and on the role of foreign seed companies are common throughout the developing world. The gains from liberalizing import policies depend on how much technology is available from regions with similar agroclimatic conditions. Greece, Italy, and Chile have benefited greatly from importing as much as half of their hybrid maize seed from the United States, which has similar growing conditions. Developing nations in the tropics cannot benefit greatly from seeds imported from temperate regions, where the most developed seed industries exist. As seed industries in developing countries mature there will be greater benefits from increasing trade with other countries in the tropics. Some of the gains from increased imports may be lost if disease or pests are imported along with the seed. Thus, to get maximum benefits from seed imports, reduced barriers to imports must be coupled with improved plant quarantine facilities.

The transfer of varieties for use in breeding programmes has been more important than seed imports as a means of transferring technology. In government research on wheat and rice, developing countries have been able to use varieties from Mexico, the Philippines, and elsewhere as parents for breeding successful high-yielding varieties. Governments and multinational corporations have adapted improved lines of maize to the tropics by breeding them with locally developed varieties. Recent studies show that tropical countries in which multinational corporations are conducting more research have a higher yield of maize than those in which they are conducting less. This indicates that multinational corporations can have a positive impact. Public and private sector research to adapt exotic varieties has large payoffs. Governments should encourage rather than restrict such research.

Governments should monitor the impact of infrastructure and price policies on the seed industry. The existence of more effective property rights apparently provides an incentive to private research on self-pollinated crops, but no estimates indicate how large that impact would be. Developing rural infrastructure would decrease transportation and communication costs and increase the productivity of the seed industry. Price and exchange rate policies that are more favourable to the agricultural sector should increase demand for improved seed. Theory and common sense support these statements although empirical studies of the impact of infrastructure and policy on the seed industry have not been conducted.

The model presented of the stages of development highlights the importance of sequencing properly the government's policies toward the seed industry.

REFERENCES

CIMMYT (Centro Internacìonal de Mejoramìento de Maìz y Trigo) (1987) *CIMMYT World Maize Facts and Trends: The Economics of Commercial Maize Seed Product in Developing Countries.* CIMMYT, Mexico City.

FAO (1985) *FAO Seed Review.* Rome.

Feder, G., Just, R.E. and Zilberman, D. (1985) Adoption of agricultural innovations in developing countries: A survey. In: *Economic Development and Cultural Change.* University of Chicago Press, Chicago.

Groosman, A.J.A., Linneman, A. and Wierma, H. (eds) (1988) *Technology Development and Changing Seed Supply Systems.* Research Report 27. Instituut vor Outwikkejings-Vraagstukken, Tilburg, The Netherlands.

Kania, E.M. and Goldberg, R.A. (1982) Pioneer Overseas Corporation. Case Study 4-583-070. Harvard Business School, Cambridge, Mass., USA.

Lynch, J.A. and Tasch, E.B. (1983) *Food Production and Public Policy in Developing Countries.* Praeger Press, New York.

Siebeck, W.E., Evenson, R.E., Lesser, W. and Primo Braga, C.T. (eds) (1990) *Strengthening Protection of Intellectual Property in Developing Countries: A Survey of the Literature.* World Bank Discussion Papers No. 112. World Bank, Washington, DC.

Tetley, K.A., Heisey, P., Ahmed, Z. and Ahmad, M. (1988) Farmers' seed sources and seed management. In: Heisey, P. (ed.) *Transferring the Gains from Wheat Breeding Research and Preventing Rust Losses in Pakistan.* Centro Internacìonal de Mejoramìento de Maìz y Trigo, Mexico City.

Unnevehr, L.J. (1986) Consumer demand for rice grain quality and returns to research for quality improvement in Southeast Asia. *American Journal of Agricultural Economics* 68, 634–641.

VI
FOOD GRAIN SUPPLY MANAGEMENT

Operating policies for food grain supply and price stabilization systems in developing countries were set out by Greupelandt and Abbott in 'Stabilization of internal markets for basic grains: implementation experience in developing countries' (*Monthly Bulletin of Agricultural Economics and Statistics* **18**(2), 1–9. FAO, Rome, 1969). They were elaborated in the FAO guide on rice marketing (1972). D.S. Tyagi *Managing India's Food Economy: Problems and Alternatives.* Sage Publications, Delhi, 1990, reviews the success of this approach in India, pointing also to the scope for cost saving with better production intelligence and targetting of supplies to the poor. See also K. Subharao *Rice Marketing System and Compulsory Levies in Andhra Pradesh! A Study of Public Intervention in Food Grain Marketing.* Allied Publishers, Bombay, 1978.

C.P. Timmer argued cogently for self targetting in 'Food prices and food policy analysis in LDCs', *Food Policy* **5**(3), 183–199, 1980. P.S. George *Public Distribution of Food Grains in Kerala: Income Distribution Implications and Effectiveness*, IFPRI, Washington, 1979, is the authority on two price systems to help low income consumers. These are reviewed in U. Kracht's 1980 paper for the World Food Council reproduced below.

Food Policy Analysis by C.P. Timmer, W.P. Falcon and S.R. Pearson, Johns Hopkins University Press, Baltimore, 1983, has been rated high by both reviewers and readers. We include their passage on the non-answers.

R. Ahmed and N. Rustagi's work on marketing margins for foodgrains, summarized below, broke new ground in comparisons between where stabilizing parastatals operated in an open market, and where they had been assigned a monopoly of wholesale transactions. Jones' paper in Part II sets out a modified role for a food marketing board in the African environment.

T.G. Pinckney *Storage, Trade and Price Policy Under Production Instability: Maize in Kenya*, IFPRI, 1989, and F. Colette, R. Ahmed and N. Chowdhury *Optimal Stock for the Public Food Grain Distribution System in Bangladesh*, IFPRI, 1991, have demonstrated how optimal stock policies could save public resources. J. von Braun *et al.*'s *Improving Food Security of the Poor: Concept, Policy, Programs*, IFPRI, 1992, is a clear compact review. See also M. Revallion *Markets and Famines*, Clarendon Press, Oxford, 1987.

28

THE NONANSWERS TO FOOD PROBLEMS

C.P. Timmer, W.P. Falcon and S.R. Pearson

It is a mistake to think that governments do not feel the urgency of the hunger of their people. The pressures to do something are great, the time for analysis is short, and political priorities and constraints usually limit the scope for intervention. Many governments respond to this environment with programmes that have great emotional and political appeal, especially if cast in a rhetoric that oversimplifies complex problems for mass consumption. Most of these programmes do not work, and some make the problem worse. The leading candidates for nonanswers are discussed below.

Eliminating the middleman

Most policymakers see the middleman as an unscrupulous rascal who buys food at low prices from disadvantaged small peasants and sells it to desperate consumers at prices so high they cannot afford bus fare to work. Surely the government can move food from farmers to consumers more fairly than that. The promise to do so, especially when food shortages are forcing food prices up, brings wild cheers from the urban population.

In its extreme form, in which the government takes over the entire food marketing function, the strategy almost never works. Consumers find that the government cannot provide food as cheaply as their corner market stall. Farmers discover that the government purchasing agent is missing when the crop needs to be sold or that payment will be delayed several months, even years. A furtive private trade springs up, reinforcing the government's view that the middlemen who conduct it are antisocial elements. Both producers and consumers, however, find they are better off dealing with then. Very quickly, the government's marketing programme becomes a visibly empty shell.

Source: Passage from *Food Policy Analysis*. Johns Hopkins University Press, Baltimore, 1983.

There is too much truth in this caricature to ignore. Significant opportunities exist for government interventions to improve food grain marketing, to the benefit of both producers and consumers. Such interventions must take account of the productive roles played by marketing agents. If the private sector is not carrying out marketing functions efficiently, and most empirical evidence says it is, the government must understand why and how to intervene to improve matters. Simply attacking or supplanting the middleman will almost never be the answer.

Crash programmes

Problems that have been allowed to develop into crises usually provoke a call for immediate and drastic action – a crash programme. By definition, such programmes are not built on an analytical understanding of the problem at hand, nor is there time to build one. The call for drastic action has political appeal precisely because of its hurried and makeshift aspect. The time is past for research, for analysis, for planning. Now is the time to act.

Food problems are extraordinarily complicated, and expedient short-run interventions frequently have devastating long-run consequences. Procuring grain at gunpoint, seizing private warehouse stocks, or placing embargoes on exports are examples of shortsighted policies. Food policy analysis attempts to identify the relationships between the short-run and the long-run effects of a policy. Failure to design interventions in the food sector consistent with long-run objectives eventually causes a policy fiasco. The greater the short-run pressures to implement a programme – any programme – the greater the probability that it will have effects just opposite to those intended. Crash programmes tend to crash.

Subsidizing farm inputs

For any output price, the incentive to increase production by more intensive use of an input can be improved by subsidizing the cost of the input. Using subsidies to lower fertilizer costs is an especially common technique for increasing the profitability of intensive agriculture while keeping food prices low. When total fertilizer use is low and the ratio of incremental grain yield to fertilizer application is high, such subsidies can be a highly cost-effective strategy relative to higher output prices or greater food imports with subsidies. Fertilizer subsidies can also speed the adoption of modern seed varieties. As fertilizer use becomes much more widespread, however, the costs of the programme rise dramatically. The production impact per unit of fertilizer subsidy drops for two reasons: declining marginal response rates, and few nonusers of fertilizer remain to be converted to users.

Many governments subsidize other inputs as well. Irrigation water is provided to farmers well below cost, frequently at no cost, over much of the world. Subsidized credit is widely used to encourage the purchase and use of modern inputs despite poor repayment records and little apparent impact on output. Furthermore, no input subsidy programme is able to encourage farmers to use more labour and provide better managerial care for the crops. All subsidies tend to distort the intensity of use of inputs from their economically optimal levels, and significant waste is a result. Since not all inputs can be equally subsidized, output price increases will have a greater impact on productivity than will input subsidies, especially in the long run. Consequently, input subsidies can keep farm profitability high and consumer prices low only for a particular stage of input use and for a short time. After that the short-run distortions significantly impede an efficient long-run growth strategy.

Direct deliveries to the poorest of the poor

The basic needs movement has focused on the essential bundle of goods and services necessary to permit human dignity for the poor. In the absence of structural reforms that enable the poor to earn incomes that allow them to purchase these necessities, strategies have been proposed that would simply deliver a basic needs package directly to the needy. Some components of the package, especially clean water, education, and health care, can be provided as public goods. But housing and especially food tend to be supplied through private markets, and these two components of the basic needs package present major difficulties for delivery mechanisms. A direct delivery system must circumvent these private markets, with all the attendant difficulties just discussed. The alternative, to use private markets as the most efficient vehicle for delivering food, raises all the complexities and dilemmas discussed as the core of this book. Direct deliveries may well work for some components of the basic needs package, but getting more food to the poor will require much more sophisticated analysis of food policy mechanisms.

Nutrition intervention projects

Traditional nutrition intervention projects, such as school lunch programmes, iron fortification, amino acid supplementation of basic cereals, or milk distribution schemes, cannot solve the problem of chronic hunger caused by poverty. Such projects can be useful because of their demonstration effect, and some can be highly cost-effective in delivering important benefits to the poor. Their failure is relative to expectations – that somehow a marginal intervention to remedy a specific micronutritional problem will significantly alter the socioeconomic context of poor peoples' lives. Food policy analysts can be supportive of effective nutrition projects without mistaking them for answers to the basic food issues they are addressing.

Food aid

Food aid has had a very mixed record in reducing hunger. Its availability on short notice is critical for famine relief. In emergencies, tens of millions of people have been kept from starving by the speedy dispatch of food from donor countries, especially from the United States. But as a vehicle for more permanent improvements in the ability of poor people to feed themselves, food aid has been a failure. Countries that relied on food aid supplies to keep prices low created serious disincentives for their farmers. In countries where the food aid displaced imports, did not distort farm prices, and had a large enough flow to affect the level of macroeconomic resources available for development, its impact was entirely mediated by the efficacy and equity of the development strategy. When this was positive, as in Korea and Taiwan, food aid helped. When it was not, as in many other countries, it did not.

Food aid can provide both the macroeconomic resources and the food required for a country to switch from an urban-biased development strategy to an agricultural strategy based on incentives to farmers and designed to increase food production and the flow of income to rural areas. The short-run food consumption problems that make this switch difficult have been repeatedly stressed. Food aid can help by providing resources in the short run to soften the squeeze on the poor. But the overall volume of food aid available to poor countries is quite limited, and bridging strategies that use food aid as a support will be available to only a few. Even where it is available, the macro food policy context will dictate whether the food aid helps or not. Food aid is not a substitute for a sensible food policy. But it can provide useful assistance in putting one in place and speeding it on its way.

ELEMENTS OF A WORKABLE FOOD POLICY

Food systems are complicated, and food policy is dependent on powerful macroeconomic policies and on the international economy. The chances of choosing the wrong path are great because no invisible hand guides policymakers, and good intentions are no guarantee of good results. Painful experience shows that uninformed policymaking usually makes matters worse. Analysis is needed to improve the poor performance of policymaking done blind – analysis that is done in the specific context of a country's own problems and resources. This book can show how to do that analysis, but it cannot show the specific results. The best the authors can offer as solutions at this stage is to reemphasize the basic themes.

Productive jobs

No practical and lasting resolution to the food policy dilemma is possible without the creation of vast numbers of productive jobs for relatively unskilled urban and rural workers. Such jobs provide two components of the answer: increased economic output to fuel economic growth, and greater earned income of the poor so that they are able to buy the essentials of a dignified life. Finding ways to create these jobs has occupied much of the development profession since serious concerns over the distribution of gains from economic growth emerged in the 1960s. One major lesson is that governments do not create these jobs very efficiently. Public sector or public enterprise approaches to unemployment cause massive overstaffing of bureaucracies and state-owned factories, stifling initiative and performance in both.

Efficient job creation is primarily a function of private or cooperative initiatives in conjunction with facilitative macroeconomic policy. Appropriate macro prices, fiscal control over budget balances, and careful attention to monetary growth foster an environment in which investment decisions create productive jobs. Macroeconomic policy is even more important in determining the real productivity of the poor who are the primary concern of food policy.

Price incentives for food production

There is no substitute for positive price incentives to the agricultural sector based on long-run opportunity costs. The pressures for such a price policy can be circumvented for a while through input subsidies and subsidized grain imports, but a poor country cannot long sustain the capacity to provide cheap food for everyone. Societies that insist on keeping it cheap will gradually distort their economies sufficiently to choke off the economic development process. The move to an incentives-oriented policy need not be immediate. Gradual increases over the course of a decade are reasonable if the government can maintain such a long time horizon.

Two critical variables that are subject to policy influence determine the level of rural price incentives compared with international opportunity costs: the foreign exchange rate and the domestic farm-gate price for food. If the government maintains an overvalued exchange rate, an additional and sometimes impossible burden is placed on domestic price policy. Discussions of domestic food price policy that do not include the role of an equilibrium exchange rate miss the most pervasive aspect of policies biased against agricultural production and rural income generation.

Removing the common biases against agricultural production in many developing countries would set the stage for a much more dynamic rural sector. For food-importing countries, however, a case can be made for going further and setting domestic food prices at a slight premium, perhaps 10% on average, over the opportunity cost of imports. Maintenance of such a price premium is justified by the second-round effects of the additional purchasing power in the countryside, where job creation for unskilled workers is most likely, by the further improvement in rural–urban income distribution, and by the added impetus to investment in future agricultural productivity. Private investment decisions are sometimes myopic, especially in the face of high and fluctuating interest rates. Agricultural investments in particular require long time horizons. A small premium on such investments in the interests of future generations can be produced with food price incentives somewhat above efficiency levels. The premium might vary by commodity, with preferred grains receiving a larger premium than staples consumed by the poor. Since the price premium will be reflected in market prices, such commodity discrimination would minimize the adverse consumption consequences for the poor.

Public investment in agricultural productivity

In no country does agriculture receive a share of public investment as large as its contribution to gross national product. Very few governments devote even half of agriculture's share in GNP to investment in the sector. While no economic law dictates that the shares be equal, public investment should be directed to projects with the largest social payoff. Following this rule would probably double agricultural investment if the projects could be prepared and administered. The bottleneck is in the preparation of sound agricultural projects. Simply going this far would significantly improve the balance of public sector investments. But as with market prices to encourage private investment, it is desirable to go somewhat further. Because of its importance to the welfare of the poor, food can be treated as a merit good for purposes of public investment and valued at a small premium over its long-run opportunity cost via imports or exports. Placing such a premium on food also addresses the food security concerns of most countries by indicating their willingness to pay a positive but not infinite price for food self-sufficiency. A premium of perhaps 10% is a suitable starting point.

Since market prices for food grains in many countries tend to be below their long-run international opportunity costs, implementing a market premium is probably an issue for the future. The use of a premium in public benefit-cost analysis of investments, however, could be started immediately. For public investment analysis, the relative premiums by commodity might well be reversed: the inferior goods consumed primarily by the poor would receive additional credit in the project appraisal while the preferred staple could be valued at its (premium) market price.

Targeted food subsidies

A food price policy that provides farmers with positive price incentives relative to the opportunity cost of food from imports, when coupled with a favourable macroeconomic development policy, will gradually increase incomes of the poor and enable them to purchase their basic needs. The food price policy dilemma arises because the poor bear the brunt of the short-run adjustments needed to implement this long-run strategy. Historical experience suggests that only targeted food subsidies are likely to ease the nutritional burdens of these adjustments. Subsidies are critical because the poor do not have the resources to purchase adequate amounts of food from the market. Targeting is essential because society does not have the resources to subsidize food for the entire population. Much of this book has been devoted to understanding the likely efficiency of various targeting mechanisms in real-world circumstances. Since the ability of bureaucrats to administer a fair means test in most of the developing world seems questionable, much of food policy analysis involves a search for more effective self-targeting mechanisms for delivering food to the poor.

No single targeting mechanism seems adequate to the task. Some combination of intersecting mechanisms will probably prove essential to effective targeting and control over food subsidy budgets. One approach is to use fair-price shops in locations accessible primarily to poor people and to sell commodities that loom large in the budgets of the poor but not in those of the middle class. Distributing food stamps good only for certain commodities in particular government shops might be another. Whatever the specific mechanism or combination of mechanisms, only a clear understanding of the food consumption patterns of the poor can provide insight into its probable effectiveness.

A policy debate focused on food prices

The emphasis throughout this book has been on the central role of food prices as the link between producers and consumers in the short run and as a significant determinant of investment decisions that connect the short run to the long run. A sensible price policy alone will not solve a society's food production problems. Price incentives exacerbate the consumption problems in the short run. However, an understanding of the positive and negative aspects of a country's food price policy illuminates most of the issues at the core of the food policy debate. With this understanding, government policymakers have gained a vantage point to survey the entire development process.

Food policy analysis can improve the quality of that debate by providing the best answers available to the tough questions that policymakers have every right to ask. How much will food production increase if food prices rise? When? Will the marketing sector be able to handle the additional supplies? Will traders siphon off all the gain? How badly will food consumers be hurt? How can they be helped? What are the implications for the budget? For the balance of payments? An analyst who can provide honest answers to these questions has learned all this book has tried to teach, and more.

29

AGRICULTURAL MARKETING AND PRICE INCENTIVES: A COMPARATIVE STUDY OF AFRICAN AND ASIAN COUNTRIES

R. Ahmed and N. Rustagi

Before presenting differences in prices across geographic regions, rural and urban centres, and various seasons in a year, definitions of different types of prices have to be clearly stated. Producer prices generally relate to prices at primary market or farmgate prices. The producer price could again be a market price or an administered price determined by government. Most producer prices used in this study are market prices except for a few East African countries where public monopoly precludes a free market price. However, if a parallel market price was available and taken into consideration, it was reflected in the average. At the terminal markets there are retail prices for domestic consumers and export prices for foreign consumers (on an f.o.b. basis). Administered prices at the retail level (such as ration prices) are not considered in any estimate of price spreads.

Quality, particularly as its relates to processing, was carefully examined before a price statistic was used in analysis. For example, farm prices are often for paddy but retail prices are for cleaned rice. In comparing retail and producer prices of rice in this example, the rice price equivalent of the paddy price was estimated by using a conversion factor and milling cost.

The estimates of marketing margin and regional and seasonal price spreads, as presented in this study, represent central tendencies or averages for a country. Proportions of marketings in various regions or markets were used as weights in such a procedure. But it must be admitted this procedure was not followed in a systematic manner in all cases because often the necessary data was not available.

Source: International Food Policy Research Institute, Washington, 1985. (Passages from a paper prepared for FAO.)

SPATIAL SPREADS IN PRICES

Two categories of spatial prices are considered. Price spreads between producer and consumer ends of a product market represent a category where the market margin is equivalent to the spread in prices at the two ends. The other category of spatial spreads reflects the differences in prices at various regional markets at a particular time. This second category of price spreads may include differences that go beyond the explanations provided by the marketing margin. Marketing margin and spatial price spreads are synonymous where the two price points are integrated by a functioning market or trade link. A spatial price spread could represent a price difference between two points having no functioning trade link between them. This is what is known as a nonintegrated market. Therefore, the regional price spreads are in some respects a partial indicator of the lack of market integration when read in combination with marketing margins.

The two types of spatial spreads in prices in foodgrain markets of selected countries are presented in Table 29.1 and Fig. 29.1. The statistics in the table as well as their graphic representation clearly show that farmers in Africa receive a smaller proportion of the price paid by final users of marketed foodgrains than farmers in Asia. In general, the average producer prices expressed as a percent of terminal market prices in selected Asian countries *vary* from 75 to 90%; the comparable figures for Africa range from 30 to 60%. Farmers in Africa receive a share of the final value of a foodgrain product that is almost half of the share received by their Asian counterparts. There are variations in the shares received by farmers for various commodities. For example, in African countries, rice offers a relatively large share of final value to producers compared to other foodgrains. This is primarily because production of rice in most African countries is concentrated in specific locations, which means that marketing of rice is not as costly as for foodgrains, which are scattered geographically. But the differences in the shares of producers among countries of a continent is moderate, although this difference is sharper among African countries than Asian countries. Farmers of Nigeria and the Sudan, the West and North African countries, appear to share about 55–60% of the price paid by final users of foodgrains, whereas farmers of Malawi, Kenya, and Tanzania, the East African countries, received only about 35–50% of the price paid by consumers.

The regional price differences within each country, as seen in Table 29.1 and Fig. 29.2 are again larger in African countries than in Asian. On average, regional prices of foodgrains in African countries differ from one another by a multiple of two to three (that is, the low price in one region could be only a third or a half the amount of the high price at another region). These figures indicate that the so-called panterritorial pricing

Table 29.1. Regional and producer–consumer price spreads in selected countries of Asia and Africa (1975–1980)

Country	Commodity	Weights (by production)	Regional spread[a] (%)	Weighted average	Producer–consumer price spread[b] (%)	Weighted average
Nigeria	Maize	14	36.60		54.5	
	Rice	6	72.89	46.1	57.0	58.9
	Sorghum	80	45.92		59.8	
Malawi	Maize	79.5	21.86	31.2	48.2	49.6
	Rice	20.5	68.20		55.1	
Tanzania	Maize	76.0	25.70		38.2	
	Rice	9.8	61.27	30.6	56.6	41.4
	Sorghum	14.2	35.47		48.1	
Kenya	Maize		30.0	30.0	42.0	42.0
Sudan	Sorghum	91.9	48.2	48.5	61.2	61.2
	Wheat	8.1	52.1			
Indonesia	Rice		71.9	71.9	84.0	84.0
India	Rice	54	69.8		82.0	
	Wheat	38	65.9	68.0	79.5	81.0
	Sorghum	8.0	63.5		80.0	
Bangladesh	Rice		75.0	75.0	79.0	79.0
Philippines	Rice	70	82.7		87.0	82.4
	Maize	30	64.2	77.3	71.5	

Source: Constructed from a large number of secondary sources.

$^a = \dfrac{\text{Lowest price}}{\text{Highest price}} \times 100$

$^b = \dfrac{\text{Producer price}}{\text{Terminal market price}} \times 100$

policies, particularly in the case of Tanzania, were often not effective.

The absolute magnitude of the regional price spread is significantly larger than the marketing margin (the producer–consumer price spread) in Africa. This implies that many markets may not be linked with one another in African countries because of high transportation costs due to poorer transport and communication infrastructure or government restrictions. In Asian countries the regional price spreads are quite close to marketing

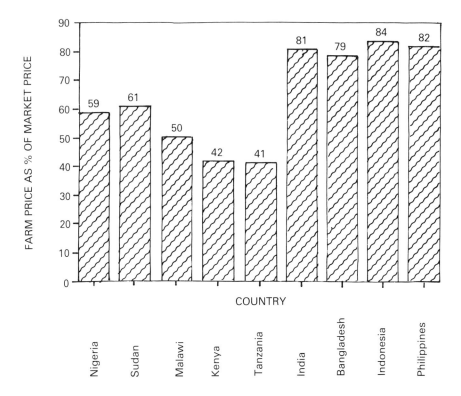

Fig. 29.1. Producer–consumer price spreads.

margins indicating that the markets scattered over various regions are probably well integrated with one another. Although it is quite unconventional to derive a conclusion on market integration from a set of data as is done here, the conventional practice of using correlation among prices as a measure of integration is often unreliable.

PUBLIC INTERVENTION AND MARKETING COSTS

As described earlier, public intervention in foodgrain marketing is widespread in all countries but there is significant difference between Asia and Africa, particularly East Africa. In the selected Asian countries private trade is not only allowed to operate side by side with public trade but also encouraged through various market development activities. These include development of market places, dissemination of price and production information, introduction of standard grades and weights, maintenance of law and order in transport channels and markets, provision of credit to traders, initiation of agricultural processing and specific storage facilities,

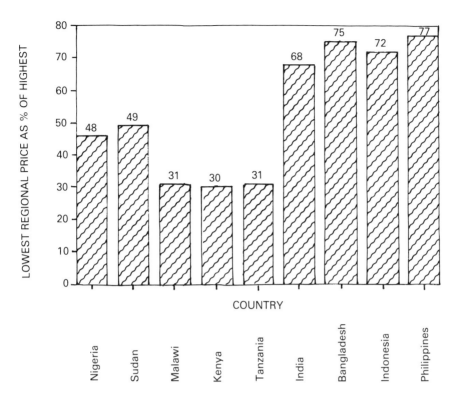

Fig. 29.2. Regional price spreads.

and provision of electricity to rural markets.

Nigeria and the Sudan resemble Asia in foodgrain marketing in that private trade is allowed to operate with minimal hindrance, although market development assistance to private trade is quite insignificant in these countries compared to their Asian counterparts. In contrast, East African countries are well known for their restrictive measures against private trade through public monopolies in foodgrains.

Public marketing affects the overall marketing margin both directly and indirectly. The direct effects arise from the relative inefficiency of government marketing compared to private trade and the inadequacy of public resources to support public marketing. The efficiency differential between public and private marketing (high public marketing costs) may not be shifted directly to producers and consumers if the government has adequate resources at its disposal to take the burden on itself, as is the case with Indonesia. But this is not generally possible for most developing countries. Data on marketing costs in public parastatals are difficult to obtain; nevertheless, it is widely known that such costs are generally much higher than comparable costs in private trade. Labour cost is the largest component of marketing cost. Labour costs in public parastatals reflect a high proportion of formally educated manpower. In private trade most workers and managers acquire on-the-job skills without much formal education; a marketing study in Kenya indicates that most of the traders in the primary market are illiterate. In the secondary markets about 16% are illiterate and about 70% have studied up to the levels between the fourth and eighth grades. It is true that the salary scales for the same level of skill could be higher in private enterprise than in public organizations, but the effect of economizing on the use of formal skills would make private trade less expensive than public trade on account of labour cost. Skilled manpower is more scarce in Africa than in Asia. Moreover, wage rates are generally two to five times higher in Africa even though average labour productivity is not higher in Africa than in Asia.

The indirect effect of public intervention on spatial price spreads, including producer–consumer price differences, is considered to be larger than the direct effect. These indirect effects can be traced to a variety of reasons related to public operation in foodgrain marketing.

Transaction costs that are imposed upon traders and farmers by the operation of public marketing and trade controls in the domestic channels are numerous and diverse. These costs generally originate from rules, regulations, and the practices of public agencies. Public marketing often operates within an environment of licensing and prohibition of private trade. Getting a license or avoiding a prohibition involves considerable explicit and implicit costs. Although systematic study of transaction costs is rare, some marketing studies in eastern Africa indicate that transaction costs to overcome trade restrictions and get legal protection are as high as 15–20% of the marketing margin. Restrictions on marketing make it difficult for private trade to avail of the opportunity of economies of scale. Instead of using trucks or railways for transport, traders use buses or small taxis to avoid movement restrictions. If public marketing is involved in any part of the marketing channel, the practice of payment by cheques to farmers and traders and the arbitrariness in assessing the quality of a product generally result in considerable transaction costs to farmers and traders.

Empirical evidence of relative efficiencies of public and private trade are rare, although public trade is widely believed to be costlier. Comparing the free trade of Benin with the marketing monopoly of the Nigerian marketing board for palm oil, it was found that the cost differences in the two systems are such that farmers end up receiving 11% less for palm oil in Nigeria than in Benin. Schmidt (1979) also indicates that storage and interregional marketing costs in Kenya are 15–25% higher in public than in private trade.

On the basis of a large amount of diverse and scattered information, an attempt was made to assess the contribution of various causal factors (including the effect of public trade) in the overall difference in marketing

Table 29.2. Shares of causal factors in differential foodgrain marketing margins between Asian and African countries

Factors	Absolute margin (points)			Shares of the factors in the difference (%)
	Asia	Africa	Difference	
Taxes	0.6	3.9	3.3	9.4
Transport and associated costs	13.8	27.5	13.7	39.1
Profit	4.0	12.6	8.6	24.5
Transaction costs (residuals)	1.6	11.0	9.4	27.0
Total	20	55	35	100.0

Source: Estimated on the basis of information from Bangladesh and Indonesia in Asia and Kenya and Malawi in Africa.

costs between selected Asian and African countries. Because of the nature of the data, this exercise is meant only as an indicative one, although the authors are convinced that the indications are close to reality. Table 29.2 shows the difference between Asian and African countries in the causal factors of marketing costs.

The figures in the table clearly point out that differences in transport and associated costs constitute the largest cause of a differential marketing margin between Asia and Africa; nevertheless, the share of residual transaction costs associated with the indirect consequences of public trade as well as the small but significant tax component in the differential marketing margin together represent a proportion almost as large as that of transport.

POLICY IMPLICATIONS

The policy implications of this analysis of marketing margins are quite clear. Improvement in infrastructural facilities is an obvious measure with large potential gains in incentives to both producers and consumers. A second major area of improvement would be public policies in agricultural marketing. While most improvements in infrastructure take a long time, reforms in public marketing policies can produce results in a relatively short time in African countries.

A strategy of infrastructural development that gives priority to areas where actual or potential growth in production is large is an obvious policy choice. To some extent, a sharply dualistic agricultural sector has exerted a natural influence in some countries of Africa so that infrastructural development has primarily been concentrated in areas where production is concentrated. Thus, Sudanese agriculture is dominated by a number of irrigation projects (including Gezira) and mechanized rainfed agriculture where most of the Sudan's infrastructural investment has taken place. This strategy has resulted in low marketing costs for produce grown there.

Public intervention in marketing seems to have been motivated partly by conditions arising from occasional (as well as locational) market failures. A number of factors are responsible for this. First, infrastructural underdevelopment tends to encourage farmers to be independent of the market. Second, large- and small-farm dualism is so sharp in Africa that large farms tend to vertically integrate marketing and production. This leaves a very thin market for the small, peasant sectors who do not have a scale for efficient marketing. The thin market is generally an unstable market. Development of an efficient exchange system, economies of scale, and service provision become very difficult and uncertain in thin markets. Ironically, public intervention in marketing designed to rectify the problems associated with market failures further accelerates the process of thinning, making the problem a complex one. Although a wholesale dismantling of public parastatals in Africa is perhaps an irresponsible prescription, few will disagree with the need for a substantive reduction and improvement in public marketing in African countries. It thus seems logical that African countries need to follow a path of gradual transition to private trading through selected public intervention in marketing, continuous efforts on market development, and a heavy commitment to properly formulated infrastructural development. The central element of this approach is that private trade should be allowed to work freely. Market development policies involving improvement in legal and physical facilities and flow of information should be another component of this process of transition. Direct public marketing may be limited to those areas where infrastructural backwardness makes agricultural income and production low and uncertain and to areas where management of security stocks for foodgrains is strategic.

REFERENCE

Schmidt, G. (1979) *Maize and Beans in Kenya: The Interaction and Effectiveness of the Informal and Formal Marketing Systems.* Occasional Paper 3, Institute for Development Studies, Nairobi.

30

ASSESSMENT OF FOOD SUBSIDY AND DIRECT DISTRIBUTION PROGRAMMES

United Nations World Food Council

Approaches to distributing food to those who would otherwise not get enough to eat invariably involve some form of explicit or implicit subsidy. Consumer and producer subsidies can have a significant impact – either positive or negative – on food production, food security, producer and consumer income distribution, and the pace and direction of overall development. This means that the choice of the right price policy, which is at the heart of subsidy decisions, has to be made within the broader food and development policy framework related to production, employment, income distribution and nutrition. A wide range of approaches has been tried in different political, social and economic country conditions. While there is no universal answer to what policies would be most effective, some appear to have greater potential or successful application than others.

Price-fixing throughout the food system

In an effort to keep food prices down, some governments, particularly in socialist countries, have exercised control by price-fixing throughout the food system, establishing retail prices at levels within the reach of low-income consumers. This does not necessarily imply government control over production and marketing, although such control simplifies price enforcement. In the case of Tanzania as presented to the Castelgandolfo Consultation in 1979, the government annually fixed producer and retail prices for major food commodities, taking into consideration domestic production and distribution cost and international commodity prices. Staple-food procurement was fully controlled by the National Milling Corporation, the Government's sole procurement agent. By law, farmers

Source: Report prepared following an export meeting at Castelgandolfo, Italy, for presentation to the sixth ministerial session at Arusha, Tanzania, 1980.

must sell all marketable surpluses to the Corporation. Similarly, distribution was in the hands of Government-controlled trading companies selling to retailers within the fixed-price structure.

However, without fiscal subsidies, even strict price control throughout the food system was often not enough to assure low-income consumers access to adequate amounts of food. The Tanzanian Government at times subsidized the National Milling Corporation's procurement and processing costs at levels that permit retail prices for wheat flour, rice and maize flour to be fixed at 25–35% below the Corporation's cost price. The precise impact of Government control and subsidies on food consumption in Tanzania is difficult to assess on the basis of the information available.

Subsidizing food production and consumption

While price-fixing throughout the food system has generally been confined to centrally-planned economies, many market economies apply price-fixing at retail level (often combined with consumer subsidies) or marketing-margin controls. Almost all countries use some form of subsidy injection at some point in the food system. The industrialized countries have a long history of heavy producer subsidizing. This has significantly helped to increase farm incomes and food production. But it has also led to structural imbalances in the food and agriculture sector, resulting in costly problems of surpluses, often with little benefit, and high cost, to consumers in developed countries and with adverse effects on developing countries' trade opportunities. Producer subsidies have been widely used around the world, but have sometimes been criticized as a potentially inefficient means of protecting agriculture and assisting consumers, and as leading to over-use of subsidized inputs and misallocation of scarce resources. By raising food production, producer subsidies can benefit consumers indirectly through greater availability of food and relatively lower prices. But they are less effective in reaching low-income consumers than are more direct measures.

Consumers generally benefit more from subsidy injections into the distribution system. These may take the form of generally subsidized and fixed retail prices for basic foods, or subsidized 'fair prices' for foods made available in rationed quantities, with the Government bearing the full or partial difference between retail prices on the one hand and procurement, processing and distribution costs on the other. This usually involves some form of public distribution or special state marketing organizations. In several Latin American countries, state marketing organizations operate with the aims of increasing the incomes of the rural poor and providing urban low-income groups with essential foods at low cost. Among these is Mexico's CONASUPO, the activities of which integrate support to agricultural production and rural development, procurement, processing, and retailing, particularly in poor urban areas. Its retail outlets sell a wide range of basic foods and reconstituted milk and have recently started to stock some basic medicines. Charging lower prices than other retail stores, they are able to have an influence on the overall price levels for basic foods. Mexico's lead has been followed by other Latin American countries: Peru, with IPSA; Ecuador, with EMPROVIT; and Brazil, with COBAL. These and other experiences indicate the food-marketing sector's potential as an important instrument to influence patterns of food production, employment, and income distribution in rural areas, while influencing the real incomes of low-income urban consumers through the price mechanism.

Complementary to specific government control, subsidy and organizational efforts within the food system to secure adequate food consumption levels for low-income groups are more general government interventions such as buffer stocks and open-market-sales operations, involving government investment and operational subsidies.

Experience with some of the major approaches to consumer-oriented food subsidies is summarized below.

General consumer food subsidies – expedient but high-cost and high-risk

Subsidizing basic foods is an expedient way of raising the food consumption levels of low-income groups, without the needs for difficult administrative operations. It frequently involves some form of foreign-exchange manipulation, which might include importing essential foodstuffs at preferential exchange rates. In a number of countries, food obtained on concessionary terms has been distributed at, or below, cost. Food export taxes have been used in a few cases – for example, in Thailand for rice exports – to keep the domestic price below world prices.

Raising the food consumption levels of the poor through general subsidies usually carries high financial and economic costs, as illustrated by Egypt's experience. The Government subsidizes about a dozen food commodities, partly through general subsidy schemes (e.g. for common bread), with no restriction on access by the better-off consumers or on quantities bought, partly through ration schemes (e.g. sugar, tea, edible oils), and partly by subsidizing foods of limited availability (e.g. frozen meat, dairy products), without rationing and on a 'first come, first served' basis. Who exactly benefits from the subsidies has yet to be assessed, but it is obvious that Government food subsidies provide income supplements for the rich as well as the poor. It also seems that a significant number of better-off consumers have managed to profiteer from commodity re-sale to those who, for one reason or another, do not have full access to subsidized foods (probably a particular problem in rural areas). Heavy general subsidies do lead to food misuse (e.g. for animal feeding) or to outright waste – an observation made in economies as diverse as those of Egypt and the Soviet Union.

As a result of Egypt's liberal consumer subsidy policy, demand for subsidized foodstuffs rose steeply over the five years from 1973 to 1978: by 50% for wheat, 80% for flour, and 150% for lentils. these increases clearly outstripped both population and income growth. Growing demand for an increasing number of subsidized foodstuffs, combined with rising international food prices has placed a heavy burden on Egypt's budget: in the 15 years from 1960 to 1975, the subsidy allocation increased about seventyfold, from LE 9 million to LE 622 million, equivalent to some 20% of public expenditure. Efforts to reduce food subsidies in 1976–77 led to public unrest and subsidies were immediately restored to previous levels.

There is the clear danger that general subsidies will depress domestic prices to the farmer. Guarding against strong disincentive effects on domestic food production can involve costly efforts to maintain higher prices to farmers and to keep consumer prices low enough to have a significant effect on the food consumption of the poor – and the difference must be borne by the entire economy, or by large-scale food aid on highly concessional terms. In addition, general subsidies are a politically sensitive issue. Once established, these programmes are extremely difficult to discontinue.

Food rationing within a dual-market mechanism – mixed experiences

Rationing schemes are one way to reduce both cost and risk. Some industrialized countries during and immediately after World War II, and Cuba after the revolution, instituted complete rationing schemes for major foods, suspending all private trading outside these schemes. However, most of the rationing schemes now in operation in developing countries function within a dual-market mechanism: an open market with commercially-determined prices and the ration-market segment, making limited amounts of free or low-price staple foods available generally or to specific target groups, often through publicly-operated or licensed outlets ('fair price' shops, ration shops).

Examples of apparently successful operations include those of Sri Lanka and the Indian State of Kerala. Both schemes have covered, or are covering, almost all urban and rural income groups, providing some 450 calories per caput per day in the form of rice and wheat, with, in Kerala, the addition of amounts of cooking oil and sugar.

Kerala's rationing scheme is estimated to have resulted in a significant net increase in calorie consumption among low-income groups and has been positively related to child nutritional status (Kumar, 1979; George, 1979). Rationing has contributed in some degree to real-income redistribution among consumers, while the State's levy procurement system (compulsory procurement, varying according to farm size) has helped improve equity in farm income distribution. State food-grain import controls have maintained high incentive prices for farmers on the open market.

In Sri Lanka, rice rationing appears to have produced significant substitution effects in low-income households, leading to more modest increases in calorie consumption as compared with Kerala. However, the incremental real income obtained from rice rationing has enabled consumers to diversify and qualitatively improve their diets and spend more on housing and clothing. Ration subsidies have contributed substantially to improved income distribution – high farm prices for paddy, through an effective voluntary Government procurement system, have benefited farm incomes and contributed to rapid growth in the rice paddy sector. The last, in turn, has been responsible for a large part of overall employment growth, and the increased expenditure resulting from higher real incomes due to ration subsidies and the scheme's employment effects is likely to have contributed to further income generation in rural areas.

Much of the relative success of the Kerala and Sri Lanka schemes appears attributable to the overall political environment; to effective procurement systems and food-grain import and export controls; to flexibility in administrative arrangements, facilitating access by the poorest to rationed grains; and to the high degree of public awareness of the ration system. These factors and the achievements in both countries also demonstrate the interrelationship between food ration subsidy schemes on the one hand, and food sector policy and agricultural and broader socio-economic development on the other. But they raise some critical issues of general relevance.

The fiscal costs of Sri Lanka's former net food subsidies have been as high as 18% of total government expenditure. At these levels, subsidy expenditures give rise to concern about their impact on Government spending for capital investment requirements. Moreover, questions have been raised about their cost-effectiveness, considered in nutritional terms; according to one estimate, for each ration-food calorie which went to increase the consumption of calorie-deficient people, another 11 either went to non-deficient population groups or were substituted for commercial purchases. Considerations of this kind influenced the decision of the Sri Lanka Government in mid-1979 to replace its traditional rationing system with a new food-coupon programme.

The proportion of the population covered by rationing schemes elsewhere in Asia has been significantly lower. Pakistan's wheat (atta) rationing scheme reaches about one-third of the population. The scheme covers mostly urban populations, where it helps keep down wage levels, and extends to some rural populations, particularly in food-grain deficit areas. Participating households are entitled to a ration equivalent to about three-quarters of estimated cereal requirements. But relatively few small farmers and landless labourers benefit from the scheme, while in urban areas all income groups have access to rationed wheat flour, which carries a public subsidy (shared by central and provincial governments) of 24% at present. In all India and Bangladesh, somewhere close to 10% of the population benefits from rationing schemes; in absolute numbers, 'close to 10%' in India means 59 million people.

Bangladesh's rice and wheat rationing scheme has aided the urban population significantly. Without it, the calorie consumption of the poorest 15% of the urban population would probably have been 15–20% lower (Ahmed, 1979). But much of the rationed food goes to less needy, middle-income people. In rural areas, with 90% of the total population, the poor, especially the landless, seem to have benefited relatively little. In a country where the majority of the population is below poverty and subsistence level, the brunt of the subsidy provided by the Government may actually be borne indirectly by the poorest, who, in addition, do not enjoy any of the benefits. It has been estimated that an extended scheme to increase the food consumption of the poorest quarter of the rural population would cost about four times as much as the present one, which has accounted for 7–13% of the total Government budget and 21–36% of net public revenue – making it an option that appears neither economically and financially nor politically feasible. Basing an extended scheme entirely on food aid would be likely to depress domestic food-grain prices by 10–27% (Ahmed, 1979).

The Government of Bangladesh is well aware of these problems. Its present efforts are directed at limiting the extent of 'statutory rationing' for the population in the major urban centres and extending its 'modified rationing' and food-for-work programmes, both intended to reach more of the population in need. Attempts are also being made to increase the proportion of less expensive wheat to rice in the statutory rationing system. And, by way of experiment, the distribution of sorghum is under way. Sorghum offers high nutritional value at low cost, and high income elasticity of demand among low-income households, thus facilitating targeting to the people in need; in addition, prospects are favourable for growing sorghum on marginal soils in Bangladesh, without competing with major food and agricultural crops. The Government also intends to test the potential of open-market-sales operations as a means of contributing to price stabilization and overall food security, although they may play only a secondary role in efforts to meet the food requirements of those most in need.

Food coupons – benefits for consumer and government

Rationing schemes call for a considerable degree of government intervention in the food-grains distribution system, but this and the problems inherent in dual-market mechanisms are obviated when some form of food-coupon scheme is used. Food coupons stimulate effective demand, and possible disincentive effects on food production are avoided. Food-coupon programmes at present operate in Sri Lanka and Colombia and are being considered by the Philippines as an element of its new food strategy. The United States of America also operates a major food-coupon programme and its experience has been included in this review, to provide insight on programme design and administrative issues.

Sri Lanka's new food-coupon programme, instituted in mid-1979, is the result of efforts to reduce Government involvement in the food distribution system, in line with its overall development policy, and to reduce the costs and increase the effectiveness of its food subsidies. From the almost universal coverage of its former rationing scheme, the Government moved to reduce the number of eligible beneficiaries by about one half. A family income of Rs.300 (about US$20) per month (for a family of five) was made the ceiling for eligibility. Eligible households were issued free food coupons in amounts varying according to household size and age composition, with higher allocations for children up to eight and from eight to twelve years old. The value of coupons was calculated in such a way as to ensure food benefits about equivalent to those of the previous rationing system. In addition, to meet rising energy costs, each eligible household receives kerosene coupons, which can also purchase major food items.

The new programme seems to offer advantages to both low-income consumers and to the government. When cashing the coupons, consumers can now choose from about 10 major food commodities, as opposed to one or two under the previous rationing programme. They can also deposit unused food coupons in a Post Office savings account – an option somewhat resembling China's ration savings system. What is not clear, however, is how consumers are shielded against inflation, unless coupons are in some way price-indexed. By limiting the number of beneficiaries, the Government has been able to reduce its subsidy expenditure by some 40%, from Rs. 2450 million to Rs. 1450 million in the current fiscal year.

A specific feature of the new programme is a complex means test for programme applicants and participants, which is re-validated every three months. The test, which requires a high degree of functional literacy, also serves to identify the economically weaker sections of the community, with the explicit objective of according them priority in employment and in moving them into income brackets above the food-coupon eligibility ceiling. This linking of food assistance to practical efforts to bring about the long-term advancement of low-income households is the programme's most important feature, elevating it beyond simple welfare to a truly equity-oriented development. This is a concept which merits the Council's consideration and support.

In the USA, a $6.9 billion food-coupon programme provides food assistance to 16 million people, allowing them to use coupons for any foods, excluding liquor and tobacco. This programme is the United States' most flexible and comprehensive income redistribution effort. Food coupons can be made available within two days of application, and participation in the programme can quickly be terminated as employment and higher income are obtained. The programme has broadly-based support among the population; it has the support of trade unions, in view of its flexibility and non-bureaucratic rapidity, and of the food industry and distributors for its demand-stimulating effects. Recently, the Government indexed coupons to food-price inflation and removed the cash contribution requirement for participants – a measure which increased programme participation among the (cash deprived) rural population by some 40%.

The US food-coupon programme is credited with having a positive effect on the food and nutrition situation of the poorest households, although there may be nutritionists who feel that the consumer, faced with the virtually unlimited choice of food, has not made optimum use, nutritionally, of the coupons. To ensure that the specific nutritional requirements of nutritionally-vulnerable groups are met, the Government operates a special programme whereby food coupons for the purchase of highly nutritious foods are issued to pregnant women and mothers of young children among low-income groups.

Nutritionally-vulnerable groups are also the target of the Colombian food-coupon programme, which operates over about half of the country. In the context of a complex food, nutrition and rural-development programme supported by the World Bank, the food-coupon programme is designed for the nutritional improvement of mothers and young children and harnesses the public health sector (distributing the coupons), the research and development, production and marketing capacity of the food industry (at present about a dozen food manufacturers have signed agreements with the Government), and the private distribution system (exchanging the coupons). Coupons can be exchanged for state-approved foods in four commodity groups:

1. Nutritionally-fortified pastas/noodles.
2. Highly nutritious vegetable mixtures.
3. Texturized vegetable protein products.
4. Nutritious biscuits.

Coupon users are required to make a cash contribution at the time of exchanging coupons. On an average, for every purchase made with coupons, coupon holders are estimated to make a cash contribution of

40%, with the coupon providing the remaining 60% of the purchase price.

Two features of the Colombian coupon programme must be noted. First, while coupons are directed at vulnerable groups as in the US special-coupon programme, recipients can exchange them for 'family foods' (like pastas), as well as for special foods for infant and maternal nutrition. Second, the nutritionally-fortified foods available under the programme cost several times as much as the basic commodities. Less expensive commodities, produced under the rural development scheme, are being considered for inclusion in the food coupon programme. An interim evaluation of Colombia's food and nutrition programme will be available during 1980.

Food-coupon programmes are essentially a form of income supplementation, and it is frequently asked why cash cannot be substituted for coupons. There are two arguments in favour of the coupon approach. First, cash transfers have proved difficult politically and are generally not supported by the tax-paying public. Second, food coupons tend to increase food consumption to a higher degree than cash transfers (probably with the exception of the very poorest consumers). According to an estimate by the US Department of Agriculture, US food coupon participants spend 50 cents of each dollar of subsidy on food; in contrast, a direct income transfer is estimated to provide only 20 cents' worth of food consumption for every dollar (Reutlinger and Selowsky, 1976).

Direct distribution schemes: food-for-work and feeding programmes

Direct food-distribution schemes have been used extensively in emergencies and for easily identifiable groups, often replacing market purchases for target groups and avoiding explicit means-testing. Food-for-work and supplementary-feeding programmes for vulnerable groups are common examples.

The effectiveness of supplementary feeding had already begun to be called into question when the Council, at its third ministerial session in Manila in 1977, requested the United Nations agencies to undertake a thorough assessment of the usefulness of these programmes. A full response to the Council's request is still awaited, but the Sub-Committee on Nutrition of the United Nations Administrative Committee on Co-ordination, after commissioning two related studies, will again examine the issue at its sixth session in Paris in February 1980.

Food-for-work programmes are among the most widespread measures used in rural food distribution schemes, mostly as short-run or emergency-relief measures. The potential advantage of these programmes over other food-subsidy programmes lies in their dual role of providing employment/

income and improved food consumption to the rural poor, while at the same time contributing to infrastructure investment. But often, these programmes have provided food in exchange for some largely *ad hoc* work, rather than developmentally-purposeful work paid for partly by food rations. When the Tanzanian participant at the Council's Castelgandolfo Consultation stressed that there were no food-for-work programmes in Tanzania, but only work-for-food programmes, he meant to emphasize more than a semantic subtlety: the focus on purposeful work contributing to development.

This development aspect is the guiding principle in the Guaranteed Employment Scheme applied in the Indian State of Maharashtra. A public-works programme offering employment to those seeking it, the programme's principal objective is to build up productive assets. Labourers are paid partly in kind, partly in cash – in kind in order to reduce their dependence on not always trustworthy food retailers; in cash to enable them to meet other basic needs. Cash transfers for these programmes are obtained through taxation of the better-off. Food-for-work programmes, designed as part of a rural-oriented development strategy, offer, in combination with international food and financial assistance, a promising mechanism for the re-distribution of food and income on a national scale.

REDUCING SUBSIDY COSTS AND IMPROVING PROGRAMME EFFECTIVENESS AND EFFICIENCY

The high costs of subsidy programmes naturally lead Governments to search for opportunities to reduce the fiscal burden. One obvious way is restricting the number of beneficiaries. But, with increased restriction, programmes become progressively more difficult to administer. Among the difficulties encountered are the effective identification of eligible beneficiaries and the avoidance of 'leakages' to non-eligible groups.

Complex means tests of the Sri Lanka type place heavy demands on both administration and programme participants and require a high degree of functional literacy. Many countries will need simpler approaches. Colombia's 'area targeting', based on a poverty-zoning map developed from easily-obtainable socioeconomic indicators, may be one practicable approach: food coupons are made available to the entire population of communities located in the major poverty zones. Targeting food benefits to the nutritionally-vulnerable groups and restricting the choice of foods to those most needed by mothers and young children, as in the United States' special-coupon scheme, is a further step towards more sharply orienting food subsidies to the food and nutrition requirements of the people in most need. Other possible approaches include the community-based identification of beneficiaries and the subsidizing of commodities primarily consumed by low-income groups, as could become the case with sorghum in Bangladesh.

Closely related to programme targeting is the question of how to reduce programme 'leakages'. These occur in several ways: at the administrative level, through diversion and misappropriation of food destined for the needy population; through fraud, such as the forging of ration cards or vouchers; and, at household level, through re-sale or barter of subsidized commodities, or simply purchase substitution, where the availability of subsidized foods frees part of the household income otherwise spent on food. Administrative and other fraudulent leakages can and must be dealt with forcefully. But leakages at household level are a more complex proposition; programme cost-effectiveness cannot be maximized at the expense of a certain amount of freedom of choice and respect for human dignity.

Reaching the rural poor has remained an unresolved problem in many countries. As much as anything, this is a problem of physical and administrative infrastructure. The inadequacy of rural roads, transport and storage infrastructure means that large segments of the population are relatively inaccessible. Distribution and marketing mechanisms are inefficient: as major difficulty encountered with many rationing schemes in rural Asia centres around the functioning and viability of ration shops. At times of high demand for rationed foods, shops are often inadequately stocked. When ration off-take is low because conditions are favourable to local production, ration shops have to fight for their survival. The low retail margins allowed to ration shops often do not permit shopkeepers to ensure adequate availability of rationed food at all times, nor provide sufficient incentive for them to do so. The design of food-subsidy and direct-distribution programmes must include measures which address the basic physical and organizational inefficiencies of rural distribution systems which jeopardize rural food security.

The need for this linkage of food subsidies with fundamental development efforts to improve rural distribution systems and local food security was strongly emphasized at the Castelgandolfo Consultation. Purposefully-designed food-for-work programmes, supported internationally with food, cash and technical assistance on a major and sustained scale, can significantly contribute to these objectives.

ADVANTAGES OF DIFFERENT FOOD-ENTITLEMENT APPROACHES

There is no right policy by which all countries can ensure food security for their people. But from this assessment a number of conclusions have emerged regarding the potential of different approaches for successful application under a wide range of different country situations.

The principal conclusion is that food-subsidy and direct-distribution programmes can be feasible and effective instruments for improving equity in the distribution of food and ensuring a degree of food security for all people. They can also be feasible ways of improving income distribution. Properly designed, they can contribute to development, rather than compete with it, by meeting consumption needs in ways which stimulate and support production programmes. However, if these subsidies are to achieve the dual objective of raising the food-consumption and real-income levels of the poor and stimulating and maintaining high levels of food production, then decisions concerning them must be taken within the framework of overall development policy, most specifically as an integral part of agricultural and food-price policy. The choice of the right price policy, reconciling consumption and production concerns, is at the heart of subsidy decisions.

More specifically, our assessment suggests that properly-managed consumer subsidies are more likely to effectively meet low-income consumption needs, while also stimulating production, than are producer subsidies by themselves. First, producer subsidies chiefly benefit those farmers with most food to sell and those consumers able to buy most, with the possible effect of worsening real-income distribution on both the producer and consumer sides. Second, even if producer subsidies can be more closely targeted to poorer farmers, their possible benefits on the consumer side remain indirect and non-selective, resulting in a relatively low cost-effectiveness in raising food consumption levels of those in need. Third, there is the risk of price reductions as a result of producer subsidies benefiting middlemen rather than being passed on to consumers. Fourth, producer subsidies rapidly become incorporated in the cost of supplies, the level of rents and the price of land. There exists the danger that the capitalized value of the stream of subsidies would be largely removed from agriculture, as costs rose with the level of returns. Fifth, there is a risk that the widespread introduction of producer subsidies in developing countries could lead to many of the structural problems of food production encountered in developed countries over the last 40 years.

While the use of producer subsidies as part of a balanced food-price policy nevertheless may be required in many countries, our assessment suggests the need for greater efforts aimed at exploiting the potential of selective or targeted consumer subsidies for raising food production through increasing demand among hungry people. The increased use of such subsidies, as adopted by some developing countries, provides a direct and efficient way of alleviating hunger, through targeting financial transfers and attaching them to food.

General consumer subsidies, on the other hand, tend to be less cost-effective in raising food consumption levels of those in need. A major difficulty often encountered with general-subsidy and fixed-price schemes is that the maintenance of an acceptably-low fixed consumer price can lead to disincentives to farmers, in turn necessitating subsidies on the producer side. Inevitably, too, such subsidies constitute an income supplement for rich as well as poor.

Targeted rationing and food-coupon schemes are more efficient when judged by social and economic criteria and can be more effective in raising food intakes than can direct income transfers. They do, however, place considerable demands on administrative capacity for actually reaching those in need and require information on the incidence of poverty and hunger. The choice between rationing and coupons is probably not one to be decided on economic grounds alone, but coupon programmes have a somewhat lower economic and social cost, since they allow consumption patterns to reflect more fully consumer choice and avoid often costly government intervention in the food distribution system.

Selective schemes can be linked with employment and income policies. This benefits both low-income groups and the public sector. In the case of Sri Lanka's new food-coupon programme, through well-directed efforts to identify employment opportunities for food-coupon recipients, the Government is able to reduce the size of its subsidies while low-income households are enabled to become self-reliant in meeting their food needs. Employment generation is also linked with direct food distribution through food-for-work programmes. The Guaranteed Employment Scheme of the Indian State of Maharashtra is an example of linking food distribution with employment and cash-income generation through tax-generated income transfers, and with development through the building up of productive assets.

Direct-distribution schemes may be particularly appropriate where low-income people are largely excluded from the wage and commercial economy, as in rural areas. Although they may run into serious logistics and administrative problems, as has been demonstrated by many food-for-work programmes and efforts to reach nutritionally-vulnerable groups, these schemes often allow target groups to be reached without the need for individual identification through difficult means tests. And their potential as a tool in rural development has not yet been fully explored.

The further development of food-subsidy and direct-distribution programmes will need to emphasize rural coverage and a better integration with other programmes aimed at rural development. General schemes may not be made specific to urban areas, but often still have the effect of directing most of the benefits to urban consumers. Targeted subsidies may be more difficult to administer in rural areas, though coupons in particular may be useful in places where food distribution centres are less accessible, and may be a means to circumvent many of the efficiency and viability problems encountered with rural ration shops. Rural hunger is often seasonal and special programmes may be required. Most targeted food-subsidy schemes are probably initiated with urban consumers in mind, though many schemes are being expanded to cover rural areas.

REFERENCES

Ahmed, R. (1979) *Foodgrain Supply, Distribution, and Consumption Policies Within a Dual Pricing Mechanism: A Case Study of Bangladesh.* IFPRI Research Report No. 8, Washington.

George, P.S. (1979) *Public Distribution of Foodgrains in Kerala: Income Distribution Implications and Effectiveness.* IFPRI Research Report No. 7, Washington.

Kumar, S.K. (1979) *Impact of Subsidized Rice on Food Consumption and Nutrition in Kerala.* IFPRI Research Report No. 5, Washington.

Reutlinger, S. and Selowsky, M. (1976) *Malnutrition and Poverty: Magnitude and Policy Options.* World Bank Staff Occasional Papers No. 23, Johns Hopkins University Press, Baltimore.

VII
EXPORTS

There has been a steady flow of advisory papers on export markets and the preparation of products to meet export requirements from the GATT/UNCTAD International Trade Centre in Geneva and the former Tropical Development and Research Institute in London (e.g. J. Joughin *The Market for Processed Tropical Fruit*, TDRI, London, 1986). FAO Commodities Division has been a continuing source of export market situation reports, e.g. *The World Market for Tropical Horticultural Products*, FAO, 1985. FAO, the United States Department of Agriculture and trade sources have also put out short and longer distance market forecasts. These can often be self defeating. Uma Lelé once pointed out that whenever the World Bank had denied loans to expand export production of a commodity on the basis of market prospects studies it had been proved wrong.

E. Reusse's study on livestock exports from Somalia, presented here, brings out the ability of traditional export channels to overcome logistic and other difficulties where there is sufficient incentive, in this case access to scarce foreign exchange.

The incorporation of small farmers of developing countries into export production/ marketing systems came to the fore in the 1970s and 1980s. Minot's study is summarized here. D. Glover ('Contract farming and small holder outgrower schemes in less developed countries'. *World Development* 12 (11/12), 1984) researched in parallel. Abbott's *Agricultural Processing for Development*, Gower, Aldershot, 1988, is a convenient source for some outstanding cases. The view that concentration on export production ran counter to rural welfare in some developing countries led to systematic studies by IFPRI and the World Bank. That by J. von Braun et al. on export production of snowpeas in Guatemala is summarized below.

S. Jaffee wrote his paper on exports from Kenya as consultant to the Department of Agriculture and Rural Development of the World Bank. It details the realities of finding and holding on to distant export markets. Studies of other successful export groups are 'Export marketing for off-season fresh produce, the case of Agrexco Israel' by B.W. Berman, M. Ben David and F. Meissner in *Marketing and Development: Toward Broader Dimensions*, JAI Press, 1988, and *The Fruit and Vegetable Export Sector of Chile: a Case Study of Institutional Cooperation* by C. Barriga et al. of the Agricultural Marketing Improvement Strategies Project of US AID, 1990. For the other side of the medal read *Constraints on Kenya's Food and Beverage Exports* by M. Schluter, Institute for Development Studies, University of Nairobi, 1984.

31

RESPONSIVENESS OF A NOMADIC LIVESTOCK ECONOMY TO A PROFITABLE EXPORT OPPORTUNITY: THE CASE OF SOMALIA

E. Reusse

In most of the developing countries the philosophy of settling nomads also has political motivations. Post-colonial state borders tended to follow divisions among tribes according to the suitability of areas for export crops or mineral exploitation while little importance was attached to livestock, the mainstay of the nomadic economy. The nomadic hinterland bordering the cultivated areas was, therefore, often attached to the territorial areas of administrative systems with little regard to the ethnic and cultural identity of nomadic tribal systems. These nomadic hinterlands, as a consequence, now extend across one or several state borders, adding to an unstable situation in many regions. The economic contribution of the nomadic systems to the national economies of such states is therefore not readily measurable and is mostly underestimated.

It has to be appreciated, however, that any such contribution is really a net value-added item since it originates from land resources that can hardly be exploited in any other way and is produced and marketed with almost no inputs from outside the system. The nomadic economy is remarkably independent of modern commercial input requirements, unlike alternative systems for marginal areas, which are heavily dependent for their exploitation on pump irrigation, fertilization and mechanization. Even the movement of the crop, i.e. the livestock offtake on the hoof, to consumer markets (domestic and abroad) is to a large extent free of purchased transport energy inputs. Thousands of trade cattle, for example, walk more than 1500 km from the western Sudan to Khartoum during the main trekking season without appreciable loss of condition. Of nearly equal length are many treks of trade cattle from nomadic range production areas in the Sahel to gathering points on the northern border of the tsetse belt for

Source: *World Animal Review* 31, 1981. FAO, Rome.

subsequent shipment by road and rail to coastal consuming centres such as Lagos and Abidjan, where Sahel beef is the basis of the meat supply.

During the present period of rapidly rising energy costs, which limit the expansion of mechanization, irrigation, chemical fertilization and road transportation, the exploitation of nomadic livestock systems is bound to remain important, or increase in importance.

It is generally known that extensive areas of Somalia and the neighbouring Ogaden areas are nomadic rangelands. In fact, 70% of the Somali population is devoted to nomadic pastoralism, contributing 60% of the GNP and 80% of the value of national exports.

While the long-distance cattle trade in west Africa and the Sudan has received moderate recognition in economic, social and technical surveys, the extraordinary flow of Somalia's nomadic livestock for export to markets on the oil-rich Arabian peninsula, with Jeddah as the dominant receiving port, has found little, and not always well-informed, publicity. Few readers would know, for example, that the small Somali port of Berbera on the Arabian Gulf was, until very recently, the world's number one livestock shipping point, handling over two million 'sheep units' per annum. Three quarters of these 'sheep units' consist of sheep and goats, the balance being cattle and camels (one head of cattle equalling five sheep units and one camel equalling eight sheep units.) It is also interesting to note that approximately half of all the livestock arriving at Jeddah, the world's number one livestock receiving port, are of Somali origin. Many readers might also be surprised to know that until the recent drastic shift in Australian sheep exports from carcass meat to live animals, Somalia was the world's major sheep and goat exporter. Even in 1976, one year after a very severe drought, Somali exports at real f.o.b. prices were valued at one-sixth of world livestock exports (US$ 340 million).

EXPORT MARKETS

Against the background of rising consumer incomes in Arabian markets, there has been active demand over the past decade for Somalia's livestock export surplus and this situation is likely to continue. Prices have been rising steadily, encouraging intensified animal husbandry, fodder production efforts and range water development, mostly on a private basis. This has been accompanied by restraint on the part of producers of their own personal meat consumption. The only two factors hampering the continuous expansion of exports have been shipping facilities, which, in recent years, appear to have been acutely short during peak export seasons, and droughts with their retarding effects on animal growth and fertility and which in severe cases cause heavy losses among flocks and herds.

Somali livestock is in a strong position on Arabian consumer markets, especially those of Saudia Arabia. Until the 1974/75 drought caused a severe set-back in Somalia's sheep/goat and camel production, its share of Saudi imports of these animals was 75 and 90% respectively. During the three years following the drought, new suppliers, especially Australia, took the opportunity to fill the gap, while the Sudan, Somalia's main competitor, came back into the market after the Government-imposed export ban (from 1975 to mid-1977) was lifted. Somalia's overall share of Saudi Arabia's livestock imports over the past three years is roughly 50%, with a slightly lower percentage in sheep and goats and a higher percentage in cattle and camels. In spite of the reduced market share, this has meant substantial increases in Somali shipments because of the very high growth rate of meat consumption in Saudi Arabia.

RANGE CAPACITY

Past assessments of Somalia's range capacity appear to have grossly underestimated this vitally important national resource. As far back as 1964, and probably much earlier, experts first reported the danger, and a little later the apparent fact, of overstocking. Against an estimated livestock population of 1.4 million cattle, 2 million camels and 7 million sheep and goats, the *FAO Somalia Livestock Development Survey Report* of 1964 referred to the 'basic problem that so far the industry has been based on expansion of the grazing area'. It continued, 'This possibility has now been exhausted'. Later reports strongly emphasized the 'degradation of the rangelands by overgrazing' and the need for 'de-stocking', some of these reports drawing on assessments of total available herbage as against the nutritional requirements per animal unit. Such reports were based on livestock population estimates that ranged from 10 to 15 million sheep and goats, 1.8 to 2.5 million cattle and 2.2 to 2.5 million camels.

The 1975 human population and livestock census undertaken in *Jilal* (the height of the dry season), immediately following the severe 1973/75 drought, which caused serious herd and flock losses over large areas of the country, arrived at figures of 25 million sheep and goats, 3.7 million cattle and 5.3 million camels; i.e. roughly 2.5 times the 1964 estimates. It was at the time of the latter estimate that it was pointed out that saturation of the range capacity had been reached. In fact, three years after the census, flocks and herds were examined that were in excellent condition, showing a marked expansion in herd or flock size. Export performance, with the exception of camels, had fully recovered from the drought-induced setback and showed a brisk upward trend while domestic consumption requirements were being met satisfactorily.

OFF-TAKE DISTRIBUTION

At the present time, approximately one third of Somalia's total livestock off-take is for export. Of the remainder, half is retained for consumption by the producers and the other half sold in the domestic markets. The latter comprises stock of lower grade and value, including old and immature animals, the trade routes for them being short compared with those for the export stock.

The direction of Somalia's livestock trade is strongly toward the Burao-Hargeisa-Berbera export triangle. Smaller numbers of stock move toward Bossasso and other minor shipping points on the north coast. The comparative advantage of shipping animals via the northern ports to both Red Sea and Gulf ports, combined with favourable climatic conditions in adjacent holding and staging areas, attract export stock from as far south as the Upper Shebele River area, and even inter-riverine areas of Somalia instead of to the much nearer ports of Mogadishu and Kismayu. The higher trekking expense is compensated for by the better prices paid in the north, on account of lower sea freight rates and the premium paid in export markets (especially Jeddah), for 'Berberi' shipments. Major export numbers of stock originate from areas in the Ogaden, especially from the Haud, where large numbers of Somali herds graze during the main export season, September to January.

The domestic market for slaughter animals draws animals mainly from nearby areas, with the exception of Mogadishu, which receives supplementary numbers from the south-eastern part of the Central, the Upper and Lower Shebele and the Inter-River Regions.

Since only male stock is permitted to be exported, the share of male stock in local slaughter is to some extent indicative of the surplus livestock produced over and above local consumption in the area concerned, i.e. the higher the share of male slaughter the lower the surplus. Typically, as reported by the 'chief butcher' at Hargeisa, post-drought years see a higher share of male slaughter stock, while in pre-drought years the share is small. Recently this share has again decreased to between 10 and 25% so far as sheep and goats are concerned.

ORGANIZATION OF MARKETING

The trader

The livestock marketing organization in Somalia is principally the result of experience and pioneering innovations undertaken by a multitude of actively competing traders over many years who, with few exceptions, have come from pastoral nomadic clans and, as a rule, still own a family share in herds or flocks. At the horizontal level, each appears to be concentrated on his own transactions, with little inclination to cooperate or work with his competitors. Vertically, cooperation appears to be versatile and efficient. As in most traditional developed trades, specialization is high. The efficiency of the system is characterized by very small price differentials (other than transport-cost-related) and the widespread dispersion of traders' activities in the buying of surplus stock. Nomads in northeastern Somalia, when asked where they normally sell their animals during the course of their wide-ranging migration, indicated several market places but stressed that there was never a problem since traders came to meet them 'everywhere'.

Marketing finance

With such a high share of export surplus production and linked to relevant markets through a competitive marketing system, many of the nomadic producers are relatively wealthy. This is shown by their ability to 'finance' traders, in some cases until the final remittance from the foreign importers is received, by permitting deferred payment for the stock they have sold. While no qualifying statement on the frequency and extent of this form of trade financing is readily available, fragmented observations indicate its widespread nature. In the Galguduud and Mudug Regions, for example, it was reported common practice for producers to wait several months until the long distance trader, who had collected, staged and trekked the animals to Burao and then sold them on credit to exporters, had received payment from the latter. Because of the essential role of the 'grey margin' (the difference between the official Minimum Export Price and the price actually obtained) in exporters' trade calculations, full payment is rarely received until the animals have in fact been re-sold in the foreign wholesale market concerned. The producer's' ability to await deferred payment raises the average price that traders are able and willing to pay. To what extent this form of financing prevails in other regions has not been ascertained, but it may be assumed that it forms a substantial portion of that part of the actual c.i.f. cost price of the animal not normally covered by the letter of credit (LC) bank advance; (on average at least 60% is financed in this way). It should be explained that LC bank advance is normally given up to 50% of the LC value that is usually calculated on the basis of the official export minimum price. To put the scope of the traditional financing transactions in perspective, it should be realized that, during the peak quarter, i.e. the three months' period preceding the Hadj (the period of the annual pilgrimage to Mecca), the private traditional production and marketing system is providing finance for up to approximately Somalos 250 million (US$40 million), while the banking system is adding another Somalos 100–150 million (US$16–24 million) to finance the preparation and transaction of the export of approximately 700 000 sheep units.

Apart from the capacity to wait for deferred payment, many nomads have become holders of deposit accounts with local bank branches. In Burao, for example, total nomad savings were estimated by a local bank manager to constitute nearly one third of total private deposits. Those earnings, however, are kept almost entirely in the form of demand deposits, i.e. they represent, for the time being, surplus cash entrusted to the bank and awaiting further decision on how to spend or invest it. More common still appears to be the entrusting of unspent cash by nomads with established *zouk* (market shop) traders at popular livestock markets, thus contributing to the financing, not only of the livestock trade, but also of the trade in those commodities typically bought by the nomadic producers. The amount of nomad investment (mostly indirectly through urban or rural settled relatives) in shops, houses or, less commonly, trucks, is said to be substantial, although probably less than the financial support of their school-age children lodged with settled relatives or foster parents at places having educational facilities. Finally, the amounts held in cash in nomad camps, formerly the dominant means of saving, might still exceed some of the above-mentioned methods of financing, saving or investment.

The nomadic producer's apparent propensity to finance, by accepting deferred payment, and otherwise to save and invest is an important factor in the efficiency of the livestock production and marketing system in Somalia. It makes him an attractive trade partner and thereby contributes to the prevalence of effective trader services at convenient trading points in the migration and stock route system. It further gives him the capacity to invest in up-grading and enlarging his herd or flock whenever the need or opportunity arises and provides the financial resources desperately needed in periods of herd and flock rehabilitation after severe droughts. Finally, it has anti-inflationary effects. These advantages may well be borne in mind when considering the adoption of purchase against cash as an attraction in Government livestock trading operations in preference to existing local practices.

Livestock movement

The movement of trade stock is principally by trekking. Practically all camels and cattle and the majority of sheep and goats are trekked. Trucking is a common alternative on certain routes for sheep and goats during periods of rapidly rising export demand, especially preceding the Hadj and during dry season periods when trekking becomes hazardous.

The costs of trucking are much higher than those of trekking. But, with the current high market value of the animals, a moderate saving on potential weight loss, the avoidance of mortality during late dry season trekking or a narrow gain on c.i.f. value through an *ad hoc* advancement of a shipping date, might more than compensate for the additional trucking expense. Typically, the nearer to the northern export points and consequently the higher the value of the animals, the more common is the use of trucks for animal transport. But, even between the Hargeisa/Burao staging areas and Berbera, trekking would probably dominate if allocation and advice of shipping facilities could be planned and executed in a more logical manner than has been the case in recent years when this area of activity, in which the Government shipping agency has assumed sole authority, appears to have been beset by hazards.

Trucking is by no means always superior to trekking in regard to the condition of the animals after the journey. During the *Gu* and *Der* seasons of good grazing, animals are reported to gain condition during the trek, suffer hardly any mortality, and usually arrive in good health on account of the gradual adaptation to climate and the grazing in the area of destination. Trucking, however, generally results in some mortality and always in loss of condition as well as sickness and susceptibility to disease resulting from the sudden change of environment. This has been demonstrated in Government transport operations between Galguduud and Burao. In fact, the call for livestock roads, appearing sporadically in sectoral development studies, needs to be critically evaluated. The real need for trucking arises only in the latter part of *Jilal*, the dry season. By then, it is possible for trucks to reach almost everywhere, crossing the sun-baked terrain on a variety of makeshift tracks and following the most convenient routing.

Markets

Livestock markets, i.e. habitual meeting places of potential sellers and buyers of livestock, have established themselves at all places of relative significance. Their location appears to be determined by the presence of major stock route junctions or terminals, major watering points, or major centres of consumption. As in other developing countries, these markets have been legalized and subjected to local tax collection. Market and, far less, slaughter fees provide the main revenue for local administrations.

The basic requirements, so far as market facilities are concerned, appear to be ample space for transactions and the provision of drinking water for market users. Little justification for investment in fencing, pens, scales or auction rings can be established since the system seems to work well in its present simple, highly flexible form.

Holding/staging operations

All along the stock route, holding, staging and re-conditioning operations take place. Animals collected during poor trekking seasons are put in the care of sedentary pastorialists until good trekking conditions commence. Trade mobs of optimum size are assembled at staging points near major stock routes. If feasible, butchers condition their purchased animals on hay and sorghum for one to three weeks before slaughter. At Hargeisa, for example, a small haymarket and a number of *zariba* stock yards (thornbush holding-areas) for 20–40 head of cattle each are attached to the livestock market, serving this kind of operation as well as providing for the assembly of small export mobs by local traders or agents.

Larger export herds and flocks are held on the open range of the highlands (*ogos*) near northern ports, especially in the Hargeisa/Burao area from which Berbera shipments are supplied. In these cases, the stock is sheltered in *zariba* enclosures during the nights and supported during the dry season by hay from commercial fodder units established by local people, until shipment is due. Further and more costly final staging operations, requiring hay to be delivered some 70–150 km, take place at Berbera.

A frequently listed item in national investment programmes for the livestock industry is 'holding grounds' for the staging and resting of export animals. However, it appears that in a nomadic range environment where there is ample space and low-cost skilled manpower to herd and tend trade animals, the need of investment in enclosed holding grounds is difficult to establish. The problems of water shortage, maintenance and the utilization of the present Government holding grounds are indicative of this. Naturally, any organization or individual would appreciate having the right to enclose a piece of rangeland as a privileged dry-season grazing reserve. The resistance of the nomadic population to permanent enclosures has, however, become acutely felt by interested parties and Government rangeland development plans, which in the past have included provisions for enclosed ranches, fodder production units and grazing reserves, are being modified to provide facilities which are independent of permanent enclosures.

Under the present system, trade animals are maintained and staged within the Burao-Hargeisa-Berbera grazing areas, and supported by loosely demarcated private and group-operated seasonal grassland reserves and hay production plots within the framework of the communal open grazing system. This system seems to have considerable potential in coping with the staging requirements of the export trade. Any assistance that can be given in order to raise the efficiency of the present system by increasing water availability, improving transport access to hay production plots, establishing grass trials and seed multiplication schemes, etc. would appear to be most effective when the system is known to be an open one in which the interests of the participating nomadic, transhumant, settled and trading populations are safeguarded by private or communal arrangement between the parties concerned.

Animal health

An efficient, low-cost marketing system depends on effective veterinary care and supervision. Livestock merchants whose trekking and/or staging operations involve prolonged possession of trade herds and flocks, usually carry their own drug and instrument kits and are well acquainted with the symptoms of those diseases common to the environment as well as with the application of prophylactic and curative treatment. They, or their qualified agents, meet trekking stock at predetermined intersections to check the condition of the animals, discuss any problem that might have arisen with the stock drovers, treat sick animals, separate those needing rest and recuperation and lodge them with local pastoralists to be added to a future trek, and supply the drovers with appropriate drugs needed for the next trekking section.

An area under particular threat from animal disease hazards, with high values of trade livestock at stake, would appear to be the Burao-Hargeisa-Berbera triangle zone. This zone, apart from being the centre of dense transhumant and semi-settled livestock rearing activities, carries large numbers of trade herds and flocks in transit from other areas of Somalia.

Since the area seasonally carries export stock of up to 300 000 sheep and goats alone, at a total f.o.b. value of approximately Somalos 150 million (US$24 million), investment in and operation of an intensified veterinary control system for the area would appear to be fully justified.

Prices and margins

At present, livestock prices in the privately-organized marketing system in Somalia are high, reflecting the booming demand in the Near East region and especially in Saudi Arabia. Trade margins, including those in the butchery trade, are however narrow.

Observations made during a recent survey in central and northern Somalia confirmed the presence of a low-cost, efficient private marketing system with narrow trade margins. The difference between the rural buying price and the central (urban slaughter or export staging) market price rarely seemed to exceed 10% or – if long distances were involved – 15%, for all types of animals.

Equally low were butchers' margins. Butchers assess animal values on a carcass basis. When looking at an animal, they envisage the amount of meat they will be able to obtain from it. Their carcass weight estimates can be surprisingly accurate. It was difficult in fact to establish a case where a butcher had paid less per kg carcass equivalent on the hoof than what he charged per kg retail, his slaughter and retailing expenses, as well as a very modest trade profit, being covered only by the sale of the by-products (hides/skins and 'offals'), which normally yield, of the total slaughter value, about 6–9% in the case of camels, 9–12% in the case of cattle and 11–14% in the case of sheep and goats. In fact, nearing the Hadj, the butchers' remuneration even appeared to become negative, involving a short period of actual financial losses in order to stay in regular business and be compensated later when retail prices tend to follow falling livestock prices, only partly and with a considerable time lag. Caution in not wanting to provoke enforcement of government control prices is also involved in this market behaviour.

Another reason for the obviously low butcher margins is probably the high participation of women in this business. At least as far as central and northern Somalia is concerned, the butchery trade in sheep and goats is nearly exclusively in the hands of women, while their participation in the cattle butchery trade seems to include most activities except the actual killing. As regards camels, the largest share of carcass preparation is still undertaken by women, but the retailing of the meat appears to be done by men. This extensive participation of women is obviously a traditional practice, stemming from labour distribution within the nomadic family. In southern Somalia, with its increased share of settled population and the reduced importance of sheep and goats, the women's role in the butchery trade appears to be somewhat less pronounced, although their presence on the slaughter floor is still common.

Seasonal price fluctuations are substantial for both livestock and meat, though less erratic and in a narrower range for meat because of the reasons stated above. Peak prices always occur toward the period of the Hadj, preceded by a three-month period of raised price levels, commencing some two to three weeks before the start of Ramadan (which is one month long and is followed by a period of two months and ten days before the Hadj). The Ramadan-Hadj season advances by about ten days each year. The raised price level during this period is strongly supported and stabilized by the continuous build-up of export trade stocks for the Arabian markets. Lower price levels occur during the three to five weeks following the Hadj, in the latter part of *Jilal* (the dry season, i.e. February–April), and during the hot months of the second and early third quarter of the year when the Kharif winds on the northern coast reduce shipping activities and meat demand for export and the domestic market drops. During the latter part of *Jilal*, prices in rural surplus producing areas are depressed on account of trekking difficulties. During drought-crisis months, i.e. when rains have failed during two preceding 'rainy' seasons (*Gu, Der*), *Jilal* prices in the affected areas may drop to the level of salvage value.

Manpower and skills

The Somalia livestock industry has grown entirely out of the private traditional sector. The growth has been a slow and steady one. Already in 1958, official records show the export of approximately 450 000 sheep and goats, 12 000 cattle and 3000 camels. These numbers have increased about fourfold for sheep and goats, sevenfold for cattle and tenfold for camels over the past 20 years, i.e. at an annual average of 8-16%. During the same period, local consumption has grown at about 2.5% per annum and marketed supplies for consumption at 3% per annum. Assuming that over the period as a whole one quarter of total production was marketed locally and a second quarter was exported, then the total (domestic and export) market volume rose at an annual rate of approximately 6.5% and total production at 4.5%. Such progress over a 20-year period is remarkable and can only have been achieved with a continuous growth of manpower and skills involved in pastoral production as well as in the trading sector. Since the share of exported animals in total production obviously rose, from about one fifth to about the one third at present (and this, under the export ban on females, means 60% of all male off-take), the achievements in the range management and livestock husbandry fields are certainly equal to those in the export trade. They are based on knowledge and care, and both are deeply entrenched in the Somali method of handling animals.

Export operations might be said to start with the meticulous care given to newly-born lambs and kids and to the early selection of young males to be raised to export standards. Young animals prior to weaning are maintained in shaded *zaribas* and suckled twice daily. After weaning of the small flock of lambs and kids the owner leads them into the nearby range where a grazing site is selected with suitable herbage and shelter from the sun before, if possible, the flock is handed over to the care of children. The animals are regularly checked for tick infestation and, in the case of the very young ones and the young males raised for export, ticks are picked off the animals by hand as often as time permits. Equal attention is given to the general behaviour and to the excreta of the animals, de-worming drugs, if available, being used whenever indicated. At night, all animals are concentrated in the *zaribas* for protection against predators and cold winds, with the herdsmen sleeping on raised bedding in the centre.

Throughout the production and marketing process, the animals' social requirements are respected. Small groups, formed by the animals themselves within the flock, are sold together whenever possible. Thus, the strain of trek or truck journey, new climatic and grazing environments, vaccinations or dipping, port handling, sea voyage and arrival at foreign shores is shared among a group of individuals well acquainted with each other and giving a necessary sense of protection. This sense of group protection is probably the main factor in keeping mortality in the trade channel at such very low levels (generally stated as 0.5–1.5%). A Somali veterinarian accompanying a shipment of 25 000 Somali sheep and goats to Abu Dhabi on an Australian livestock carrier informed the writer of the surprise and even disbelief of the Australian captain at the low number of deaths among the animals.

An indispensable factor in the movement and maintenance of trade animals from the point of first sale until slaughter or arrival in export markets is the availability of experienced and loyal herdsmen and drovers at moderate daily wages (Somalos 20 in 1978–US$3.20). The number of animals in the trade channels during the peak export period is about 1.5 million sheep units. With herds and flocks averaging about 200 units in size, each under the care of a herdsman or drover, the number of men involved would number some 8000 to 10 000.

CONSTRAINTS AND ACHIEVEMENTS

The continuous increase of Somali livestock exports over past years to the unprecedented level of approximately 2.4 million sheep units in 1980 has been achieved despite the adverse effects of the Somali-Ethiopian war (on the production side) and the great increase in Australian live sheep exports to the Near East (on the marketing side).

The effects of the competition from Australian supplies has been aggravated by new port regulations at Jeddah, which subject the short-distance Berbera-Jeddah trade to the same shipping requirements as those applied under Australian export regulations, thereby substantially reducing the former transport cost advantage enjoyed by Somali exports.

The achievement of the Somali livestock export sector under such a complex set of adverse factors is remarkable and unique in the history of world-wide livestock export performance. That it has been achieved by a nomadic system of production and marketing should give additional encouragement to those responsible for the re-evaluation of animal production systems in marginal zones now being carried out.

32

CONTRACT FARMING AND ITS IMPACT ON SMALL FARMERS IN LESS DEVELOPED COUNTRIES

N.W. Minot

The most serious constraints on small farm production relate to problems of access to production resources (inputs, services, and information) and access to markets. First, small farmers often lack the production and marketing information necessary, particularly for new crops and varieties, and obtaining such information is difficult. Second, even with sufficient information regarding a profitable investment, small farmers may lack the financial reserves necessary, and the availability of external credit is limited by the lack of collateral. Third, small farmers operating near subsistence are probably more risk averse than larger farmers. They understandably tend to assure themselves a minimum supply of food before expanding commercial production for an uncertain market. And fourth, public intervention has been ineffective and even counterproductive in relieving these constraints. In the case of both credit and fertilizer, government efforts to subsidize inputs have led to unreliable supplies and rationing, generally favouring large farmers. Additionally, public extension efforts and policies to promote mechanized agriculture have had more impact on large farmers than small.

Thus, it is clear that, in the interest of both efficiency and equity, it would be useful to investigate the institutional mechanisms which (i) facilitate small farmers access to credit, technical assistance, and inputs and (ii) reduce the uncertainty in marketing their output. To the degree that such mechanisms are developed and policy biases reduced, small farmers will be able to raise their incomes by producing these high-value crops.

Source: Michigan State University International Development Paper No. 31, Department of Agricultural Economics, MSU, East Lansing, 1986.

POTENTIAL OF CONTRACT FARMING

One institutional form which deals with many of these constraints in an integrated manner is that of contract farming. Contract farming may be defined as agricultural production carried out according to an agreement between farmers and a buyer which places conditions on the production and marketing of the commodity. It is also called 'core-satellite' or 'outgrower' production. One variation occurs when the contracting firm also operates a large farm, or 'nucleus estate', which is used to supplement the supply of raw materials from outgrowers.

In the United States, contract farming accounts for an estimated 17% of crop and livestock production, playing a particularly important role in vegetables for processing, sugarbeets, seed crops, poultry, and fluid-grade milk. In the less developed countries, such figures are not available but contracting is used in the production of tobacco, bananas, tea, oil palm, sugar, rubber, poultry, milk, and many fruits and vegetables.

The buyer, frequently a processing and/or exporting firm, finds it profitable to contract growers to assure reliable supply of the commodity. In order to obtain sufficient supplies of the right quality and at the right time, the firm often provides technical assistance and inputs to the farmer as well.

In general, the buyer has an incentive to reduce the cost of production and raise the quality since the willingness of farmers to join and remain in the scheme is dependent on the farm-level profitability. The advantage to the farmer is that the market for the commodity is relatively assured, and, in many cases, the farmer is provided access to technical assistance, production inputs and services, and production credit. Thus, the farmer is able to produce higher-value commodities and improve productivity, thus raising farm income. Furthermore, these services are often provided wholly or largely by private firms, thus saving scarce public resources.

On the other hand, contracts cannot cover all contingencies and enforcement may be costly. There is a short-term incentive for opportunistic behaviour or outright violation of the contract on both sides. After the farmer has planted the crop, the buyer may use various pretexts to force down the effective buying price. Alternatively, the farmer may use the inputs and technical assistance, but avoid repayment by marketing the commodity elsewhere. Even without opportunistic behaviour, contracting involves some costs which must be justified by improvements in market coordination. As discussed later, it is appropriate only under certain conditions. For example, there appears to be little economic incentive to produce grains and other staple food crops under contract. Thus, contract farming should not be considered the key to raising world food production, but rather one strategy for income generation, useful in specific circumstances.

Contract farming is best seen as a form of vertical coordination since it contributes to the harmonization between adjacent stages in the commodity marketing channel with respect to the quantity, quality, timing, and location of supply and demand. As such, a contractual relationship between grower and buyer may be seen as an alternative to (i) an open market relationship and (ii) the organization of production and marketing functions within the same firm, such as plantation-processing plant complexes.

Spot markets are efficient when the good is homogeneous, there are many small buyers and sellers, and there is perfect information. However agricultural markets sometimes suffer from wide variation in product quality, monopsony, and various information problems. In such cases, 'tighter' forms of coordination such as contracting and vertical integration may be favoured.

Vertically integrated plantations are favoured when special quality requirements must be met, when local growers are unfamiliar with the production technology, when supplies must be carefully scheduled, and when other kinds of marketing information are more available to the processor than local growers. On the other hand, if the economies of scale are large for one marketing function, say processing, but small for agricultural production, then integrated production is not very efficient. In this case, contractual relations may provide some of the coordination mechanisms but allow production to be carried out at a more efficient scale.

Contract farming accounts for around 22% of the value of agricultural production in the United States and a smaller but growing proportion of the agricultural product in less developed countries. This expansion is based largely on the growth in domestic urban demand for processed and high-quality goods and the growth in exports.

The extent of contract farming varies greatly among agricultural commodities, being the greatest with high-value perishable commodities which are processed such as vegetables, fruits, milk, poultry, tobacco, and many traditional tropical exports. Conversely, it is least common for basic food grains which are not processed.

The extent of contract farming also varies among countries. The published cases of contract farming tend to be concentrated in Mexico, Central America, Kenya, and Thailand, though there are examples in a wide range of less developed countries.

Many contracts involve the provision by the buyer of agricultural inputs to the farmer. Seed and fertilizer are the most common, but other chemicals, pest control services, machinery hire, and harvesting services may also be supplied.

The cost of these are subtracted from the crop payment at harvest, either implicitly through a lower crop price or explicitly. Although production contracts are often associated with the pre-planting determination of crop prices, in fact, formula prices and negotiated prices are probably common.

Contract farming often involves technical assistance as well. The extension and supervision effort is generally quite intensive with farmer–agent ratios of less than 200:1. Many of the more successful schemes incorporate locally-hired paraprofessionals at the field level to take advantage of their knowledge of local farming patterns, constraints, and cultural factors. These may work as salaried employees or on a commission basis.

However, there is great variation in the contract provisions, the size of farmers contracted, the type of technical assistance and services provided to growers, and the bargaining relationship between the buyer and the growers. For any one scheme, these variables depend primarily on the commodity produced and its final market, and to a lesser degree on the existing land tenure system, the technology of processing, the policy environment, and other factors. Thus, it is useful to examine the patterns of contract farming by commodity.

Although bananas are often produced by integrated plantation/export companies, the use of independent contracted growers has expanded since the 1950s, partly under political pressure and partly to reduce labour costs. Generally, the contract growers are large farmers due to economies of scale in production, but there are examples of successful cooperatives of small farmers which produce under contract with substantial support from the company.

Tobacco, being very labour-intensive and requiring careful husbandry, is, in many cases, produced by thousands of small growers under contract. Firms like the British American Tobacco Company insist on plots smaller than one hectare to maintain quality. Technical assistance is intensive, and focuses on both production practices and some processing, such as curing and baling.

Rubber and oil palm are produced on plantations, by small holders (contracted or not), and in nucleus estate/smallholder schemes. Many of the latter are public settlement projects involving several thousand outgrowers per scheme. Due to the perishability and bulkiness of oil palm, fruit, production must be clustered around the processing plant and deliveries carefully scheduled. Rubber is less perishable but vertical coordination serves to ensure high quality and transfer technology regarding local-level processing. There is a trend toward greater farm-level processing, assisted by the contracting entities.

For sugarcane, there are economies of scale in production and even more so in milling. Each mill may be supplied by one or several plantations and sometimes small outgrowers or contracted tenants. Again, perishability and bulkiness of the cane requires geographically concentrated production and careful scheduling.

Although tea is viewed as the ideal plantation crop, there are examples of smallholder contract production, particularly in Africa. The best-known scheme is the Kenya Tea Development Authority with its 140 000 growers. Other such schemes have been attempted though none have succeeded on the same scale. Intensive technical assistance, careful grading, and rapid assembly are keys to successful tea contracting.

Poultry (broiler) production is increasingly carried out under contract. A firm contracts growers, providing them with chicks, feed, veterinary services, and credit. This system can lead to concentration of ownership and displacement of small producers, though it can reduce poultry prices significantly. The growers tend to be large, with tens of thousands of birds.

Milk production is often contracted because of the need to assure regular supplies of the perishable commodity. There are some examples of firms providing technical assistance and inputs as part of the contract though it is not clear how common this is. Milk pricing controls often constrain the development of improved marketing systems with small-holders.

Fruit and vegetable production is frequently contracted, particularly when the commodity is for processing or for export. This is because of the labour-intensivity of production and the importance of quality control. In addition, fruits and vegetables for processing must be carefully scheduled to assure stable supply to the plant, while exports must reach the market during a seasonal period of shortage. There are numerous examples in Mexico, Central America, northern Africa, Thailand, and Taiwan, among others. They involve contract production of tomatoes, peppers, melons, green beans, and cucumbers. Leakage to fresh markets may be a problem for the buyer, depending on the product, and contract violation occurs on the part of the buyer as well. However, successful schemes are generally based on institutional innovation and mutual trust between contracting parties.

CONCLUSIONS

Contract farming is generally successful in supplying credit, inputs, technical information, and market information to growers. In doing so, it transfers production technology to the growers as well as providing, in many cases, a more secure market outlet. In almost all cases for which the data are available, the implementation of contract farming schemes has resulted in significantly higher incomes for participating growers. Furthermore, there is often a long waiting list of growers interested in participating.

The literature supporting contract farming tends to overlook several limitations. Perhaps the most important one is that the impact of contract farming, though intense, tends to be relatively narrow. Even in Kenya, with a wide range of such schemes, only 12% of the smallholders are contract growers. While there is probably scope for expansion of contract farming as an institution, it is probably not appropriate for the production of basic food grains. Thus, this system should not be considered the basis for an entire rural development strategy, but rather an important component in improving agricultural production and marketing.

Another limitation is that greater incomes are not always translated into improvements in standards of living broadly defined to include nutrition, education, and health. Although the impact of higher incomes is generally positive, some schemes involve a shift from subsistence food production to commercial production. Although not harmful in itself, this is sometimes combined with poor choices regarding food purchases and/or inequitable distribution within the household. Men are assumed to be the growers and heads of household by the company and receive the crop payment. In cultures where men and women have separate budgets and spending responsibilities, such as in much of Africa, women are generally responsible for care and feeding of the family. Thus, payment to the men may bias household purchases away from food and health related items. In several cases, income from the schemes has made possible excessive consumption of alcohol. However, it should be noted that this problem can occur with any income-generating project and is not a problem unique to contract farming.

The literature critical of agribusiness argues that contract farming is simply a method of obtaining cheap labour and of 'transferring' risk to growers. This is an oversimplification and unnecessarily pessimistic for several reasons:

1. Lower implicit wages are only one reason for contracting as opposed to plantation production: improved labour productivity, dispersion of production zones, and reduced investment risk are also factors. To the extent that the lower cost of labour is an incentive, this tendency has a positive equity impact since the firm is (indirectly) employing precisely those workers who have the poorest alternative employment opportunities.

2. With regard to 'transferring' risk, it is important to note that the distribution of risk is not a zero-sum game. A contract may reduce (or increase) risk for either or both parties, depending on the details of the case and the alternative to which it is compared. For example, fixed price contracts reduce gross revenue variability for the grower (compared to spot sales) when market price variation is due to shifts in demand rather than in supply. On the other hand, they reduce buyer risk (compared to spot purchases) when market raw material prices are unrelated to final product prices. These are not mutually exclusive. In any case, one would expect any increase in risk bearing by growers to be compensated by greater average returns; otherwise, the grower would be reluctant to contract.

Contracting firms are also depicted as abusing their monopsony position and violating contract provision in order to maintain their large profit margins. It is true that the literature provides numerous examples of conflicts between buyer and growers, often related to the quality control and trading practices. However, this argument ignores the fact that such behaviour is generally not sustainable for annual crops grown by owner-operators: growers will withdraw from the scheme within a year or two if they find the returns insufficient. In fact, when parallel markets for the commodity exist, leakage is generally a serious problem unless the company pays a price superior to the market price. Long-lasting contract farming schemes almost always involve a deliberate attempt by the company to develop mutual trust between itself and its growers. Finally, it should be noted that high profits are not a foregone conclusion: the literature provides abundant examples of marginally profitable schemes and outright failures.

Another debated issue is the size holding preferred by contracting firms. Critics of agribusiness argue that firms tend to contract only large farmers, whereas proponents emphasize the cases where small holders are contracted. In fact, the tendency depends greatly on the commodity. Sugar-cane, bananas, and poultry appear to have significant economies of scale, but tobacco, tea, and many vegetables are generally contracted out to quite small farmers. For each commodity, however, there is a range, as indicated by successful smallholder schemes producing sugarcane and bananas. Contracted small farmers require expensive outreach efforts, for credit, training, and inputs. But their advantage over larger farmers is greater motivation, lower implicit labour costs, and less risk that they will come to market their own produce.

The geographic patterns of contract farming also vary depending on the commodity. Bulky crops such as sugarcane and oil palm tend to be produced quite close to the processing plant, whereas higher-value commodities such as tobacco, tea, and vegetables tend to be more dispersed. It is not clear whether central or remote locations are preferred. Central locations offer better access and lower outreach costs, but land and labour are more expensive and leakage is more likely to be a problem.

HYPOTHESES

Contract farming for the domestic market is more likely where there are large, relatively high-income urban areas.

Contract farming for the export market is favoured by proximity to these markets and good transportation networks.

Contract farming is favoured where price controls and market regulation are minimal. For export commodities, contract farming is favoured when the currency is not overvalued relative to foreign currencies. The existence or possibility of government market regulation and government investment in productive or commercial enterprises acts as an inhibitor to the establishment of private contract farming schemes.

Contract farming is more likely to be accepted by growers if they have alternative market outlets for the commodity and if there is no nucleus estate also supplying the raw material since this will improve their bargaining position and reduce the risk of monopsony abuse.

Contract farming is more likely to be accepted by buyers if growers do not have alternative market outlets for the commodity and if there is a nucleus estate to supplement contracted production. However, fruit and vegetable canneries, particularly publicly-financed ones, are often based on unrealistically low estimates of alternative market prices. Successful adaptation requires effective enforcement of contracts and/or accommodation to market prices.

Fixed-price contracts are more acceptable to growers when yields are relatively stable, such as in areas with dependable rainfall or irrigation, but the market is unstable or unpredictable. Fixed-price contracts are more desirable by buyers when the buyer exports, buys from a remote market, or sells to a remote market. In other words, buyers are reluctant to offer fixed price contracts where raw material and final good prices are closely related.

The problem of arbitrary or variable rejection rates is most likely to occur in the first few years of a scheme. After several years, growers will have better knowledge of expected net return, after learning about standards and company procedures in applying them. Furthermore, after the first few years, buyers will either refine grading practices or, if net returns to growers are insufficient, be forced out of business. With inexperienced growers, detailed instruction and supervision may be necessary and growers should be informed of their labour and financial obligations. With experienced growers, it is often better to provide economic incentives for performance and let growers decide the best way to achieve it. Again, after the first few seasons, growers will have better knowledge of obligations and company practices, and thus of the net return relative to alternative crops.

Contract growers will be more concentrated around a processing or transportation centre if the commodity has a low value:bulk ratio, the commodity is fragile, timing is very important, or there are large economies of scale in processing. Thus, small economies of scale in processing and high value:bulk ratios (tea) allow dispersed production, while large economies of scale and low value:bulk ratios (sugarcane) require concentrated production.

The returns of land are increased to the extent that contract farming raises the income-producing potential of the land. This may occur through the introduction of a new crop or yield-increasing technology. In addition, this can be the result of the construction of a processing plant in the region, particularly if it is costly to transport the raw material and supplies must be obtained close to the plant. In this case, land owners or those with traditional usufruct rights benefit, but not necessarily renters, sharecroppers, or hired labourers.

The returns to labour are increased to the extent that contract farming improves labour-productivity through the introduction of new technology or production practices or through the creation of specialized farm-level capital. If the new skills are limited to the contracted growers, it is likely that the return to their labour will increase. If labour productivity is improved for anyone working on the farm, it may result in greater use of hired labour. This effect is more likely to benefit hired labourers (through employment generation) and other landless farmers.

There is widespread use of informal contractual relations between growers and buyers which is not reported in the literature because it does not involve large firms. This contracting is probably most common with perishable and specialty crops. These contracts are likely to employ formula prices rather than fixed prices.

Contract farming has the greatest development impact when the crop is labour-intensive, when it does not completely displace home food production, and when intrahousehold distribution is equitable.

There is scope for expansion of the extent of contract farming, mainly within the commodity groups identified here, through policy reform.

33

NONTRADITIONAL EXPORT CROPS IN TRADITIONAL SMALLHOLDER AGRICULTURE: EFFECTS ON PRODUCTION, CONSUMPTION AND NUTRITION IN GUATEMALA

J. von Braun, D. Hotchkiss and M. Immink

The adoption of new technology and new crops and their related effects on increased market integration are cornerstones of employment-oriented agricultural development. Increased foreign exchange problems and deteriorating prices of traditional export commodities lead agricultural policymakers to seek diversification in export crop production. As nontraditional crops, export vegetables appear to be a promising option due to their high labour intensity and expanding demand in industrialized countries. This study deals with a case of export vegetable production and its effects on food production, employment, consumption, and nutrition in Guatemala.

The focus of this study is the recent introduction of labour-intensive vegetable production for export into the traditional small-farm sector in the Western Highlands. It can be hypothesized that the growth of this new export crop sector within the smallholder sector will have very different effects for the poor than growth in Guatemala's traditional large-scale agricultural export sector. The study area is well known for its problem of poverty and malnutrition. The population in the area consists mostly of Maya-Indians.

The research is based upon two detailed rural household surveys (400 families) in the Western Highlands of Guatemala which were undertaken in 1983 and 1985. The sample is divided into two groups of households – those who produce the new export vegetables (snowpeas, broccoli, cauliflower, and parsley), under a cooperative scheme, and those who do not. Also, differences in duration of participation in the export crop scheme –

Source: International Food Policy Research Institute, Washington, 1987. (Selected passages.)

the cooperative *Cuatro Pinos* – characterize the subsample of the export crop growers (one to seven years). The sample is drawn randomly in a structured way from a census undertaken in six villages of the study area in 1983.

COMMERCIALIZATION OF GUATEMALA'S DUALISTIC AGRICULTURE

Guatemala's agriculture has shifted away from food production to agro-industrial crops. Food crops covered 58% of the country's crop area in 1950 as compared to 37% in 1979. Small farms decreased their basic food crop area from 97 to 87% also in this period.

Export orientation in agricultural production has a long tradition in Guatemala and has resulted in a highly dualistic agricultural sector. The traditional export crops (coffee, cotton, sugarcane) are grown in the large-scale 'modern' sector in the favourable areas of the country. Land ownership in Guatemala is extremely skewed. Two percent of the farms hold 67% of the agricultural land. Sixty percent of Guatemala's farms fall in the group below 0.7 hectares, about the average farm size in the sample for this study.

THE ACTORS AND PLAYERS IN THE COMMERCIALIZATION PROCESS

Commercialization of traditional agriculture in a market economy is driven by changes in incentives or technology, but the process is to a large extent also influenced by actors and players. The smallholder cooperative *Cuatro Pinos* in the Western Highlands, which produces the export vegetables led to the dynamic expansion of export vegetable cultivation in this case. The actors were:

1. a multinational company that provided know-how and initially organized the export channel;
2. development organizations that provided seed money to open up the export channel from Guatemala; a nongovernmental organization that stimulated the formation of the cooperative, provided training, and was instrumental for its sustainability in the beginning;
3. local farmers who formed the cooperative that organized the vegetable production and domestic handling and which later moved toward independent handling of the export marketing; and
4. public institutions in Guatemala that provided know-how on agricultural technology and farm-level credit.

Export vegetable production did not end up in the small farms as a planned undertaking but moved there in several steps. Originally, the multinational grew the crops on large-scale units managed by the company itself. It moved then to medium-sized farms (20–30 hectares) in the form of contract growing, and from there to the smallest farmers (average 0.7 hectares) in the Highlands. The crops' characteristics which apparently have negative returns to scale in production and management led to this trial-and-error development path in which the actors and institutions mentioned above interacted and responded to economic incentives.

The new export vegetables were rapidly adopted by the smallest farmers. Our model analysis shows that in the early phase of adoption, small farmers with somewhat bigger holdings (1–2 hectares) and households who had no reasonably secured off-farm income source showed a significantly higher probability to join the scheme.

RISKS OF THE NEW CROPS

The new export vegetables have certain risks for the small farmers. Such risks can possibly result from crop failures, from price collapses on the export market, and from a breakdown of the marketing institutions. Relative production variability of the new crops is not higher than in the traditional crops but because the new crops are much more input-intensive, the potential absolute income loss is higher than for the traditional crops.

The price variability of the new crops – especially of snowpeas – is extreme. Within-year prices in 1985 fluctuated between 0.10 and 2.00 Quetzal per pound, but farmers can partially cope with this by spreading the growing seasons and having a long harvest period (12 weeks).

The export channels of fresh commodities from developing countries are risk prone not only in a technical sense. Government action on exchange rate manipulation is a factor too. This study exemplifies that the exchange rate policy introduced in 1985 in Guatemala implicitly taxes the export vegetables from the small producers by about 25%. The effect of this tax is adverse for employment in the small farm export sector.

Recently, in addition to the individual multinational company and the cooperative, many more other traders have handled the export channel. Also, local processing and freezing of fresh produce has been initiated. These developments reduce the risk of a sudden collapse of the marketing channel.

PROFITABILITY OF THE NEW CROPS

Nontraditional export crops are substantially more profitable to farmers than traditional crops. Net returns (gross margins) per unit of land of snowpeas – the most important new crop – are on average five times than those of maize – the most important traditional crop. Returns of the new crops per unit of family labour were about twice as high as maize and 60% higher than traditional vegetables produced for local markets in 1985. The input cost for snowpeas, however, are on average about 1600 Quetzales per hectare while they are about 430 for traditional vegetables and 120 for maize. Short-term financing of inputs poses a problem to small farmers and indicates the importance of rural credit. A functioning rural credit and banking scheme is necessary and desirable for input financing and would also provide a savings opportunity and enables farmers to cope with short-term liquidity problems that may destabilize food security due to production failures and other risks discussed above.

EFFECTS FOR LAND USE AND FOR THE LAND MARKET

Farm households which are not members of the export crop cooperative grow the traditional subsistence crops (maize and beans) on 78% of their land, whereas participants in the scheme grow them on 52% of their land. The smallest farms of coop members allocate the highest shares of land to the new export crops. For instance, farms below 0.25 hectares of coop members allocate 45% of their land to the new crops. The other side of this coin is that these smallest farms grow only 38% of their land with maize and beans, while those of nonmembers of similar size grow 81% of their land with these subsistence crops.

A model to explain the area allocation confirms that the new export crops have not only replaced traditional vegetables but reduced the area allotted to the traditional subsistence crop (maize). Yet, this may hardly have had price effects for maize as the study region was a net importer of staple food before. The degree of self-sufficiency in the six communities is only 50.4%.

The new export vegetables are grown on plots of better land quality. Increasingly land is purchased and rented by farmers that grow new crops. Land prices and rental values have reportedly increased substantially in the communities partly due to the more profitable production opportunities and partly due to generally increased pressure on land because of population growth. Landowners not participating in the export crop scheme benefit indirectly through increased land values. Adversely affected are households that rented land and did not join the scheme.

EFFECTS ON EMPLOYMENT AND DIVISION OF LABOUR

Nontraditional export crops created local employment directly on farms and indirectly through forward and backward linkages and multiplier effects resulting from increased income spent locally. Labour input in agriculture increased in the export vegetable producing farms by 45%. About half of this increase is covered by family labour and half by hired labour.

There is a shift of labour from traditional to new crops and a partial substitution between family labour and hired labour. Total labour input is cut back in maize by 13%, in beans by 43%, and in traditional vegetables by 29%, but hired labour input into maize even increased somewhat, as family labour in the traditional subsistence crop was partly displaced by hired labour.

A substantial share of the increased family labour is from women. It is 44% of the increase in the farms below 0.5 hectares and 32% in the farms above one hectare. While women's share in total family labour decreases with increasing farm size, men's share remains stable and children's share increases. Women's labour input into the subsistence crop (maize) is low (9%) as compared to traditional vegetables (25%) and the new export vegetables (31%). Yet most of the field labour provided by the family is from men (in maize, 83%; in snowpeas, 59%; in traditional vegetables, 61%).

Combining farm level employment to the roughly estimated employment created through the input supply and output marketing yields an overall 21% increase in agricultural employment in the six communities. This reduces off-farm work and interregional migration of farm household members. From export vegetable producers' households, 0.7 persons per household worked away for an average of 2.3 months while in other farm households 0.9 persons worked away for an average of 4.2 months.

A more detailed model analysis controlling for relevant factors in household characteristics confirms that off-farm earnings are substantially reduced with increased farm income and participation in the export crop scheme.

EFFECTS ON SUBSISTENCE FOOD PRODUCTION

Farm households in the Western Highlands have a strong desire for food security based on own-produced maize. This is quite understandable in view of unstable food markets, insecure off-farm employment, and low levels of institutional provisions for food security in crisis situations in Guatemala. Practically all export crop producers maintain some maize production (94%).

Despite smaller areas allocated to traditional subsistence crops by export crop producers, the great majority tend to have higher amounts of maize available (per capita) for consumption from own produce than other farmers of same farm size because coop member's maize and beans yields are 30% higher on average than nonmembers' yields. A combination of factors is responsible for the increase in yields. Fertilizer inputs are increased and cropping practices are more labour-intensive (more weeding labour).

Our production function analysis shows that new export vegetables have favourable yield effects over and above the increased input use, and this is not an effect of a possible self-selection bias of more efficient farmers who became export crop producers. One reason is the positive effect of snowpeas on soil fertility (nitrogen fixation). A second more hypothetical reason is that export crop producers improved their crop management of subsistence crops once new export crops were introduced.

Analysis with the help of a consistent farm household model based on the survey data shows that with new export crops, the opportunity cost of maize produced for own consumption increases drastically. While the shadow cost of maize for own consumption in a typical farm was 0.49 Quetzal per kilogram without the new crops as an alternative, it increases to 1.16 Quetzal when competition with the new crops is introduced. Assuming the household wants to maintain its earlier degree of self-sufficiency, the difference between the shadow cost and the actual market price (0.29 Quetzal in 1985) may be interpreted as an 'insurance premium' which farmers are willing to pay for the degree of self-sufficiency they actually maintain. This 'insurance premium' is relatively high. It can be brought down further by technological change in the subsistence food production. Farmers adopting the nontraditionals have a strong incentive to increase productivity in subsistence crops which brings down the 'insurance premium' and this is what they actually did.

The Highland farmers were able to adopt the balanced strategy of achieving joint growth in new export crops and staple foods because of indigenous know-how about yield increasing measures, input supply channels, and hired labour. In contrast to the case of export crop promotion in the large-scale sector in the dualistic Guatemalan setting, positive spillover effects into the traditional food crops were obtained in the case of commercialization within the small-farm sector. Yet, much scope for further improvement of productivity and yields in maize and beans remains to enhance the process. Research based technological change has so far not substantially impacted on productivity of the two main traditional food crops in the Western Highlands – maize and beans.

EFFECTS ON CHANGE IN INCOME

The export crop production scheme led to increased income in the participants' households. This increase between the two surveys (1983–1985) was most pronounced in the group of new adopters in which expenditures –

used as an income proxy – increased by 38% above the average nominal increase in the total survey population. The income gains were highest among the adopters in the smallest farms. The new export crops had a favourable effect of moving the poorest upward on the income scale. As for the export crop producers among the poorest one-third of the households, only 38% stayed in this tercile from 1983 to 1985, compared to 55% for the lowest tercile of the other households.

EFFECTS ON EXPENDITURE PATTERNS

Export crop producers spend on average 64% of their total expenditures on food compared to 66.8% among the other households (including own-produced food consumed by households). The difference can be mainly explained by Engel's Law. The relatively lower budget share to food among export crop producers is due to relatively lower expenditures on almost all foods but meat, fish, and eggs. Nevertheless the absolute per capita budget spent on food is on average 18% higher in export crop producing households. Proportion of total expenditures spent on food decreases with increased income, but as expected in a poor population, this decrease is not very rapid. A 10% increase in total income decreases the budget share to food by only 1.4%.

At the same income levels, export crop-producing farm households spend less of additional income on food. While nonmembers in the lowest quartile on the income scale spend 61% of additional income on food, coop members in the same income class spend 53%. Income from new export crops is mainly male-controlled. Holding income constant suggests a negative effect of an increased household income share from (male-controlled) new export crops for the budget share spent on food. Among the nonmember households, a similar relationship for food expenditures is observed in case of male-earned nonagricultural income.

EFFECTS ON CALORIE CONSUMPTION

More than half of calories available in households come from maize but this share decreases with increasing income. Accordingly, there is a higher price per calorie and more diverse diet with rising income in both, the export crop producing and other households. Export crop producers spend about 10% more per calorie.

Additional income increases calorie acquisition significantly but at decreasing rates at the margin. At the sample mean a 10% increase in income increases calories in the household by 3.1%. Joining the export crop scheme apparently reduces the positive income effect on calorie consumption, but does not eliminate it. Coop households in the lowest half of the income scale increase their calorie consumption by 2.8% with a 10% increase in income while noncoop households increase theirs by 4.4% which is consistent with above-mentioned results of the analysis on budget allocation.

EFFECTS ON NUTRITION

Despite the relative closeness to Guatemala City and to infrastructure of some of the sample villages, the general nutritional status in the study area is as poor as in other more remote rural Guatemalan regions where there is comparable data.

The nutritional situation of the sample population in general, as measured by anthropometric standards of children, has hardly improved between 1983 and 1985. While slight improvements are recognized in the top income tercile, prevalence of underweightness and stunting actually increased in the bottom and middle income tercile. Guatemala was going through a severe economic crisis during these years that affected the rural areas in the form of reduced overall employment in the nonagricultural sector.

The effect of relative increased income for nutritional improvement appears small in the short run. This is not surprising given the general low levels of income in the study populations. Two types of regression models are specified to evaluate the effects of increased commercialization on nutrition. A first set of models attempts to trace the income effects of the new crops via its impact on food availability and household's health and sanitary environment. Significant positive relationships between increased food availability and nutritional improvement are established especially with respect to the short-term indicators (weight/height). Thus, as found in the consumption analysis, nutritional improvements resulted from the addition of the new crops to real income and through that to calorie availability – as found in the consumption analysis.

A second set of models takes a short-cut approach to the income–nutrition relationship as some of the dynamic complexities of the food intake and health-related connections to nutritional status were probably not captured in the cross-sectional and short-term longitudinal analysis which the data permit. It is found that increased income leads to significant nutritional improvement, but decreasing so at the margin. The hypothesis that an increased share of male-controlled income – be it off-farm income or income from the export crops – would lead to adverse nutritional effects (holding total income constant) does not find support by this analysis. A higher share of especially women-controlled nonagricultural income, however, tends to add more to nutritional improvement than does men's nonagricultural income. The effect for nutrition of an increased share of export crop income is not significantly different from zero over and above the total income effect. Once the income level, income source, and composition are controlled for in the models, still a significant positive effect of membership in the cooperative for nutrition is found. This can be attributed to the social support programmes launched by the coop, such as food and nutrition education. Twenty percent of coop profits are spent on education and social services programmes. This outcome underlines the importance and potentials to further enhance income effects of commercialization for nutritional improvement through such accompanying programmes.

The nutrition-related analysis also suggests that it may be misleading to draw inference for the welfare effects of nontraditional crops in the households just from the income and expenditure or consumption analysis.

1. The production and income-related analysis concluded with favourable effects of the nontraditional crops for food crop productivity, employment, income growth, and income distribution.
2. The expenditure and food consumption analysis found that incremental income earned from the nontraditional crops tends to be relatively less spent on food than other income, and this is also reflected, although to a lesser extent, in calorie availability. Thus, food expenditures and consumption increased relatively less than expected. Yet, still substantial absolute increase in food expenditures and food availability in households (calorie) due to the scheme was found.
3. The nutrition-related analysis shows that the increased food consumption and income in the nontraditional-producing households improved nutritional status. The effect was further enhanced by the food and nutrition programme of the coop-making households more efficient in achieving nutritional improvement.

Especially in the late 1970s and early 1980s, steps to alleviate poverty and improve living conditions in Guatemala were constrained by the economic and political environment. The case study shows, however, that with appropriate access to resources and markets and effective assistance in institution building at the community level, the poor in the Western Highlands can substantially improve their income and welfare. That this also rapidly translates into nutritional improvement of children requires appropriate health and nutrition-oriented social infrastructure.

No general automation between growth and social articulation of the poor which could lead to development can be postulated based on this study. This case study demonstrates the existence of niches in the system that provide this potential and also highlights the specific conditions under which it may work. Central to these conditions are the diseconomies to scale in producing the export vegetable and the ecological conditions in the

Western Highlands along with the labour market situation.

The sustainability and expansion potentials of the programme depends on the functioning of the marketing channels, domestically and internationally, both for inputs and outputs. Policy dialogue has to focus on maintaining and improving this. The implicit taxation of small farmers through overvalued exchange rates for outputs is a matter of concern. The riskiness of the new crops due to potential disturbances in the marketing chain is to be a matter of constant attention.

34

KENYA'S HORTICULTURAL EXPORT MARKETING: A TRANSACTION COST PERSPECTIVE

S. Jaffee

Despite the quantitative significance of Sub-Saharan Africa's horticultural trade and despite the evident need for many African countries to diversify their exports away from a narrow basket of beverage crops and industrial raw materials which face weak medium-term market prospects, the region's horticultural export experience has been the subject of little social science research. Research which has been done has focused on production patterns, contractual and other linkages between producers and processors/exporters, and government programmes and other interventions. While most of this work emphasizes the constraints of marketing on horticultural export development, there has been virtually no research examining the marketing of Africa's products. For example, what markets and consumer segments have African exporters targeted? What have been the implications of alternative marketing arrangements for the performance of African horticultural export sectors? How are African supply countries and firms perceived in overseas markets?

This paper examines these issues through the lens of Kenya's horticultural export experience. Among African countries, Kenya has developed one of the most successful horticultural export sectors, covering a broad range of fresh and processed products and achieving double-digit rates of growth in trade volume and value over the past two decades. Discounting the trade of South Africa, Kenya accounted for more than one-third of Sub-Saharan Africa exports of fresh vegetables in 1988, one-half of its exports of processed vegetables, two-thirds of its exports of processed fruit products, and over 80% of its exports in cut flowers. In this paper we will argue that considerable part of Kenya's success in this line of trade can be attributed to the very effective inter- and intra-company

---

Source: Paper presented at the Workshop on the Globalization of the Fresh Fruit and Vegetable System at the University of California, Santa Cruz, December 6–9, 1991.

linkages which were developed between Kenyan (and joint venture) exporters and their overseas trading partners.

FACTORS INFLUENCING TRANSACTION COSTS

What factors contribute to the level of transaction costs in any particular trading context? Physical distance may prove a major barrier to information-gathering, communications, and the monitoring of transactor activity. There is the extent to which legal provisions are transparent, enforceable, and equitable. In the literature of transaction cost economics, these variables would be subsumed under the operating condition of 'uncertainty'. In this literature, it is argued that the level of transaction costs is associated with three dimensions of the trading environment (Anderson, 1988):

1. *Asset specificity*: the extent to which the physical and other assets required for production and exchange are durable and specialized for a particular product or trading relationship.
2. *Uncertainty*: The overall degree of uncertainty surrounding the exchange.
3. *Competitive market structure*: particularly the number of alternative buyers and sellers available.

For any particular production and trading operation, individuals may undertake either generalized or specialized investments. Certain types of plant, equipment, materials, and knowledge have potentially generalized use across a broad range of products or trades. Other assets are highly specialized for a particular product or trade outlet and have little or no alternative use or value. Making investments in specialized assets exposes the investor to potentially severe bargaining and contractual enforcement problems as it will be 'locked-in' to particular production or trading activities and be exposed to pressures from trading partners to improve their own terms of trade. Examples of asset-specificity in agriculture include crops with extended gestation periods (e.g. fruit trees), large-scale specialized processing and postharvest facilities, and use of highly specialized production inputs and technical knowledge.

In any particular trading context, the degree of uncertainty may vary – uncertainty regarding the availability of supplies or market outlets, the quality of the products on offer, the timing of supply and demand, the trading terms being offered, interventions by governments, and so on. Such uncertainties tend to be more pronounced in agriculture than in industry because of the important influence of changing weather conditions, the seasonality of production and/or trade, and the perishability of many agricultural commodities.

Transaction costs will also be influenced by the prevailing market struc-

Table 34.1. Competitive and ownership structure of Kenya's horticultural export trade, 1985–1986

	Fresh fruit & vegetables	Cut flowers	Processed fruit & vegetables	Combined
Number of firms	95	14	16	125
Concentration of trade				
3 Leading Firms	49 %	90 %	95 %	
6 Leading Firms	67 %	98 %	99 %	
Export shares for foreign vs. local firms				
Foreign owned (MNCs)	0 %	58 %	91 %	54 %
Private locally owned	97 %	38 %	9 %	43 %
Kenyan Asian	81 %	0 %	7 %	30 %
Kenyan European	9 %	35 %	2 %	10 %
Kenyan African	7 %	3 %	0 %	3 %
Local cooperative	0 %	1 %	0 %	1 %
Parastatal	3 %	3 %	0 %	2 %
Total	100 %	100 %	100 %	100 %

Source: Author's field research, 1985/86.

ture, especially the numbers of alternative buyers and sellers. The existence of few alternative buyers and sellers may be expected to result in higher search costs, lower screening costs, and considerably higher bargaining and enforcement costs. Where there are relatively few alternative trading partners, one might expect (i) less complete disclosure of interests to trade and less disclosure of product information, (ii) better opportunities for strategic bargaining, and (iii) more transaction enforcement problems since threats to terminate trade and deal with competitors would be less credible.

The competitive and ownership structure of Kenya's horticultural trade in the mid-1980s is summarized in Table 34.1. The table indicates that while a large number of firms are active in horticultural processing and export trade, only a few firms accounted for the majority of trade. For example, three firms accounted for 95% of Kenya's processed fruit and vegetable exports, while three producer/exporters accounted for 90% of the country's cut flower sales. The fresh fruit and vegetable export trade is much more competitive with nearly one hundred licensed exporters. Nevertheless, only three firms accounted for one-half of this trade and six firms accounted for two-thirds of trade. The vast majority of fresh produce exporters conduct trade on a very small scale (e.g. less than 5-10 tonnes per week), typically on a seasonal or intermittent basis. In the years since the data below were gathered, only the cut flower trade has experienced any noticeable reduction in concentration.

Table 34.1 indicates the dominant role played in this trade by foreign-owned companies and companies owned by members of Kenya's small Asian and European communities. Foreign-owned (and joint venture) companies account for nearly all of Kenya's processed fruit and vegetable exports and a majority of its cut flower exports. Companies owned and managed by Kenyan Asians account for 81% of the country's fresh fruit and vegetable exports. None of the leading firms had many years of experience in horticultural production and/or domestic trade before entering the export market. Many have made complementary investments in farming or freight forwarding since developing their export trades.

Despite direct investments and several other interventions by the Kenyan Government designed to 'Kenyanize' control over this trade, the proportion of trade accounted for by Kenyan African firms has not increased since the early 1970s and was only about 5% (including parastatals) in the mid-1980s. While many African-owned firms have entered the trade, the majority have gone out of business, with most of the others remaining as small-scale, part-time exporters whose bargaining power *vis-à-vis* growers, air-freight suppliers, and foreign market importers has been very weak.

Despite the close proximity of Kenya to the Middle East and the large expansion in import demand in that region for many products exported by Kenya, the large majority of Kenya's horticultural export trade has been directed to Western Europe and especially to the United Kingdom. This is a reflection of Kenya's close economic and political ties with the United Kingdom during both the colonial and post-colonial periods. Many Kenyan traders have received business or academic training in the UK or have close social or familial ties to that country. Especially for Kenya's small-to-medium scale fresh produce exporters, informational barriers to trade have been much lower in the UK than elsewhere. Other important factors have been the investments made by European firms in Kenya's horticultural sector (geared toward supplying their core markets) and the relatively extensive air traffic connections between Kenya and parts of Western Europe.

PRODUCT AND MARKET CHARACTERISTICS AND EXPECTED CONTRACTUAL FORMS

Table 34.2 provides ratings for asset-specificity, uncertainty, and market structure variables for the major commodities in Kenya's horticultural export trade. Its contents are based on the author's interviews of processors

Table 34.2. Asset specificity, uncertainty, and market structure for major Kenyan products

	Asset spec.	Perish-ability	Imp. seas.	Imp. conc.	Market segm.	Brand differ.	Expected type of contract
Processed products							
Canned pineapple	High	Low	Low	High	Med	High	Vertical integration
Canned green beans	Med	Low	Low	High	Med	High	Long-term contract
Dehydrated vegs.	Med	Low	Low	High	High	Low	Long-term contract
Passion fr. juice	Med	Low	Low	High	Low	Low	Long-term contract
Fresh fruit and veg.							
Off-season fr. & veg.	Low	M/H	M/H	Med	Med	Low	Long-term contract
Asian vegetables	Low	Med	Low	Med	Med	Low	Spot mrt. or L-T ctr.
Trop./sub-trop. fruit	Low	L/M	Med	Low	Med	Med	Spot mrt. or L-T ctr.
Flowers and seeds							
Cut flowers	L/M	High	M/H	Low	Med	Med	Long-term contract
Flower/veg. seeds	L/M	L/M	Med	High	High	High	L-t contract or V.I.

Asset specificity: A qualitative judgement based upon the need for specialized grading, storage and transport facilities and the (potential) scale economies and actual product specificity of processing facilities.
Perishability: Period maintaining quality before deterioration under practiced storage and transport conditions. High – less than one week. Medium – 1-3 weeks. Low – more than 3 weeks.
Import market seasonality: High – 2/3 or more of imports (from outside the region) are done in three months or less. Medium – 2/3 of imports are made in 4 – 6 months. Low – 2/3 of imports made over seven or more months.
Import market concentration: High – very few market entry points with largest four firms accounting for 50% or more of trade. Medium – oligopolistic core, but competitive fringe providing market access. Low – decentralized trade with largest four firms accounting for 25% or less of trade.
Market segmentation: High – multiple user/consumer groups with major quality/price differences. Medium – several different user/consumer groups with moderate quality/price differences. Lcw – predominance of single user/consumer group.
Brand name differentiation: High – major importance of brand name promotion and labelling. Medium – some branding occurs but is not the norm. Low – virtual absence of brand name promotion.

and traders, together with published market surveys for particular commodities or commodity groups.

For each of Kenya's major processed fruit and vegetable products, processing facilities involve at least moderate degrees of product specificity, this being potentially most risky for pineapple canning where there are also potentially large economies of scale (and thus very high sunk costs in plant and equipment). Neither product perishability nor trade seasonality is a problem in the trade of these products. For each product (group) there exists high levels of concentration at the import stage throughout Western Europe. In the European market for canned pineapple and in the French market for canned green beans, a very limited number of major firms vie for increased market shares through brand name product differentiation. Product differentiation by brand is not important for dehydrated vegetables and passion fruit concentrate since the bulk of supplies serve as intermediate raw materials for further processing. For three of the four products, market segmentation exists with alternative outlets among industrial, service, and direct consumer sectors.

Kenya's trade in 'Asian' vegetables entails little investment in post-harvest facilities (e.g. for grading and cooling) which are specialized for this range of commodities. Moderate levels of uncertainty surround this trade, as most 'Asian' vegetables experience quality deterioration after 10–20 days under practiced storage and handling procedures. For the majority of 'Asian' vegetables there is little seasonality in trade and prices: they are consumed year-round by the UK's 1.5 million Asian residents, and as there is very limited European production of such commodities, the 'Asian' vegetable trade in the UK features at the import level – five firms handled 65–75% in 1985 (but no one with over 20%).

Market segmentation in this trade consists of the differential product mixes required by the different South Asian immigrant communities. There is no brand name promotion in this market. The core problem in supplying this market lies in attaining a rough balance in the quantities supplied of the broad range of vegetables, since these vegetables are normally traded (and sometimes consumed) in combination.

Kenya's trade in temperate 'off-season' fruit and vegetables (e.g. french beans, melons, strawberries, courgettes, cherry tomatoes) is associated with low levels of asset-specificity, yet moderate to high levels of uncertainty. Several individual commodities are highly perishable and can be profitably traded by Kenya only during narrow mid-winter periods. European markets for these commodities tend to feature an oligopolistic core at import and wholesaler/distributor stages, yet a large competitive fringe which provides many additional, if less stable, points of market access. Product differentiation by brand is relatively uncommon. Several alternative market segments and types of distribution channels exist, although multiple chain supermarkets are generally becoming the most prominent final outlet for these commodities. Most of the tropical and subtropical fruits (e.g. mango, avocado, passion fruit, pineapple) which Kenya exports feature a slower rate of perishability and a lower degree of trade and price seasonality than for the temperate commodities discussed above. Trading uncertainty is thus of lower magnitude. While trade in these commodities features somewhat lower degrees of concentration in many European markets, brand name promotion has become increasingly important in this trade, following the examples set by Israeli and South African marketing companies.

Kenya's trade in cut flowers is associated with low-to-moderate levels of asset specificity (in grading and storage facilities), but relatively high levels of uncertainty given the rapid perishability of the commodities and the narrow market 'window' which Kenya has for some types of flowers. The West European trade in cut flowers is highly competitive at import and wholesale levels, providing many potential points of market access. Product differentiation by brand is uncommon, although trade values at import and subsequent stages are influenced by the reputations for quality held by growers and traders. While the market is not segmented at the import stage, it is segmented further downstream, especially with the distinction between institutional buyers and individual consumers.

For Kenyan producers and traders of horticultural seeds and seedlings there are low-to-moderate levels of asset-specificity (e.g. for seed cleaning and other post-harvest equipment) and moderate levels of uncertainty associated with product perishability and trade seasonality. Potentially more problematic is market access and limited bargaining power in a highly concentrated European market. Product differentiation by brand and market segmentation are both significant in the European horticultural seed market.

Hence, from a transaction cost perspective, we would expect only a small proportion of Kenya's horticultural export trade to be conducted on a spot market basis. Only 'Asian' vegetables and selected tropical and subtropical fruits would appear to have the technical, post-harvest, and market structure and demand characteristics which would enable exporters (and foreign market importers) to conduct much of their trade on a short-term, consignment basis without facing excessive risks and transaction costs. While transaction cost economics tends to place greatest emphasis on the problem of asset-specificity, in this particular context, problems of commodity perishability, seasonality of trade, continuous market access, and market penetration may be more serious barriers to profitable trade. The above analysis suggests that long-term contracts are likely to be the most efficient mode for organizing much of Kenya's agricultural trade.

Kenya's exports of processed fruit and vegetables are almost entirely governed by intra-firm trade or long-term contractual arrangements. Only a tiny share of Kenya's processed horticultural product exports are handled on a consignment basis or any other form which constitutes arms-length trade. While exports of canned green beans and dehydrated vegetables were technically governed by annual or longer-term marketing contracts, all or nearly all of this trade was conducted through European firms which also played important management and technical roles in the Kenyan operation. In the case of green beans, export sales were made exclusively to the French firm Saupiquet which maintained a long-term technical, management, and marketing contract with the Kenyan firm Njoro Canners. For dehydrated vegetables, some 90% of sales were made to the West German firm Bruckner Werke which held a small ownership stake in the Kenyan firm (i.e. Pan African Vegetable Products Ltd.) and maintained a long-term technical and marketing contract with it.

While 90% of passion fruit juice sales were made through Passi Co. (Switz), the majority shareholder of the leading Kenyan producer of this product (i.e. Kenya Fruit Processors Ltd.), Kenyan exports of canned pineapple products have been handled exclusively by a UK-based affiliate of Del Monte which is also the majority owner of Kenya Canners Ltd., the major Kenyan producer. Such a high degree of functional and/or ownership integration between the Kenyan exporter and foreign market importers was not expected by our analysis of asset specificity, uncertainty, and market structure conditions, with the exception of canned pineapple products.

While actual product shipments have been made from Kenya to more than one country or final buyer, in each of the above cases practically all marketing activities have been orchestrated by the Europe-based affiliate and practically all exports are invoiced to such affiliates. In each case, it has been the Europe-based affiliate and not the Kenyan firm which has been responsible for short- and long-term decisions regarding product lines, sales volumes, delivery times, and market destinations. Market research and market development activities have been left to the European affiliates as has been the determination of distribution channels for the product and the negotiation of sales or agency agreements with individual distributors.

Several industry-wide patterns should be highlighted. The first of these is the considerable importance of integrated exporter–foreign importer operations. Approximately one-fourth of Kenya's trade is conducted on this basis. Unlike in the case of processed products where trade between subsidiaries of MNCs was important, in Kenya's fresh produce trade 'vertical integration' involves trade between members of the same family. Many exporters, both Asian and African, have family members resident in the UK or elsewhere with whom trade is conducted. Some of these relatives have established formal trading companies; others trade fruit and vegetables only as a side-line activity.

A second (and related) pattern is the strong degree of personalization of trading relationships. While nearly one-half of (the volume) of Kenya's trade is conducted on a consignment basis, very little such trade occurs between totally unassociated companies. The bulk of Kenya's trade is conducted through highly personalized, continuous trading relationships, even though the specific volume and terms of trade may change from transaction to transaction. Only when sales are made by small-scale, part-time exporters or when established exporters initially attempt to penetrate a new market, are transactions undertaken between previously unassociated companies.

The predominance of personalized, continuous trading relationships in Kenya's fresh produce trade can be linked to potential problems of 'agency' and of buyer (or seller) opportunism when highly perishable and quality variable commodities are traded over long distances in a market which features multiple prices for very similar commodities. While the payment of a sales commission is supposed to offer an overseas agent an incentive to sell an exporter's product at the highest possible price, the agent's incentives are normally influenced by additional factors plus the agent may seek to take advantage of his favourable access to information. Given the asymmetric distribution of information, the agent may misreport actual sales results, misinform the exporter about overall market conditions, wrongly claim that produce quality has deteriorated in transit, and/ or delay or withhold payments owed to the exporter. These have all been major problems faced by Kenyan exporters.

A third industry-level pattern is the major differences in the export marketing arrangements employed by the larger, first-tier exporters who trade on a year-round basis and those of the many small-scale, part-time exporters. A majority of the leading firms (e.g. those whose annual exports in the mid-1980s exceeded 1000 tonnes) conduct 75% or more of their trade on the basis of either longterm contracts or intra-family linkages. Many of these leading firms have been closely involved in horticultural production (either on their own arms or through intensive outgrower arrangements) and/or domestic produce marketing and have begun to invest in modern postharvest storage and handling facilities. These investments have led them to be especially keen on developing long-term trading arrangements and, at the same time, provided them with improved capacity to meet long-term commitments. Their larger scale of operations, their broad range of products, and their trade continuity has led selected European and Middle Eastern buyers or brokers to cultivate long-term trading ties with them. In contrast, most of Kenya's small-scale exporters conduct the bulk of their trade on a consignment basis, negotiating prices, commissions, quantities and other trade terms for each transaction.

Turning to individual commodities, for both avocado and mango, practically all of Kenya's export sales are made on a consignment basis with the f.o.b. value normally determined by wholesale market prices in the destination countries. For both commodities, short and longer-term uncertainty about crop availability and high variability in product quality have prevented most Kenyan exporters from making and fulfilling forward supply commitments. Partly as a result of Kenya's supply problems, the import demand for Kenyan avocado and mango in Europe and the Middle East has been highly variable, with the Kenyan products used to supplement or fill seasonal gaps in the deliveries of more preferred international suppliers. Trade in such products has thus been highly risky, leading few firms to invest in proper grading, storage, and handling facilities.

In sharpest contrast are the export marketing arrangements for strawberries. During the mid-1980s, Kenya's exports were conducted almost exclusively through seasonal sales contracts with a limited number of major European buyers at fixed prices. Kenya's major producer of strawberries has targeted the mid-winter European market, when supplies from alternative sources are limited and premium prices are offered by retail chain stores and catering distributors. Several buyers have offered technical support in production and postharvest quality control.

The contractual arrangements governing exports of french beans are varied. Such diversity stems in part from the multiplicity of markets and distribution channels through which the Kenyan product is sold. In the main market outlets for Kenya's product (e.g. the UK, France, and Belgium), supplies are channelled to supermarket chains, to the catering industry, and to independent greengrocers, each of which has different procurement needs and utilizes different types of importers or wholesalers to serve their needs. This is reflected in the varied purchasing or agency arrangements employed at the import level. Different competitive conditions among individual markets also contributes to the diversity of sales arrangements. For example, in France and Belgium, french bean supplies from West Africa compete directly with the Kenyan product and there is thus stronger resistance to seasonal and fixed price commitments than in the UK where such competition is minimal. Most Kenyan exporters who handle relatively large volumes of french beans (e.g. 10–30 tonnes per week) have sought to cultivate long-term trading links with major European importer/wholesalers so as to service multiple-chain retailers.

The 'Asian' vegetable export trade (to the UK) is based entirely on fixed price sales, normally made in the context of extended seasonal agreements or intra-family operations. The typical trading arrangement is for the importer to have a standing order for some two dozen different vegetables, stating required volumes, number of weekly consignments, and prices. Sales prices tend to remain stable over long periods, unless there are significant exchange rate or freight cost changes. Given the vast range of commodities involved in this trade, the maintenance of steady prices helps to economize on bargaining costs. The importer/distributors involved in this trade are not the large-scale, diversified organizations increasingly prominent in the UK's fruit and vegetable trade, but small and medium-scale trading companies, owned and managed by individuals of South Asian ethnic origin, who are specialized in supplying goods to the UK's South Asian and other ethnic minority communities.

As in the case of processed products, a predominant proportion of Kenya's flower and horticultural seed exports are governed by intra-firm trade or long-term contracts. Similarly, the bulk of this trade is based upon exclusive or near exclusive trading relationships, with the Kenyan suppliers playing little or no role in the actual marketing of its products overseas. Pure spot market sales, involving consignments going direct to the Dutch flower auctions, accounted for only a tiny share of Kenya's trade through the mid-1980s.

While the European cut flower market is both highly competitive and dynamic, the marketing strategy adopted by Kenyan exporters has allowed little flexibility, such as shifting supplies among different buyers or agents in response to short-term market conditions. Several factors have led Kenyan exporters to shun such a strategy and instead conduct their trade on a long-term basis with only one or very few buyers or agents.

PERFORMANCE IMPLICATIONS OF ALTERNATIVE MARKETING ARRANGEMENTS

In this section, we indicate selected performance implications of alternative export marketing arrangements. Given the predominance of long-term contractual ties and vertical integration in the trade, most attention is given to the apparent benefits and costs associated with such trading arrangements.

Market access and penetration

Long-term and exclusive trading ties with major Europe-based manufacturers and trading companies have provided Kenyan and joint venture firms with immediate and continued access to markets, even those which are highly concentrated at the import and wholesale levels. This pattern contrasts with the market access problems periodically experienced by Kenya's processing and fresh produce exporting companies during the 1950s and 1960s when dealing through unaffiliated importers. This also contrasts with the current problems of European market access faced by Kenya's smaller processing firms which lack ties with major European firms. Small-scale, African-owned fresh produce traders who lack long-term trading linkages with European firms have also experienced periodic problems in finding a buyer for their produce.

Long-term trading relationships with major Europe-based firms have also facilitated the deeper penetration of Kenyan products within the targeted import markets than might have been possible if reliance was made on smaller importer/distributors. The wide trade and industrial contacts of the major importers has enabled Kenyan products to penetrate the most important market segments. Kenyan and joint venture firms have also benefited from the brand name recognition of their trading partners. As a result of these factors, Kenyan exports of such products as canned pineapples, canned green beans, dehydrated vegetables, and carnations were increased even during periods when market demand was stagnant or declining. For each of these products, Kenya's market share in Europe was increased during periods of market downturns.

In several cases, shifts from arms-length to vertically integrated trade enabled exporters to better penetrate their targeted markets. One prominent case is that of Kenya Horticultural Exporters Ltd., the country's leading fresh produce exporter. When KHE initiated its trade to the UK in the 1960s, sales were made via two commission agents based in London. Such sales arrangements provided little market development for the firm's product line as the distribution channels employed were very narrow. After one of the firm's senior partners emigrated to the UK (in 1973) and set up a marketing company, a wide range of new distribution channels were tapped throughout the country, generating higher profits and enabling KHE to greatly expand its trade. KHE's UK affiliate also assisted the former in penetrating European continental markets by making initial trade contacts and providing logistical support.

Another example is that of the Oserian Development Corporation which successfully boosted its trade (and profitability) after shifting from reliance upon a commission agent to utilizing its own Holland-based trading subsidiary. Having a physical presence in the market enabled Oserian to better service many of the regional Dutch flower auctions and not only the large Aalsmeer market. The firm's increased confidence in marketing led it to invest in production and postharvest facilities to expand output.

Set against the above advantages, the reliance on single Europe-based firms to perform all or most marketing functions has at times prevented Kenyan firms from tapping potentially lucrative market outlets. For example, the joint venture firms producing dehydrated vegetables and passion fruit juice were each periodically approached by potential buyers in countries not serviced by their Europe-based partners. Such enquiries were either not followed up or the potential buyers were referred to the European firm. Sulmac's long-term contract with Florimex greatly impaired the former's ability to supply other buyers and markets other than Germany. For several years prior to its forward integration of trade, Oserian was poorly informed by its exclusive commission agent about its sales opportunities both within Holland and elsewhere in Europe.

Information flows and transaction costs

Vertical integration and long-term exclusive trading arrangements have provided some Kenyan firms with access to a continuous stream of technical and market information, enabling them to better match their product mix, quality, packaging, etc. to the tastes and requirements of consumers and end-users. Continuous access to technical and market information has been especially important for Kenya's major flower producer/exporters given the dynamic technical changes occurring in this industry, the lack of flower research in Kenya, and the frequent changes in market conditions. The provision of technical information has also been critical to the operations of Kenya Canners and Njoro Canners, enabling them to overcome production problems and to raise the quality of their final products.

Long-term and exclusive trading ties with major Europe-based manufacturers and traders have reduced the transaction costs faced by Kenyan processors and flower and seed exporters. In such trading relationships, the exporter has not needed to identify the range of potential 'downstream' customers or end-users in different countries, negotiate individual sales terms with each, and enforce the ensuing agreements. Instead, the exporter utilizes a single contract and a single partner to make its sales, relying on this better informed and better positioned partner to coordinate subsequent sales and distribution operations. In addition, the close trading ties have enabled Kenyan firms to procure production inputs at lower cost and on a more reliable basis.

In some cases, however, information flows from the exclusive buyer or agent have been either incomplete or distorted, undermining the Kenyan firms' ability to plan and execute production and sales. In the case of PAVP, Bruckner Werke provided insufficient market information to enable managers of the Kenyan firm to make independent judgements about the appropriate mix of dehydrated vegetables and to enable the firm to properly negotiate sales prices with the buyer. PAVP faced a continuous problem of asymmetric information. Bruckner Werke also did not provide adequate technical information to permit the efficient operation of the processing factory. When Oserian sold its flowers through an exclusive commission agent, it received only limited (and perhaps distorted) market and sales information. Allegations of false sales reporting led Oserian to undertake a costly lawsuit in Holland.

Fresh produce exporters who trade at least partly with overseas family members report that such family affiliates are more inclined to give accurate and complete market information than unaffiliated firms and are also in a position to pass on valuable information regarding the competence and creditworthiness of other importers. Family affiliates are more amenable to making pre-payments for produce (to enable the exporter to overcome cash-flow problems) and supplying production inputs than are unaffiliated firms. Family affiliates have also been used to track down delinquent buyers to obtain compensation for the Kenyan firms. Losses due to sudden exchange rate changes are readily compensated for in intra-family and other long-term trading relationships.

In contrast, firms which export fresh produce primarily on a consignment basis complain strongly about buyer or agent opportunism in the form of false quality claims, sales misreporting, and delayed payments. For many small-scale exporters, quality claims or delayed payments have been sufficient factors to bring about bankruptcy or a termination of business due to an inability to pay farmers for delivered produce. Probably more than one hundred small-scale exporters have failed and withdrawn from trade within their first or second seasons over the past decade.

European importers of tropical and 'off-season' fruits and vegetables complain about the high transaction costs associated with trading with many Kenyan suppliers. In a survey by Hormann and Will (1987) of fifty importers in six West European countries, Kenyan exporters were evaluated less favourably than exporters from several Mediterranean (e.g. Israel, Morocco) and other middle-income countries (e.g. South Africa, Brazil) in terms of (i) their overall familiarity with the financial and other administrative requirements of the trade, (ii) their understanding of market requirements, (iii) their continuity of deliveries and observation of delivery dates, and (iv) their ability and willingness to settle complaints. My own survey of UK fruit and vegetable importers revealed strongly held views that the fragmentation of Kenya's export trade and the emergence of many small-scale, part-time importers in the UK was fostering increased opportunistic behaviour on both sides and was undermining long-term trading relationships.

The above comments should be put in perspective. The Hormann and Will (1987) survey found Kenyan suppliers to be far better perceived than those of any other country in East or West Africa, with the possible exception of the Ivory Coast. Countries such as Mali, Cameroon, Burkina Faso, and Ethiopia were rated as 'less satisfactory', not only in terms of exporter knowledge, exporter reliability, and transaction costs, but also in terms of product quality.

Pricing and the distribution of benefits

With intra-company and intra-family trade or in the context of long-term contracts, there is wide scope for firms to manipulate export and production inputs import prices in such a way that their profits from trade are concentrated abroad rather than repatriated to Kenya. Transfer pricing may be used in order to reduce a company's local tax burden, reduce its dividend payments to local shareholders, manage exchange rate risk,

Table 34.3. Distribution of sales revenues between Kenya and overseas distribution components

	Canned[a] pineapple	Canned[b] french beans	Fresh[c] french beans	Cut[d] flowers
F.O.B. Kenya	55.0%	40.9%	25.6%	19.2%
Freight/handling	8.0	7.5	21.3	12.8
Importers costs + profits	22.0	29.3	12.7	6.4
Wholesalers costs + profits	N.A.	N.A.	10.4	11.6
Retailer costs + profits	15.0	15.8	30.0	50.0
Other costs	0.0	6.5	0.0	0.0
Total	100.0	100.0	100.0	100.0

[a] UK market, sold in supermarkets (1987). As reported by Del Monte International.
[b] French market, sold in supermarkets (1985). Calculated from data provided by Njoro Canners and Marketing in Europe (October 1986).
[c] UK market, sold by green grocer (April 1985). Based on interviews with UK and Kenyan traders.
[d] Dutch market, sold by specialist florist. Based on information provided by Oserian Development Corporation and trade data.

enable it to by-pass profit repatriation regulations, transfer financial assets abroad, and/or enable it to finance imported inputs without having to go through the Kenyan Central Bank to obtain foreign exchange allocations.

Allegations of financial improprieties and the non-repatriation of export earnings by fresh produce exporters have been made periodically by prominent government officials and by the national press. Concern about this problem resulted in the implementation of a system of 'minimum export prices' and close scrutiny of all export transactions by officials of the Horticultural Crops Development Authority. In debates within the Kenyan Government regarding the future organization of the fresh produce trade, advocates for the nationalization of trade pointed to the alleged widespread practice of transfer pricing and the non-repatriation of foreign exchange. The sums involved, however, probably pale in significance compared to the lost earnings for Kenya and non-repatriated profits occurring in the fruit and vegetable processing and cut flower industries.

Table 34.3 indicates that only one-fifth to one-fourth of the final sales value for french beans and cut flowers (which together account for more than three-quarters of fresh horticultural export earnings) accrues to Kenya. While air-freight and handling costs account for a significant share of costs, it is large retail trade margins which account for the bulk of the f.o.b.–retail price spread. Such large retail trade margins are probably more a reflection of high levels of wastage due to product perishability than of oligopoly, although this should be examined. Importer and wholesaler gross margins are not substantial and generally result in net profit margins of only a few percentage points. Even by integrating forward into overseas market import/distribution and practicing transfer pricing, a Kenyan firm's gross share of final value would still only be 38% for french beans and 26% for cut flowers.

For canned fruit and vegetable products, a far higher proportion of final retail value accrues to Kenya as export earnings. This is due to the lower unit transport costs and lower retail margins for such products. Still, Table 34.3 indicates that both Del Monte International (UK) and Saupiquet (France) captured a substantial share of the final sales value through profits at the import/wholesale level. For Kenya's canned pineapple, Del Monte's distribution costs and profits accounted for 22% of final sales value, while for Kenya's canned french beans, Saupiquet's distribution costs and profits accounted for 29.3% of final sales value. While both firms do incur storage, transport, and advertising costs, such gross margins on non-perishable products are quite high and can be viewed in part as foregone profits for the Kenyan partners or affiliates of such firms.

REFERENCES

Anderson, E. (1988) Transaction costs as determinants of opportunism in integrated and independent sales forces. *Journal of Economic Behavior and Organization.*

Hormann, D. and Will, M. (1987) *The Market for Selected Fresh Tropical Fruits from Kenya in Western European Countries.* Working Paper 57. Institute for Horticultural Economics, Hannover.

INDEX

Abbott, J.C. xiii, 1, 146, 158, 160, 323
Aid 1–2, 52, 89, 91, 176, 211–214, 304–305
  food aid 328
Anthropological studies 57–58, 93
Arbitrage 152, 156–158
Assembly 81, 230–231
Auction sales 137–138, 398

Bananas 372
Bangladesh 91, 302–303, 319–320, 346–347
Bank, banking 211, 281, 286–287
  letter of credit 22, 362
Bargaining power 12–13, 73, 285
Bates, R.H. 2
Bauer, P. xii, 1
Berg, E. 36
Beverage crops 239–241
Biotechnology 122–144, 306
Birla Institute of Scientific Research xii
Bonded warehouse 160
Botswana 69
Brand, branding 177–178, 383–385, 393, 399
Brazil 30–31, 75, 254–262
Buffer stocks 203, 237, 248–250, 301, 305
Bulk handling 201, 204–205
By-products 175, 366

Cameroon 297
Canning 180, 385, 400–403
Capital 100–101, 103, 112–113, 142–143, 179–180, 218, 233, 279–281, 318–319
Cassava 106–107
Chile 237
China 265
Cocoa 19, 246–248
Coffee 137–138, 297
Colombia 252, 286, 348–349
Commission agents 116, 231, 282, 398–400
Communications 156, 209, 275–276
Comparative advantage 298, 360
Competition 5–13, 36–37, 54–56, 104–107, 119, 125, 138, 176–177, 294–295, 299–301
Compton, J.A.F. 223
Consignment sales 282, 386–401
Consumers 45–46, 177, 210, 219, 234, 244, 342–348, 358
  preferences 171, 176, 312
Contract(s) 264, 394–398
  contract farming 73, 287, 315, 369–377
Cooperatives 73–74, 78, 90, 94, 110, 115, 127–143, 224, 265–267, 282, 286, 297, 329
Coulter, J. 223
Credit 30, 38, 72, 84, 88, 99–101, 105–111, 113–116, 128, 141,

198–199, 218, 284–288, 327, 381
Creupelandt, H. 146, 158, 323
Crop reporting 149, 156
Cyprus xii

Demand 171–173, 177, 190–195, 295, 312–313, 344, 358–359
Drying, dryers 206, 218, 315

Ecuador 282
Egypt 286, 343–344
Employment 67, 69, 179, 329, 350, 353, 374, 381–382
Entry barriers 79, 105, 113–114, 148, 233–235
Epstein, T.S. 93
Ethnic factors 131, 135, 150, 210, 299, 391
Exchange rates 117–118, 251–252, 296, 380, 387
Exploitation 14, 87–88, 102
Export(s) 18–24, 43–44, 47, 49–50, 89–91, 95–97, 145, 170–171, 244, 291–292, 370, 378–403
    versus food crops 90, 297–299, 380–383

FAO xii, 81, 169, 239, 273, 308, 355
Fair price shops 331
Fast food 120
Fertilizers 279–289, 298, 326–327
Fibre crops 239–241, 89
Financing 67–68, 98, 107–110, 119–122, 151, 210–221, 232–233, 279–289, 361–362
Fish 69, 77, 81, 115–116
Flowers 394
Food coupons 331, 347–349
Food crops 47–49, 61, 75, 144–162, 197–207, 225–235, 274–275, 292, 295–301, 325–340, 372
Food-for-work 349–350
Food security 217, 305, 331, 382–383
Food subsidy programmes 66, 241–242, 331, 341–354

Fruit and vegetables 69, 80, 142, 175, 206–207, 219, 292, 373, 378–403
Futures markets 252, 305

Geographical studies 57–58
George, P.S. 323
Ghana 76, 79–80, 83, 112
Government facilitating services 28, 72–73, 38, 180, 223
    intervention xii, 3–23, 38, 41–50, 99, 102, 109, 137, 145, 317–322, 325–328, 336–339, 362, 391
    marketing support unit xi, 84–85, 156, 212, 265, 273–275
    policy 41–50, 83–87, 89–92, 97, 104, 107–111, 131–134, 155–157, 180, 204, 209–215, 273–274, 283, 290–317, 322, 339–340, 341, 377
    socialist policies 45, 124, 131, 134, 273, 331
Grades, grading 18–21, 28, 109, 220–221, 395
Group action 114, 214–215, 264
Guatemala 90, 378–387

Harriss, B. xii, 1–2
Horizontal coordination 27–30, 129–130
Hunter, G. 94
Hyden, G. 94

IFAD 2, 67, 88, 91
IFDC 277
IFPRI 277
Incentives 50, 54, 60, 68, 92, 143, 174, 179, 218–220, 303, 322, 329–330, 333–340, 370
Indebtedness 5–6, 72, 93, 179
India xi–xii, 69, 75–76, 84, 90, 112, 125, 138–140, 182–196, 224
Indonesia 81
Input supply 44, 67, 88–89, 99, 106,

116–117, 277–322, 326–327, 371, 400
Inspection 14, 21–23
Institutional issues 4, 55–56, 83–85, 215, 223–274, 280, 285, 288–289, 369–373, 386
ITC xii, 355
Ivory Coast 50

Jaffee, S. 355
Jiggins, J. 93
Jones, W.O. xii, 35–36

Kenya 47, 68, 75, 83, 136–138, 159, 265, 285, 291, 338, 373, 388–403
Kerala 344–345
Kinship factors 101–103, 129–130, 133, 391–400
Kissinger, H. 165
Kola nuts 97–98
Korea 140–142, 265
Kracht, U. 323

Lele, U. 94, 147, 355
Liberalization 85–92, 220, 225–235, 275, 300–303
Lipton, M. 1, 66
Livestock 172, 357–368
Loan repayment 288–289
Location 178, 182–207, 217
Losses 216–219, 368

Maharastra 66, 350, 353
Malaysia 245–248, 252
Management 121, 125–126, 130–131, 162, 171–180, 270
Market(s) 6–7, 37–38, 41–49, 61, 67, 81–83, 88, 146–147, 161, 165, 206, 212–213, 264–265, 363
   information 28, 142, 149–157, 221, 235, 263–265, 268–272, 275, 400–401
   regulation 167, 224–225
   research 345

transparency 268–269
Marketing
   boards *see* Parastatal(s)
   bibliographies xiii
   channels 6–10, 25–29, 177, 360–365
   costs 4–5, 18–21, 30, 35, 73, 182–183, 208, 220, 291, 313–316, 336–339
   definition xi
   efficiency 1, 18–19, 35, 55, 107–111
   enterprises 11–23, 69–76, 93–164, 388–403
   extension 91, 108, 221, 262–267
   infrastructure 28, 36–38, 53, 77–83, 149, 155, 165–221, 280, 304, 340
   intermediaries 3–6, 8–9, 104–111, 131, 226–235, 325–326
   investment 196–206, 279–282, 314
   legislation xi, 72, 224
   margins 3, 5, 8, 107–110, 131, 316, 323–336, 365–366, 403
   research 51–61, 63
   services 10–11, 35, 48
   structure, conduct, performance 34–38, 55, 59–60, 148–154, 391–393, 398–403
   training 53, 61–63, 162, 254–264, 280–281
Mellor, J. xii
Mexico 285, 342–343
Michigan State University xii, 1–2, 30
Middlemen *see* Marketing, intermediaries
Milk 136–140, 179, 373
Monopoly 6–14, 52, 146, 242–243, 294–295, 320–321, 376

Nepal 265–266
Nigeria 21, 43, 47, 151–154, 201, 337
Nutrition 327–328, 347, 384–386

OECD 1

Oil seeds 19, 47, 140, 182–196, 372

Packing, packaging 84, 172, 177–178, 279
Pakistan 312–313, 316
Papua New Guinea 245–248, 265
Parastatal(s) 42–50, 52, 60, 74–76, 95–97, 144–162, 197–199, 206–207, 226–230, 237–251, 291–294, 300–301, 308, 316–322, 342–343
    efficiency criteria 273–274
Peru 242–243
Philippines 87, 90
Postharvest management 165, 216–223, 315–316
Poultry 106, 109, 373
Poverty 65–69, 327, 331, 341–354
Price, pricing 3–27, 33, 35, 41–50, 61, 73, 150–157, 197, 220, 229–232, 274–275, 329–330, 333–340, 365–367, 401–403
    fixing 341–342
    pan seasonal pricing 220, 242–244, 251
    pan territorial pricing 17–20, 75, 88, 209, 220, 242–244, 251, 275, 295, 303–304, 334–335
    stabilization 73, 75, 108–109, 158–160, 223, 236–253
Private trade(rs) 53, 60, 72–73, 85–87, 95–111, 147–148, 160, 198–199, 226–227, 252–253, 284–285, 292–296, 300–305, 320–322, 337–338, 360–368
Privatization 85–91
Processing 24, 44, 80–81, 84, 105, 108, 122–124, 135, 137–140, 167–196, 200, 204–205, 370–371, 393–395, 403

Quality control 65–69, 327, 331, 341–354

Rationing 343–346, 350–353

Raw material supply 170, 173–175, 185
Restrictive practices 5
Retailer, retailing 16, 207, 231, 271, 282–285, 313
    training 257–261
Reusse, E. 355
Riley, H.M. xii, 2
Risk 28, 197–199, 374, 380
Roads 164, 208, 212–213, 304, 363
Rubber 246, 252, 372

Sales promotion 177–178, 316
Savings and development 25–33, 362
Seasonality 79, 173, 218, 243, 353, 366
Seeds 301, 306–322
Self help 165, 208–215
Senegal 46, 73, 86, 225–235
Siamwalla, A. 93
Slater, C. 1
Slaughterhouse 168, 175, 178, 366
Smallholders in marketing 69–83, 110–111, 128, 135, 219, 340, 369, 380
Somalia 357–368
Spatial equilibrium 177–178, 195–196
Spinks, R. 94
Sri Lanka 287, 345, 353
Standards *see* Grades
Storage 29, 38, 78–81, 84, 91, 158–159, 199–205, 218–221, 251–252, 279–281
    cold storage 80, 176–178, 206, 217, 394
    on farm 78, 91, 221–223
Structural adjustment 2, 65, 85–87, 210, 223, 275
Subbarao, K. 323
Subsidies 294, 311–313, 326–327, 337, 340
Sudan 89, 91
Sugar 106, 372
Supermarket(s) 30, 393, 397
Sustainability 91–92, 214, 304, 387
Systems approach 56–57, 62

Tanzania 135, 243–244, 273–275, 287, 326, 341–342

Targetting 242, 269, 331, 350–353
Taxes, taxation 43, 61, 237, 245–246, 296, 363, 387
Thailand 87, 105–112, 272
Timmer, C.P. 148
Tobacco 107, 372
Transaction costs 338–339, 388–403
Transnationals 73, 90, 93–94, 117–125, 310–314, 321, 379
Transport 2, 17–18, 28–29, 37, 78, 108, 178, 195–196, 200–205, 208–209, 214, 220, 234, 275, 294, 313–316, 357–360, 391

UK National Resources Institute xiii, 355
Urban influence 43, 67, 204, 325, 346, 354
USAID xii, 30, 62, 302

Value added 167, 172
van den Laan, H.L. 93

Vertical coordination 25–30, 55, 129–130, 360–361, 370–371, 398–400
von Braun, J. 323

Weights and measures 12–15, 83, 300
Wholesalers 99, 112–113, 120, 151, 159, 207, 227–234, 280–282, 302
  training 259–260
Women in marketing 85, 93, 112–116, 143, 366, 382
World Bank xiii, 2, 165, 212, 224, 273–274, 304

Yamey, B.S. xii, 1
Yemen 219

Zaire 212
Zambia 45, 113, 265, 303